ILLUSTRATIONS

    Sunflower, globe artichoke and Scottish thistle   *frontispiece*

1. *Beans*. Butter bean, soya, pea bean, Mexican black, runner bean and jack bean   *facing page* 48
2. *Leaves and whole plant*. Comfrey, cowslip, celandine and violet   62
3. *Fruit I*. Breadfruit and leaf, olive, pandan, akee and okra   66
4. *Leaves*. Brussels sprouts, seakale, Chinese cabbage, water cress, mustard and rocket   72
5. *Cacti*. Prickly pear, beavertail and sahuaro   76
6. *Swollen leaf-bases*. Celeriac, dasheen, garlic, kohl-rabi, Florence fennel, onion and leeks   94
7. *Fruit II*. (Vines) Snake gourd, water melon, balsam pear, gourd, chayotes, summer crookneck and squash   118
8. *Bulbs*. Martagon lily, Florentina, Solomon's seal, dog's tooth violet, wild tulip, tulip and lily   126
9. *Palms*. Male and female spadices of sugar palm, sago, male and female spadices of palmyra, coconut   164
10. *Reeds and rushes*. Seaside clubrush, sweet cypress, sea arrowgrass, water manna grass, reedmace and common bulrush   196
11. *Seaweeds*. Blade kelp, bladder-wrack, purple laver and carragheen   204
12. *Roots and tubers*. Sweet potato, yam bean, arrowroot, cassava, orchis, Jerusalem artichoke, ulloco and ysaño   230
13. *Water plants I*. Lotus   246
14. *Water plants II*. Arrowleaf, water plantain and fringed water lily   252
15. *Nuts*. Moreton Bay chestnut, pine nut, cashew, Queensland nut and Brazil nut   276
16. *Grain*. Two-rowed barley, Japanese millet, oats and rice   292
17. *Aromatics and mustard*. Dill, cumin, coriander, fennel, juniper, borage, thyme, rue, fenugreek, rosemary, black and white mustard   324
18. *Spices I*. Cinnamon, vanilla pod, cardamom, ginger and turmeric   330
19. *Spices II*. Cloves, pepper, allspice, red pepper, chilli, nutmeg and mace   336

ACKNOWLEDGEMENTS

I am indebted to the following authors, their executors or publishers, for permission to quote from their works:

Columbia University Press for poems from *The Manyoshu* (New York and London, 1965) and from *Su Tung-P'O* (*trans*. Burton Watson, New York and London, 1965);

Seamus Heaney, Faber and Faber Ltd and Oxford University Press Inc., for sections I and III of 'At a Potato Digging' from *Death of a Naturalist* by Seamus Heaney;

The Society of Authors as literary representative of the Estate of A. E. Housman, Jonathan Cape Ltd and Holt, Rinehart and Winston Inc., for 'Sinner's Rue' from *Collected Poems* by A. E. Housman;

Indiana University Press for S. N. Pant's 'Fruits of the Earth' from *Modern Hindi Poetry* (ed. V. N. Misra, Bloomington and London, 1965);

Alan Jackson for 'Loss' from *Penguin Modern Poets 12*;

Jon Silkin, Chatto and Windus Ltd, and Wesleyan University Press for 'Dandelion' and part of 'Peony' from *Poems New and Selected* by Jon Silkin;

Mrs Myfanwy Thomas and Faber and Faber Ltd for 'Swedes' and 'Old Man' from *Collected Poems* by Edward Thomas;

UNESCO and John Day Co., for poems from *Anthology of Korean Poetry* (*trans*. Peter H. Lee, New York, 1964);

Peter Viereck for 'To a Sinister Potato' from *Terror and Decorum* (New York, 1948, to be reprinted in 1973 by Greenwood Press, Westport, Conn.). The slightly revised version printed in *New and Selected Poems* (Bobbs-Merrill Co., New York, 1967);

Richard Wilbur, Faber and Faber Ltd, and Harcourt Brace Jovanovich Inc., for 'Potato' from *The Beautiful Changes and Other Poems* (New York, 1947), reprinted in *Poems 1943–1956* (*London*) by Richard Wilbur.

In addition it might be mentioned that my version of the *Tsurezuregusa* is taken from the French edition, *Les Heures Oisives* by Urabe Kenko (*trans*. R. P. Saveur Candau, Paris, 1968, copyright Unesco). There is also an American translation by Donald Keene, *Essays in Idleness* (Columbia University Press, 1967).

To provide a bibliography for a book as eclectic, and unique in its particular field, as this would not be particularly helpful. It owes a little to a whole range of publications including dictionaries and encyclopaedias, nineteenth-century enquiries into the history of vegetables and edibility of exotic plants by French enthusiasts and scientists, nineteenth-century and early twentieth century works on folklore and dialect, a limited number of works dealing with single crops or with the food resources of a particular area, and so on.

INTRODUCTION

Language being what it is, the definition of a vegetable will vary according to the speaker. For my purposes it is any plant, a part or all of which is edible. Certain omissions have had to be made in order to make the plan of this book at all manageable. There are, for instance, several hundred edible fungi, the treatment of which is best left to an expert. Fruits also provide enough matter for a separate work, which means that the date and the coconut are given no more than an incidental glance, despite the fact that they form the staple food of some peoples. Certain fruits, however – beans, tomatoes, egg-plants – are traditionally considered vegetables, as opposed to a dessert, and so are included.

It is impossible to be all-inclusive in this sphere, and I am conscious of several omissions (of roses, for instance), although of nothing greatly significant. Many plants are included whose use is now limited to more or less primitive national groups, or which are of only historical interest, although popular enough in their day. There was a time when any plant which was not irredeemably poisonous, too foul to the taste or too tough, was eaten. So it is still where food is difficult to come by. It has been my intention to treat all the important foods and staples of the world, avoiding as far as possible a narrow Western viewpoint. Language is the limitation and not geography, otherwise the Kalmuk Tartar and the Hottentot, the South American *peon* and the poor Asian peasant, could all find here their favourite foods.

The chief part of this book, then, contains a wide selection of vegetables together with the herbs and condiments to which they are related. The plan is often capricious. In several cases all the edible relations of a particular family, however insignificant, are included under one heading. At other times the family is divided up in various ways. The lily family, for instance, has a large section devoted to it; asparagus, and those relations treated like it, has a section to itself, as have the members of the onion family, although these are also part of the lily family. Similarly cruciferous plants are divided into mustards, cabbages (and other leafy members), and turnips (and other rooted members). Two lesser divisions of the book deal with certain nuts and grains, and with herbs, spices and condiments. It is impossible to avoid a certain amount of overlapping but reference to the indices will help the reader avoid confuison.

The kind of information provided is in part the result of personal predelections, in part a result of meeting people in search of information and finding out what they want to know. Both my wife and I have been approached by people in markets and delicatessens and asked how we cooked the strange vegetables we were buying. Our response is two-fold: my part is to say what the plants are and give a brief outline of how they are used; my wife's is to provide more detailed culinary information and recipes. She is at present working on a cookery book of her own.

Far from being exclusively culinary, my interests are suited to the locust mentality and the grasshopper mind. Dealt with are the place of particular plants in history and in the life of man; their place of origin, spread and present habitat; the uses to which they have been put (not all of them alimentary) and the little accidents that have happened to them on the way. The literary associations are treated also, although with a certain amount of restraint, intrinsic interest and worth being the criteria. Folk-lore and folk-wisdom – proverbs, legends and ceremonial – are also given some attention.

The origin of plant names often makes an instructive study and is here dealt with quite extensively: that barley is ultimately connected with the verbs 'be' and 'bear' (cf. beer) tells us much about its importance to the Germanic tribes who named it; and that wheat is the 'white grain' bears evidence of its late introduction to the tribes who thus distinguish it. Other plants go under a multiplicity of names – pretty and amusing, grotesque and obscure – the story of whose origin is often worth knowing.

It is from the multiplicity of these popular names that the botanist's generic Latin names are supposed to rescue us. A name like *fat hen* may be applied to a number of goosefoots and their relations, depending on which continent, and in which region of the continent, it is used, but the Latin name helps avoid confusion. It is trying, however, to find a constant task of botanical reclassification going on whereby the generic name is changed not once but twice or thrice in the course of a decade or two. The case with palms, thistles and watercress might be instanced here. As great a confusion as that over the popular names could be the result.

Worse crimes are committed in the name of science against clarity of description and the Latin language. When Linnaeus first attempted an encyclopaedic classification of plants he used several latinisations of Greek names, and names from other languages, resulting in imperfections whose possible barbarism our ears have, perhaps, accepted now (although words like *heterospathe* and *ptychosperma* are not such as one would delight in hearing very often). On the whole his aim was to be as explicit as possible, giving as the family name the accepted Latin or latinised name, the second part being descriptive of the plant's distinguishing characteristic. Thus the potato, *Solanum tuberosum*, is a member of a family which includes the nightshade, and its most interesting part is the tuberous root. Some of its close and equally edible relatives, many of which go under the name *papas criolas* in their continent of origin, have of late been saddled with such jaw-breaking and uninformative names as *S. antipovickii*, *S. kesselbrenneri*, and *S. juzepczukii*. We know what to look for in *S. mamilliferum*, where to look for *S. chacoense*, and what are the Indian names for *S. ajanhuiri* and *S. phureja*. Is the only distinguishing characteristic of these others that botanists of Slavic extraction first identified them? It rather seems as if there were a most unscholarly cult of personalities among the botanists. Even more objectionable is their habit of naming new plant-families after distinguished colleagues or even themselves. Admittedly the men after whom *Dillenia, Lewisia, Torreya, Cogeswellia* and *Bultmannia* are named might be worthy of immortalisation, but not in this way. It is only a step off the practice of breeders of hybrid roses selling the privilege of having one named after him to the highest bidder.

There are congresses held in which the procedures for naming plants are discussed and formalised. In view of the above objections, one might make the following suggestions which ought to be taken up at some time. Firstly that already established names be more respected; if it is necessary to change them then this should happen once and once only. Perhaps a ten years' notice of the desire to change a name might be made mandatory before a final decision is taken. Secondly, names of persons should be outlawed in Latin descriptions of new species, except as a purely temporary measure; in no case should they become the official names. Thirdly, subject to the same limitations as above, no genetic name should be based on a personal surname. Lastly, everything should be done to remedy the present cacophonous situation.

Beside the obvious interest of the undertaking there is a serious reason behind this cataloguing of foodstuffs, from the best-known to the most far-flung and unlikely. Such knowledge might make the difference if it were ever a question of survival for us. It is shocking to think that even in a country like Britain certain classes of people, especially the old and those out of work, are very likely to suffer from malnutrition either because they cannot afford the right kind of foods or because they do not know which these are.

For various reasons Western civilisation is now one in which meat plays a great part in the diet of the majority. At one time it was practically the preserve of the rich and vegetables figured more or less largely in the dishes eaten by the poor. Rising prosperity, and a corresponding dependence on meat, has forced down the variety of vegetables readily obtainable, and at the same time a large number of nourishing plants are looked upon as no more than troublesome weeds to be poisoned by the way. Knowledge of their usefulness is dying at a time when it might very well be most important to remember them.

The decay of knowledge is not limited only to the use of vegetables. Many of the humbler meat dishes once popular are now disdained. Who, among the English, now knows which snails and frogs are edible? We no longer sit down to 'four and twenty blackbirds baked in a pie' or to sparrow pie. And even of those few animals that are eaten, many parts that are more or less edible and nutritious are often ignored: brains, hearts, necks, trotters and hooves. They have their uses of course, but the important fact is that the generality of people have been allowed to forget all but the most popular (or popularised) items of nourishment in an increasingly regimented industrial society. The result is that even in countries where the standard of living is high there is an understratum allowed to go hungry or undernourished through ignorance.

For the anthropologist the first signs of agriculture are evidence of an early advance towards civilisation. Yet many have been the wandering tribes who have subsisted, and still attempt to do so in isolated parts of the world, on roots, fruits, nuts and bark. In Greek and Roman legend the idealised golden age was one in which men were satisfied with acorns for their sustenance before men had learned to wound the earth with their ploughs or to sow for gain. But agriculture was not so much a manifestation of a fall from grace as the means by which that fall came about. No one would be so churlish as to condemn the man who first attempted to raise a crop rather than rely upon the uncertainties of a nomadic existence, the chance finding of food and the horrible struggle to stay alive during the barren winter months. Give a man land, however, and sooner or later he is sure to abuse his position.

The will to power through economic manipulation is a result of overproduction and the consequent search for an outlet for the excess product. Finally one person, or a group, assumes control of all the land and the means of growth, forcing all others into narrow specialisations not necessarily of their own choosing and exploiting their labour before allowing them to eat. In more sophisticated times the only advance has been that money replaces food as a more convenient medium of exchange, otherwise man remains as capricious, ruthless and greedy as he has always been. Many, sometimes the majority, are still in need and still go hungry in the midst of plenty.

A good half of the world, comprising those who do know everything there is to know about the art of surviving on a minimum diet, go permanently hungry and grossly undernourished. If they are not yet aware of some of the obscenities in western society it will not take them much longer to find out. Ours is a system in which the more fortunate eat too much and either ignore or waste a good quarter of the country's food potential; in which farmers are paid not to produce food-crops beyond the needs of capitalist society and in which excess produce is destroyed; which kills almost anything not of capital value needlessly and makes a sport of the arts of hunting; in which the less fortunate are allowed what the authorities are pleased to describe as a subsistence-level income, to make the most practical use of which they have never been instructed, and are encouraged only by the spectacle of those even more unfortunate in the so-called underdeveloped countries.

The problems of not enough food and too many people, and the imbalance of prosperity, must be solved. If left too late *nemesis* (or *karma*) will take over the responsibility. A number of ends are foreseeable, to indulge in surmises about which is a nice exercise in pessimism and masochism. We are going to need all the knowledge we can find merely to *survive*: by encouraging self-sufficiency now, a greater respect for our hidden resources and less reliable upon a market whose guidance is steadily passing outside the control of the average individual, we can worry less where our next meal is coming from, both nationally and individually, and think more about where other people's meals are to come from.

International co-operation and mutual assitance between classes, putting into practice the fine preaching about the brotherhood of man (instead of mutual throat-slitting in support of private interpretations of what the phrase means) is probably as Utopian a hope as the other solution, three acres and a cow for everyone. In order to bear the colossal sums needed for exploration of the planets, we are told, the nations must pool their resources. No nation can afford these sums by itself, especially when stockpiling arms at the same time, seemingly in the belief that only overkill or slow starvation can be the answer to the world's population problem. But man has no place in the planets, let alone in outer space, which he will only litter or poison with his industrial detritus; and he can have no place there until he has learned to live at peace on (and with) his own planet, to feed its people and provide real education rather than encapsulated propaganda.

YANN LOVELOCK
Sheffield 1971

CAUTION

In the course of this work several poisonous plants are mentioned as edible after some preliminary preparation. It has been thought as well to reinforce the cautions in the book with this short note. Generally the taste of such a plant is either so bitter or burning that it is simply inedible unless the preparations mentioned are made. However, it is as well to treat such plants with caution.

Other plants are poisonous to a certain degree, or under certain conditions, and care should also be taken over these. One might ennumerate some of the conditions under which some of these plants might cause trouble. A plant should not be too old, or be left uncooked for too long. Potatoes that have sprouted can be very poisonous (as can green potatoes). Some plants are only safe after being subjected to heat (the roots of the cuckoo pint) or to another method of cooking (nettles cease to sting after being cooked in a very little water). Care should be taken that only the parts mentioned of some plants are eaten: the leaves and stems of the potato and the tomato are toxic. Very large amounts of some vegetables, or prolonged indulgence in them, can sometimes be dangerous. Some sheep which had each eaten about 25 lb of chickweed (Stellaria media) died of indigestion; a lesser amount is quite fit for human consumption when cooked. Again, some Germans who had eaten such common spinach substitutes as white goosefoot (Chenopodium album) and common orach (Atriplex patula) several times for some weeks were admitted to hospital with symptoms of poisoning. Here again, the plants are safely edible if not over-indulged-in.

In some cases, because we are not used to certain foods and consequently have not built up an immunity to them, as it were, they are liable to cause some internal distress. It is as well to remember this if experimenting with foods eaten by more primitive peoples, or foods once popular but now out of cultivation. One might take, as a minor example, various reactions to members of the onion tribe (genus Allium). The eating of wild species has been known to cause some cattle distress, and an unpleasant flavour is imparted to the milk of cows who eat them. The eating of garden onions has been known to cause symptoms of poisoning in American cattle. English people are not used to garlic, and are liable to internal upset upon first exposure to it, especially if it is uncooked. The Latin poet Horace accused his patron of trying to poison him after he had been given it at a feast, giving it as his opinion in an aside that countryfolk, whom he knew to use it, must have tough stomachs!

For the more cautious I add the names of a few other plants that it is well to beware of. They are drawn from Pamela North's *Poisonous Plants and Fungi* (Blandford Press, London, 1967), and *Britain's Poisonous Plants* (Min. Ag. Bulletin no. 161), to which reference should be made for fuller details, and for other plants under suspicion only. The majority of these plants are also common in Europe and other continents, especially those which have had contact with Britain. The order of mention is that in which they appear in the book.

1. *Sunflower* (Helianthus tuberosus). This has been recorded as poisoning pigs in Estonia and is therefore suspect. If the roots did this, then the pigs probably ate too many; they tend to be rather indigestible.

2. *Bryony* (Bryonia dioica). Everybody knows the roots and berries of this are poisonous. The stems and leaves have been known to do animals harm.

3. *Lesser celandine* (Ranunculus ficaria). The sap is burning and poisonous, but the root is non-toxic and has been eaten raw in salads. The leaves (probably only the young leaves) were once used to counteract scurvy.

4. *Marsh marigold* (Caltha palustris). The sap is burning and poisonous; if eaten at all, the parts specified should be used only when very young. In case of poisoning by this, the lesser celandine, or any member of the buttercup family, administer raw eggs and sugar in skimmed milk.

5. *Bracken* (Pteridium aquilinum). The very young shoots are moderately harmless if not over-indulged in. They caused me no ill-effects.

6. *Horsetails* (genus Equisetum). The whole plant is known to be poisonous to animals in large amounts.

7. *Horseradish* (Cochleria armoracia). The plant-parts above ground are poisonous. So is the root but, in the small quantities we take it, it is quite safe.

8. *Solomon's seal* (genus Polygonatum). The berries are poisonous and the plant is regarded with suspicion in Europe.

9. *Tulip bulbs* (genus Tulipa) poisoned cattle fed on them in times of scarcity during the Second World War. This might have come about through an excess of strange food.

10. *Charlock* (Sinapis arvensis). The seeds are poisonous, as are those of *white mustard* (Sinapis alba), otherwise grown, like the above, for fodder or as a pot-herb or salad.

11. *Vetch* (Vicia sativa). This sometimes causes poisoning in animals when large amounts are eaten. It was once mixed with *Indian pea* (genus Lathyrus), also poisonous, especially the seeds. There was once some dispute over this. It would be best not to over-indulge.

12. *Lupin* (genus Lupinus). The seeds of wild species are poisonous, apparently more toxic some seasons than others. It should be noted that those who used to eat them would first steep them in water before cooking. Other species still are specially cultivated for their edible seeds.

13. *Broom* (Sarothamus scoparius). The plant has earned a bad reputation from its old association with witches. Only the seeds contain poison, and in very small quantities.

14. *Poppy* (genus Papaver). Animals have had symptoms of poisoning after eating large amounts. The symptoms are described as 'agreeable mental excitement and increased physical activity' followed by drowsiness and deep sleep. Coffee, alcohol and some other drugs, 'poisonous' in overdoses, have similar affects.

15. *Sorrels* (genus Rumex) and *wood sorrels* (genus Oxalis) have caused cattle distress when taken in large quantities. Small amounts, cooked or in salads, do no harm.

16. *Acorns* (genus Quercus) have poisoned cattle, and even, occasionally, pigs. In the past people used to avoid taking too much of poisonous species by drying them and mixing the flour with other substances.

17. *Beech nuts* (Fagus sylvatica). The poison is water-soluble.

SECTION 1

Leaves, stems, roots and fruits

Akee (*Blighia sapida*)

This tree, which is cultivated for its fatty fruit, comes originally from tropical West Africa and has been introduced into the West Indies. The fruit bursts open when ripe, and becomes dangerous to eat if over-exposed to the air, no less than if it is gathered when under-ripe. It is, therefore, best picked in the morning when newly burst. Akees are accounted good boiled or fried, and are very popular, in the West Indies especially. In Jamaica they are often cooked with bacon for breakfast and are said to look rather like scrambled egg. The fruits are obtainable tinned from delicatessens, but they are at their best only when picked absolutely fresh.

Amaranths (*genus Amaranthus*)

There are about 500 species of the amaranth in various parts of the world. Its probable place of origin is India where it is still grown as a vegetable, although others have had a very ancient history of use in the Americas. The leaves are also eaten on certain islands in the Indian Ocean and in Africa. In England the plant was once used as an asparagus substitute, while in America it was important as a grain crop: the flour obtained from the seeds is till the principal nourishment of some Indians there. The plants are cultivated in Mexico and Guatemala, where they have been in use since 5000–3000 B.C. Montezuma, the Aztec emperor, is said to have received 200,000 bushels of amaranth seed yearly as tribute. But the growing of it was considerably supressed by the Spaniards because of its importance in Mexican religious ceremonies.

Indeed, the amaranth has frequently been connected with religious and quasi-religious ceremonies. The name of the plant was originally *amarantos*, signifying 'unfading' in Greek. The 'anth' ending comes from confusing the name with those of other plants with similar endings, such as polyanthus and chrysanthemum ($\dot{\alpha}\nu\theta o\varsigma$ is the Greek for a flower). The flower came by its name because of the Greek belief that it was immortal; thus it became connected with various mystery cults, such as that of the many-breasted Asian earth-goddess Artemis at Ephesus (the 'Diana of the Ephesians' mentioned in the Bible, Acts 19), to whom it was sacred. For similar reasons it was used to decorate religious statues and tombs. In 1653 Queen Christina of Sweden set up the Amaranter Order of knighthood, obviously bearing in mind the plant's mystical associations. It is the ceremony of initiation into this that is supposed to have survived in the Order of the Amaranth within the Eastern Star Chapter of North American Masons, the largest masonic women's organisation in America.

Perhaps the best known amaranth is *love-lies-bleeding* (A. caudatus) which has been grown as a grain crop in the Andes since at least the times of the Incas, for which reason it goes by the name *Inca wheat*. Its native names include *quihuicha* and *quinoa*[1]. The plant also grows in tropical Africa and Asia, where the leaves are used as a salad and pot-herb. Its popular name refers to the red ornamental flowers, and also embodies an old confusion over the proper spelling of the Latin name, once supposed to have been *amor-anthus* (love flower). The arresting colour gave the French reason to call the plant *passevelours* (beyond velvet), a name by which it used to pass in England, together with such local variations as *velvet* or *passe-flower*. The hanging flower-clusters (referred to as *cats-tails* cf. the French *queue de rerard* and German *fuchs-schwarz*) suggested to some the warp-end of a weaver's web (known as a thrum), and thus it was called *thrum-wort*. Other names are *Joseph's coat* (of many colours, but stained red with blood to make his father think he was dead, see Genesis 37) and *Gabon spinach* in Africa.

Wordsworth devotes two very late poems to this plant, characteristically displaying more ingenuity than art. In the first he treats some classical associations with its name thus:

> *You call it 'love-lies-bleeding' – so you may,*
> *Though the red flower, not prostrate, only droops,*
> *As we have seen it here from day to day,*
> *From month to month, life passing not away:*
> *A flower how rich in sadness! Even thus stoops . . .*
> *Thus leans, with hanging brow and body bent*
> *Earthward in uncomplaining languishment,*
> *The Dying Gladiator, So, sad flower . . .*
> *So drooped Adonis, bathed in sanguine dew*
> *Of his death-wound, when he from innocent air*
> *The gentlest breath of resignation drew.*

The companion poem concentrates more upon the theme of its undying nature and upon the folk-elements that go into its christening.

There are several other amaranths grown for grain by the Indians of North and South America, some of which also serve as vegetables both in America and Asia. Amongst these are included the very important A. leucocarpus; A. diacanthus; *redroot amaranth* (A. retroflexus); *slim amaranth* or *bloody princess-feather* (A. hybridus); *prostrate amaranth* (A. blitoides); *tumbleweed amaranth* (A. graecicans); *Torrey amaranth* (A.

[1] The latter name properly belongs to Chenopodium quinoa (*see under* Spinach), which is used for the same purpose.

torreyi); and *quelite* (A. palmeri). The leaves of some of these are also used in salads and as animal forage. In Australia the aborigines eat the seeds of the *large-flowered amaranth* (A. grandiflorus). *African spinach* (A. gangeticus) is another important species common to the American, African and Asian tropics, cultivated both for its seeds and as a vegetable. As a vegetable the plants in their seedling stage are cooked like spinach, especially in such countries as India, Malaya, China and Japan, and go by such alternative names as *bush* or *spin greens* and *green leaf*. The flowers may also be eaten. In Asia the *kiery* (A. frumentaceus) is the species grown principally for its seeds, the flour from which serves as the principal nourishment of several peoples in India, especially in the Himalayan region.

Several more species of Asian origin serve principally as pot-herbs or greens. Among these are the *melancholy amaranth* (A. melancholicus) and A. mangostanus, both of which are especially popular in Japan, where they serve as a hot-weather vegetable, the leaves being boiled or salted. Others include the Javan A. polystachys, the Indian A. oleraceus and the *green amaranth* or *pig-weed* (A. viridis) common to all the tropics. *Florimer* or *flower-gentle* (A. tricolor) is also much used in India and Ceylon. Its principal name is a corruption of *floramor*, another play on the idea of 'love-flower' (*amor-anthus*).

Three species are used on islands of the Indian Ocean such as Reunion and Mauritius: *African spinach* (A. gangeticus), *sad amaranth* (A. tristis) and the *princess-feather* or *Baldar herb* (A. hypochondriacus). These go collectively under the name *brède de Malabar* in the local creole dialect, *brède* being a word of Portugese origin signifying the leaves of amaranths and other spinach-substitutes. In Europe *wild blite* (A. blitum) was once used similarly. The tender leaves and stems of the *spiny* or *thorny amaranth* (A. spinosus) are gathered by peasants in India and Ceylon under the name *kantanatya*; they are used as a pot-herb in Madagascar, where they are known as *anampatsa*. In tropical America they figure in such native dishes as *calulu* and are known in the West Indies as *prickly calulu*. But, as there are better-tasting plants which also go by the name of the dish in which they figure, these are not gathered unless the others are not available. In Brazil the *yellow amaranth* (A. flavus) is used in much the same way and goes by a corruption of the name, *caruru*.

Several relatives of the amaranth are also to be found in the tropics and are put to use by various peoples. The *prickly chaff-flower* or *bitter bloodleaf* (Achyranthes aspera) grows in Africa, Asia and Australia. In Chad the plant is reduced to ash and used for salt, while the branches serve as toothbrushes in southern Arabia. The Asian poor eat the leaves as a vegetable. Those of the closely related Aerva lanata are eaten like spinach in East Africa, and the young shoots are eaten in Indian curries. The whole plant is eaten during times of famine. The leafy shoots of Deeringia amaranthoides are eaten with rice in Indonesia; those of the *feather cockscomb* (Celosia argentea) by others among the Asian poor, who also gather the leaves of Allmania nodiflora and the young stems and leaves of the *tree-flowered joyweed* (Alternanthera triandra). Another species of the latter (A. sessilis) is eaten in Malaysia, Indonesia and the Congo in soups, with rice or like spinach. The *Tom Thumb joyweed* (A. amoena) and *alligator joyweed* (A. philoxeroides) are similarly used in these, and other, parts of the tropics.

Angelica (genus Angelica)

It was probably Syria that this umbelliferous plant came from originally. Many species have been used since prehistoric times. The plant has been regarded highly for its medicinal and magic properties, especially in counteracting poison and plague and warding-off evil. Its name derives either from these properties, or from the sweet smell of its roots, formerly called 'the root of the Holy Ghost'. One species is still called *Holy Ghost* (A. sylvestris) while another is known as *archangel* (A. archangelica). With regard to the old beliefs concerning angelica one might look at the sixteenth century translation of Du Bartas by Sylvester.

> *Contagious aire ingendring pestilence*
> *Infects not those that in their mouths have ta'en*
> *Angelica, that happy counterbane*
> *Sent down from heav'n by some celestial scout*
> *As well the name and nature both avowt.*

The plant is still widely used for a variety of purposes, especially in the making of sweetmeats and alcoholic drinks. The midribs of the leaves may be boiled as a kind of celery substitute. Laplanders eat the stalks and Icelanders both the roots and the stems, raw with butter. In Finland the stems are cooked and an infusion made from the leaves, while in Norway a bread is made from its roots. The Ainus eat the stalks of the *Japanese edible angelica* (A. edulis), and in America the leaves of the *hairy angelica* (A. villosa) are recommended as a cure for smoking. *Great angelica* (Archangelica or Coelopleurum gmelini) is to be found in North America and northeastern Asia; this is used like celery by the Eskimos.

Wild or *wood longwort* (A. sylvestris) goes under a variety of intriguing names in this country. *Jeelico* is merely a corruption of its name, while it shares the name *kecks* in common with many other umbelliferous plants.

Rheumatism plant points out one of its medicinal uses; *ground ash* may be a form of *ground ache* (parsley); *Jack-jump-about* is said to refer to the action of its ripe seed-pods blown about in the wind, but why it should be called *whey-bucket* I don't know. Children have found many uses for it and thus given it some of its names. *Water-squirt* is a name it shares with wild chervil, from the practice of blowing water through its hollow stems – a form of primitive water-pistol. *Ait-skeiters* is a Scottish name meaning oat-shooters; here the stalks, sometimes simply called *skytes*, are used like a pea-shooter. They may also be formed into a kind of pipe, and from this they gain the name of *trumpet kecks*.

The name *kecks* or *kex*, may suggest yet another use for the stalks if it should stem from the Gothic word *keek* or *kike*, to peep, peer or spy. We find it in the Middle English *kiken*, which has become the dialect word *keek* of today, and is also present in the Dutch *kijken*. Nor need this refer merely to a childish game of peering at each other through the stem. In early medieval iconography it is common to find St Gregory portrayed with the scribe Petrus peering at him through a hole in a curtain in which has been inserted some kind of hollowed stem. It may well be that this and similar umbelliferous plants contributed in this way to the primitive beginnings of espionage.

Lastly there is the related *wild angelica* or *elder* (Aegopodium podagraria) whose leaves may be boiled as a vegetable and used in soups and salads. The likeness of its flowers to those of the elder have given it such alternative names as *bishop's, dwarf* or *ground elder* and *dog eller*. In Northern Ireland it is also known as *farmer's* or *garden plague*, for obvious reasons. Gerarde names it after himself, commenting 'Herb Gerard with the rootes stamped, and laid upon members that are troubled or vexed with the gout, swageth the paine, and taketh away the swelling and inflamation thereof, which occasioned the Germains to give it the name *Podagraria* bicause of his vertues in curing the gout'. One of the French popular names for the plant also happens to be *herbe aux goutteux*.

Artichokes

1. Cardoon and globe artichoke (*genus Cynara*)

The original artichoke is a member of the thistle family which comes originally from the Mediterranean region where it is grown widely. In ancient times the heart as well as the leaf-bases were eaten. The species of oldest use was the *cardoon* (C. cardunculus), *prickly artichoke* or *little thistle*, whose name derives from the French *chardon* (a thistle). In

Australia this goes under the name of *weed artichoke* owing to its having run wild and become a considerable nuisance. Much the same thing has happened on the South American pampas, in the Californian region, and elsewhere. Nowadays we also eat the *French* or *globe artichoke* (C. scolymus).

The Greeks mentioned the artichoke often but it remained for a long time a luxury vegetable. In an Italian novel of the fifteenth century it is mentioned as an aphrodisiac 'dear to Venus'. Perhaps it was this fed the great sexual appetite of Catherine of Medici, a black magician and instigator of the St Bartholomew's Eve Massacre, who was very fond of it; so too was England's much-married Henry VIII. During medieval times it was grown mostly in Spain and Italy, where they may have learnt of its use from the Arabs. It is rich in iron and iodine and may be boiled, baked, fried or stuffed. The young stems and leaves are occasionally used as a pot-herb, blanched and dried.

The name winds its devious way from an Arabic beginning as *harshaf* or *kharchiof*, whence it travels eastwards to become *kinghir* in Persian, *kunjir* in Urdu. In the Algerian lingua-franca it is *carchouf*; *al-kharshofa* in Spanish Arabic; *alcachofa* in Spanish; *alcachofra* in Portugese; *alcarcil* in Andalusia; *iscarzoffa* in Sardinia. These names may have had some influence on European versions of the name which seem closer to the original Arabic: *carciofo* (Italian); *carcioffa* (Naples); *carciofano* (Roman); *carxofa* (Catalan); *carchoflo* (Languedoc). North Italian forms shift the initial 'c' into the middle of the word so that we get such forms as *arcicioffo*, *arcicciocco*, which lead naturally into the early English *archychock*. Somehow the 'c' now becomes a 't' as in *articioco* (Venice); *artichioca* (Genoa); *articactus* (Late Latin, looking back to the Greek name, κακτος), *artichaou* (Provençal); *artijaou* (Limousin); *artichaut* (French). The English *artichoke* (cf. the German *artischoke*) again bears much in common with the northern maritime Italian forms and ignores the French, with its hints of cabbage (*chou*) altogether.

2. Jerusalem artichoke and sunflower (*Helianthus tuberosus*)
The *Jerusalem*, or *root*, *artichoke* is really the tuber of a sunflower which, like all its genus, comes originally from North America. The 'Jerusalem' part of the name is really a corruption of the Italian *girasole* (turning to the sun). The root is like a knobbly potato with a pungent taste which the original American settlers were constantly comparing to that of the artichoke. It can therefore be seen that the vegetable goes under a gigantic misnomer, really having very little to do with an artichoke and absolutely nothing with Jerusalem. The suggested name of *sun-root* would really be much better. It was introduced to France in the seventeenth century by the French settlers in Canada who, because the climate was found inimical to

European crops, were forced to learn from the Huron Indians the use of edible roots, various of which came to be known as earth-nuts, pears or apples (*noix, poires, pommes de terre*).

The rest of Europe seems to have taken up this artichoke at about the same time. In England, where it was introduced in 1616, it was first used as a sweetmeat, but an over-indulgence in the making of these by the pastry cooks lead to its swiftly losing popularity. It is the sort of vegetable that needs to be used sparingly. It contain inulin, which makes it less easy to digest than the potato, before which its European popularity declined. Unlike many other vegetables, its flavour is said to improve with the frost. In England it used to figure in a soup called Palestine, a quibble on the name Jerusalem. It became a kind of Lenten fare in Holland, where there are many diverse ways of cooking it. Some say it is delicious fried. The leaves, also, have been used as a spinach substitute.

The cultivation of the artichoke has now spread into Asia and Africa. In the early twentieth century another species, the *ten-petalled sunflower* (H. decapetalus), was introduced to France. The root looked the same but had a slightly different taste that was compared to salsify. Although it had no great culinary success it was recommended as an animal fodder. *Canada potato* was an old name for the root artichoke after its introduction from that country; it was also known as *Canadian apple* or *truffle*, or simply as *Canada*. Various other members of the same family are still used by North American Indians. These include the tubers of the *oblong-leaved sunflower* (H. dronicoides) in the central U.S., and those of *Maximilian's sunflower* (H. maximiliani), eaten by the Sioux and other prairie tribes. The fruits of the *giant sunflower* (H. giganteus) are added to bread-flour by the Choctaw Indians. The seeds of the *common sunflower* (H. annuus) were once an important food for the Indians of the western U.S. who have been using them since about two to three thousand years ago. Flour was ground from the seed kernels, and the receptacles (the bulging seedcases at the top of the stem) were eaten like globe artichoke, as they still are in some parts. The Tartars, for instance, now eat them in this way.

The sunflower was introduced into Europe from the New Mexico area by the Spanish in 1510. Peter the Great took it back to Russia in the early eighteenth century, and it was there that the practice of chewing the seed, raw, roasted or salted arose. It was during this period also that the seeds were first used for making oil. Nowadays they are also fed to livestock, especially poultry and other birds. The plant, which often attains such a height that it is capable of hiding a horseman riding through a field of them, is now grown for ornament and cultivated as a crop all over the world, but especially in Russia and its satelite countries, South Africa, Argentina and Uruguay, and to a certain extent in the U.S. Kansas is known as the sunflower state because so many are grown there. Anyone who has travelled in the south of France will have noticed the fields of sunflowers there which so excited Van Gogh with their colours during his discovery of the Midi in the summer of 1888. His series of paintings of the flower date from August of that year, and were originally intended as decorations for his room in Arles.

3. Chinese artichoke (*Stachys sieboldii*)

The *Chinese*, or *Japanese*, *artichoke*, which occasionally goes under its Japanese name *chorogi*, is yet another artichoke-tasting root. It has been used since times immemorial in China and was first brought to France in 1882. Cultivated by Pailleux and Bois, who first put it on sale in 1887, it was immediately taken up with great success, not only in France but all over Europe, Britain included. It is related to many European plants of the woundwort family, especially *marsh woundwort* or *betony* (S. palustris). The root of this was once used by poor people, especially in times of famine, and there were attempts to cultivate it.

Several of this plant's alternative names testify to the use of the roots either by cattle or humans. It shares the name *runch* with members of the mustard family, and its Scottish name of *swine's maskert* contains a form of *mosscort* (moss = marsh + *cort*, a Somerset corruption of carrot). In Ireland it was known as *sheep's brisken*, formed from the Gaelic *briosclan*, *brisgean*, names of edible roots eaten by the poor. *Swine's arnit* or *arnut* alludes to yet another edible root (*see under* Pignut). Other names include the Irish *rough weed*, the Scottish *opopanewort*, *hedge nettle* in the U.S. and *dead-nettle* in Yorkshire, and *clown's*, *husbandman's* or *ploughman's all-heal* or *woundwort*, alluding to its curative properties. *Wood betony* (Pedicularis lanata) belongs to the speedwell family. Its roots have been commended as a food in some parts of northern Canada. The alternative name of *louse-wort* springs from a common prejudice that it gives fleas to any animal that eats it; on the contrary, however, it probably kills the fleas.

Sweet betony (Stachys officinalis or betonica, formerly Betonica officinalis) looks like the red dead-nettle (*see under* Nettles) of which it is a relative; the name *Harry nettle*, and those of the marsh woundwort, underline this. Its leaves have a bitter faintly aromatic taste and, owing to their great reputation in the Middle Ages, summed up in the name *all-heal*, were much used as a pot-herb and in salads, if the alternative name *wood salad* is anything to go by. The Italian faith in the plant was such that they had a proverb. 'Sell your coat and buy betony'; again, wishing to speak highly of a person, they would say 'He has more virtues than betony'. Antonius Musa, physician to the Roman emperor Augustus, is said to have devoted to a whole treatise to the plant, specifying it against forty-eight diseases. It was supposed to be good for the digestion; eaten before drinking it hindered drunkenness, and taken after it cleared the head. A powder made from the plant refreshed those overwearied with labour; mixed with honey it was taken for coughs and colds, or taken in wine it was given for toothache and the bites of venomous beasts. Betony was also worn round the neck as a protection against the power of evil spirits and was planted in churchyards for this purpose. The flowers are supposed to symbolise surprise. An alternative name for the plant is *St Bride's comb*. The name betony was given, according to Pliny, in reference to the Vettones, a Spanish tribe credited with its discovery.

Asparagus (*genus Asparagus*)

Asparagus is one of those curious vegetables, like spinach, which have innumerable possible substitutes, few of which are now used, and which itself has lost much of its former popularity. It may have been used by the Ancient Egyptians; something like it is depicted on murals dating from the Memphis dynasties of the third millenium B.C. To the Greeks it may have been medicinal rather than a plant for eating, although their word for it, practically the same as ours, seems also to have been applied to such diverse vegetables as cabbage, lettuce, mallows and beets, according to Galen. It was, however, eaten by the Romans, who extended the meaning of the word to such substitutes as butcher's broom (see below).

After the fall of the Roman empire the cultivated variety seems only to have survived in Syria, Egypt and Spain; its cultivation was consequently taken up by the Arabs who conquered these countries, and the plant re-entered Europe at the end of the Middle Ages. There is mention of it in France in 1469, and in England in 1538. During the nineteenth century it was particularly widely grown and had a considerable aristocratic following. Two tinned varieties are now available – white, which may be eaten uncooked, and green, which needs cooking.

Asparagus is really the young shoot of a tuberous root, picked before it becomes too bitter in the case of the wild species, which may be used as well as the cultivated variety. The Romans used both kinds, and wild asparagus is still sometimes used in country districts of Europe. The shoot, ideally, should be cooked as soon after gathering as possible. Suetonius, speaking of a precipitate action of Caesar Augustus, says it was performed quicker than the cooking of asparagus, '*citius quam asparagi coquantur*'. A Greek myth describes asparagus as the fruit of a ram's horn thrust into the ground; in later times many jokes centred upon the connection between asparagus and the cuckold, popularly supposed to wear horns.

The plant is a member of the lily family, and closely allied to the lily-of-

34 / Asparagus and substitutes

the-valley; there are about 120 species. The name appears under various guises in England. In the sixteenth to seventeenth centuries it was *sperage*, deriving from the French *esperage*; the contraction of the Latin name to *sparagus* led to the version *sparrow-grass* commonly used in the seventeenth to eighteenth centuries. The cultivated variety is A. officinalis; there is a subvariety, *seaside asparagus* (A. officinalis maritima), which has been grown since time immemorial on the shores of the Atlantic and the Mediterranean. Cultivated asparagus and *smilax asparagus* (A. asparagoides) have both been taken to Australia whence they have now escaped from cultivation and run wild.

Southern Europe seems to provide a particularly rich number of wild species, many of which are still eaten. *Spiky asparagus* (A. horridus) is distinguished by its long pointed shoots. There are also the *South European asparagus* (A. acutifolius) and *slender leaved asparagus* (A. tenuifolius). *White asparagus* (A. allus) grows on the high plateaux of North Africa and is sold in Algerian markets as *wild asparagus*; in Madeira it is known as *garden-hedge asparagus*. Several in the family are inedible, such as *bitter asparagus* (A. scaber) and *wild African asparagus* (A. africanus), a native of South Africa, whose leaves are added to ointments used by native women to stimulate the growth of hair. With others it is only the roots that are eaten, as with the Far Eastern *shiny asparagus* (A. lucidus). These are sometimes cooked with sugar in Japan, while the large fleshy roots of A. sarmentosus are candied in China, although boiled as a vegetable in Ceylon. The large roots of the *Ethiopian asparagus* (A. abyssinicus) are fried and eaten in Ethiopia and along the Red Sea shore, in West Africa those of A. pauli-guilielmi are boiled. The latter improves with cultivation and might well be made a commercial crop in the tropical African area. Poor Asians eat both root and shoot of the *sicklethorn asparagus* (A. falcatus); the roots also appear in cooling medicines and native remedies for venereal diseases. Only the fruits of the *rooted asparagus* (A. racemosus) are eaten by the Golo people of the Sudan.

ASPARAGUS SUBSTITUTES (*a*) LILY FAMILY

1. Dragon plant (genus *Dracaena*)
These are of tropical origin and were once more widespread; before the ice-ages they were to be found growing in Europe. That principally used is the *asparagus bush* (D. manii) of tropical Africa, whose young shoots are consumed both by natives and whites. The *tikouka* (D. or Cordyline australis) is alternatively known as *palm lily* and *New Zealand cabbage palm*; the terminal bud is eaten in the same way as palm cabbage in New Zealand. Other members of the family are used for the making of a dye

(known as dragon's blood). The strangest, however, was the *Canary dragon tree* (D. draco), destroyed by a storm in the Canaries in 1868. This was estimated to be about 6000 years old and was an object of worship to the original inhabitants of the island who had built a sanctuary inside its enormous trunk, which was at that time 15 feet wide. Others of great antiquity may still be found on the islands.

2. Scrambling lily *(Geitonoplesium cymosum)*
The young shoots of this plant, endemic to Australia and also known as *Australian shepherd's joy*, make an acceptable substitute on that continent.

3. Star of Bethlehem *(genus Ornithogalum)*
For many centuries the young flower-clusters of the *Pyrennean Star of Bethlehem* (O. pyrenaicum) have been eaten in the manner of asparagus. Some centuries ago they were sold in country markets in western England under the name *Bath, French, Prussian* or *wild asparagus*. The bulb of the *Narbonne Star of Bethlehem* (O. narbonensis) is edible, as is that of the *common Star of Bethlehem* (O. umbellatum), which is eaten in Palestine but referred to slightingly as *dog's onion* in England. The flowers of the latter have been baked in bread; these open with the sunrise but close at evening. Popular names for the plant refer confusedly to this: *eleven o'clock lady* (cf. the French *dame d'onze heures*); *Betty-go-to-bed-at-noon*; *wake-at-noon*; *sleepy Dick*; *peep o'day*. The appearance of the flowers are referred to in the name *summer snowflake*.

4. Butcher's broom *(Ruscus aculeatus)*
The plant is of Mediterranean origin but is now to be found growing in many parts of northern Europe. It resembles asparagus both in the use to which its young shoots are put and in that the seed has been used for a coffee substitute. *Jew's myrtle*, one of it's popular names, was given in the belief that Jesus' crown of thorns was made from this plant. Other names include references to its prickly nature: *prickly box* or *box holly*; *knee holly*; *shepherd's* or *wild myrtle*; *pettigrue*.

(*b*) PARASITES

1. Broomrape *(genus Orobanche)*
Many species of this plant, generally parasitic on the roots of various crops and much hated by farmers, grow in Europe and Asia. It is often to be found growing from the roots of broom, and its scaly root is popularly supposed to resemble a turnip (called a *rape* in some parts). This explains its English name. The stems of O. cruenta are most frequently mentioned

as being eaten like asparagus, but those of the commonest European species, O. major, although they are slightly astringent, have also been recommended. The only pleasant name ever borne by this plant seems to have been in a restricted locality of Yorkshire, where it was known as *Our Lady of New Chapel's flower*. Most others signalise the farmer's detestation: *strangle tare* (which is the meaning of *orobanche*, the Latin name of Greek derivation), *choke fitch, kill herb, devil's* or *hell root*. A popular name in some parts of Italy designates it as the *wolf plant*; as such it was once proscribed in Tuscany by public edict.

In North America the underground stems of *Californian broomrape* (O. californica) are eaten by the Indians of California and Nevada, and the entire plant of the *cancer root* (O. fasciculata) by the Indians of Nevada and Utah. In the southwestern U.S. the yellow roots of *Louisiana broomrape* (O. ludoviciana) are roasted on coals by various Indian tribes; the Californian Indians also use the roots of *tuberous broomrape* (O. tuberosa). The roots of North African relatives (genus Phelypaea) are eaten in times of hardship by the Tuareg.

2. Indian pipe *(genus Monotropa)*
This plant grows from decaying vegetable matter, and members of the family can be found in the forests of North America, Europe and the East. Among those whose stems may be eaten like asparagus are *yellow bird's nest* (M. hypopystis), alternatively known as *false beech drops* and *pine sap*, and the *one-flowered wax plant* (M. uniflora). It has no green about it; a white flower surmounts an upright white, cold, clammy stem, peering over the mould. This curious appearance has given the plant many of its popular names: *ghost flower, corpse plant, American ice plant, fit plant, ovaova*.

(*c*) OTHERS

1. Bryony
Red-berried or *white bryony* (Bryonia alba or dioica) is an evil-smelling member of the cucumber family whose poisonous berries were once used medicinally. The shoots, however, may be eaten like asparagus. The berries are variously known as *snake, fellon* or *murrain berries*; and the equally poisonous root is castigated as *mandrake* and *devil's turnip* (cf. the French *navet du diable*). General names for the plant include *wood, wild, white* or *Isle of Wight vine, grapewort, hedge grape* and *white hop*. *Black bryony* (Tamus communis) naturally shares several names with the above, and is known in addition as *adder's poison, snake's food, serpent's meat, Lady's seal, bead-bind, oxberry, rowberry, rueberry*. The berries are equally

poisonous as may be imagined, but in classical times the shoots were actually preferred to the true asparagus, and continued to be eaten in some places (as in Tuscany, under the name *tamoro*) for many centuries.

2. Hops *(Humulus lupulus)*

The hop is a member of the same family as nettles (one name for nettles used to be *hop-tops*). The young shoots, when about 3–4 inches long, may be eaten like asparagus. Nowadays hops are used almost exclusively in the making of beer, for which they have been cultivated for over a thousand years. There are records of hop-gardens in eighth- and ninth-century England and Germany. The plant grows wild in northern Europe and Asia, and is cultivated in these continents as well as in North America. The male and female flowers are produced on separate plants, the male being known as the *seeder*. Others names for hops include *bines, Jack hop, willow wolf*, and *poor man's beer*.

The hop bloom is the birthday flower for 7 April, symbolising injustice, pride and passion. Perhaps one of these associations is borne out by the proverb 'when the hop grows high it must have a pole'. It is certainly a chancy thing to grow: 'hops make or break'. Furthermore

> *Till St James' Day [25 July] be come and gone*
> *You may have hops and you may have none.*

If you are successful then there's nothing that will arrest a good hop-harvest. 'As fast (or thick) as hops' was once a widespread popular simile. One of the indications of this coming good fortune, apparently, is 'plenty of ladybirds, plenty of hops'. There is probably no truth in the rhyme which came to prominence in Tudor times that

> *Turkeys, carps, hoppes, piccarell and beer*
> *Came into England all in one year.*

In some versions of this one finds 'heresy' among the number, in reference to Henry VIII's defiance of Rome and the entry of Reformation ideas into England. This is patently false since, as we have seen, hops have been in England much longer than that:

> *Malt does more than Milton can*
> *To justify God's ways to man!*

3. Elder *(genus Sambucus)*

The young shoots of the elder may be eaten like asparagus. They are thus used in North America, where the most widespread species is the *sweet* or *autumn-flowering elder* (S. canadensis). The fruits of these are much esteemed, being bigger and sweeter than the commoner European species S. nigra, and are put to a variety of uses, amongst which they act as an adulterant to port wine. In Europe a large body of folklore has grown up round the tree, whose pith may also be added to soups. It is the birthday plant for 27 March, standing for zeal and compassion. In the druidic alphabet, however, it represents the thirteenth consonant (R) and corresponds to the barren Celtic month falling 26 Nov.–22 Dec. Part of Jesus' cross was supposed to be formed from its wood, and according to some traditions it was from this that Judas hanged himself. Langland calls it the Judas tree in *Piers Plowman*. And the *Jew's ear fungus* (Exidia auriculajudae) is so called because it is generally to be found growing on the elder.

In Ireland witches were supposed to ride through the air on eldersticks, while in England it was believed that to burn logs of the wood was an invitation to the devil to enter the house, for which reason it was known as *devil's wood*. This probably stems from Germanic folklore. Germans would raise their hats in the tree's presence and in Denmark they used to say it was under the protection of the 'eldermother', whose leave had to be asked before its flowers were gathered. The wood must not be used for household furniture, and the baby in a cradle made from its wood would be strangled by the goddess.

Avens (*genus Geum*)

The origin of the name avens is obscure. Versions such as *avance* and *avantia* appear in early French and Latin documents; there is an Irish name *minarta* which probably has no connection. The plants belong to the rose family and were once very popular in medieval cookery. That most often used was the *common, city, wood* or *yellow avens* (G. urbanum) whose roots are agreeably aromatic and known as *clove roots* (cf. the German *nägleinwurz*). The plant's alternative name is *herb Bennet*, deriving from the Latin *herba benedicta* (blessed herb, cf. the French *herbe bénite* and *benoîte*). This comes about because of the plant's magical reputation. 'Where the root is in the house', comments one writer, 'the devil can do nothing, and flies from it: wherefore it is blessed above all other herbs'. A man bearing the root about was also safe from venomous beasts.

The avens flower is a yellow five-pointed star which has gained it such names as *star of the earth, north* or *east*. This distinguishes it from the purple bell of the *drooping, nodding* or *water avens* (G. rivale); from this the latter flower has gained the names *purple avens, Egyptian, soldiers' buttons, fairies' bath*, *granny's cap* or *old woman's bonnet*. The plant is to be found growing by water, often at high altitudes, all over the northern hemisphere. The roots were once used to flavour ale and preserve it from turning sour.

Avocado pear   (*genus Persea*)

The green or purple fruits of this evergreen member of the laurel family partly replaces meat in the diets of some Central Americans. They have a nutty flavour and a high oil, protein, iron and vitamin content; when ripe they have a buttery consistency which has given them such names as *poor man's*, *soldier's* and *midshipman's butter*. The name derives from the Mexican Nahuatl Indian *ahuacatl*, which became *ahuacate* or *aguacate* when it first entered Spanish, but seems later to have been confused with *bocado* (titbit) and corrupted into *avocado*. An early English name, corrupting the original Spanish, was *alligator pear*. There are three main species, of which the chief (P. americana) grows from Mexico into the Andean states and bears fruits weighing several pounds. P. schiediana grows in Guatemala and P. drymifolia, whose fruits are not much bigger than plums, grows in Mexico. These were widely cultivated before Columbus; shortly before the Conquest the Incas are said to have carried them from Ecuador into Peru. The Spaniards do not mention them until the sixteenth century. Attempts have been made to grow avocados in Australia, South Africa, the Pacific, India and Malaysia, and in 1900 cultivation was taken up in the U.S.

Bamboos and canes

The name bamboo derives from the Malay. We generally think of it in terms of swishy jointed canes, but in the East, where the plants are abundant and have a certain value as food as well, a great deal of lore has grown up about it. Bamboo is symbolic of long and abundant life, straightforwardness, open-mindedness, and yielding but enduring strength; of refined culture and fastidiousness; gracefulness, tranquility, gentleness, modesty and peace. It bears off adversity and protects against defilement. Associated with the tiger in Chinese art, and the sparrow in Japanese, it denotes safety. The plant appears at wedding feasts in Japan as the emblem of fidelity, but the cut has to be hidden by leaves since the sight of it brings severed love to mind.

In China a bamboo staff is carried when mourning for one's father. The culms are thrown into the fire, where they explode and frighten off evil spirits (fire-crackers in the New Year's Day celebrations perform the same function). A sprig of it is used for divination, but to the Japanese it denotes madness. Bamboo also appears amongst their New Year's Day decorations, a reminder of the three friends of man, the bamboo, plum and pine, which flourish in the cold season despite adverse conditions, a constant inspiration to man that he should not falter in the face of misfortune. The three friends also represent the founders of the three principal religious systems in China: Buddha, Confucius and Lao Tzu. Bamboo and plum by themselves are emblematic of two-fold happiness, and therefore husband and wife. In Japan the bamboo and cane stand for longevity and happiness, and a bamboo grove both for the everyday world and a family of princes.

In South-East Asia particularly, the bamboo has played a great part in the musical awakening of the people. It has been suggested that a system of pivoted bamboo pipes used in irrigating the fields and springing back into position with a pronounced clunk was what originated both the love of percussive sounds and the idea for several instruments: the *angklung* rattle in which two or three notched vertical tubes tuned in octave slide to and fro in the grooves of a rectangular frame and strike its rim; tubular xylophones and vertical drums or the tube-zither made from a piece of bamboo with strings slit from its own rind and held away by tiny bridges. Then there are pan-pipes and primitive mouth-organs (*not* harmonicas), and various pipes and flutes. As a Chinese ode of the ninth century B.C. expresses it,

> *Heaven enlightens the people*
> *When the bamboo flute responds to the porcelain whistle.*

In the Far East a collection of bamboo instruments represents a gathering of people; and the sounds they make are suggestive of rushing water.

The seeds of many bamboos are used for food, but it is principally the young asparagus-like shoots or buds which are eaten. Generally they are gathered when about a foot long, and are often earthed over to keep them tender and fit for eating. Although there is an all-year-round supply, those collected in the autumn are considered the best. The Japanese *take-noto* (young shoots) may be eaten whole or cut in half, and pickled in vinegar. These were once exported to China in a dried form which expanded to normal size when placed in boiling water. The shoots are now exported tinned to many countries and are fairly easy to obtain.

The following are the species most often eaten:
1. *Genus Bambusa*
    (B. arundinacea) *Spiny bamboo*. India.
    (B. cornuta) *Horned bamboo*. Java.
    (B. multiplex) *Hedge bamboo*. Tropical Asia.
    (B. oldhami) *Oldham's bamboo*. Tropical Asia; of Chinese origin.
    (B. spinosa) *Thorny bamboo*. East Indian area.
    (B. tulda) *Tulda*. East Indian area.
    (B. vulgaris) *Common bamboo*. Tropical Asia.
2. *Genus Dendrocalamus*
    (D. hamiltonii var. edulis) Himalayas to Burma.
    (D. latifolius) China and Japan.
3. *Genus Giganthochloa*
    (G. ater) Tropical Asia.
    (G. verticillata) *Giant grass*. Java. The buds have a particularly delicate flavour and are often kept in vinegar before cooking.
4. *Genus Phyllostachys*
    (P. bambusoides) *Japanese timber bamboo*. China and Japan.
    (P. edulis) *Moso*. China and Japan. The buds are to be found sold in the markets.
    (P. riga) *Black bamboo* or *whangee cane*. China and Japan.

Many more than this are used, of course, among the poor and in the remoter areas. Other cane-like members of the grass family, to which the bamboos belong, have similar uses. The young shoots of the American *large* or *southern cane* (Arundinaria gigantea) are cooked like asparagus, and the seeds were used both by the Indians and the early settlers as a wheat substitute. The *woody grass* (Melocalamus compactifolius) of tropical Asia bears large mealy seeds which are edible and rather like chestnuts. The grass-joints (culms) of these are used as shoes by the peasants. Some of the wild sugar-canes (genus Saccharum) also provide edible shoots, like those of the *wild-sweetcane* (S. spontaneum) in Indonesia, a plant otherwise of little use and low in sucrose.

Bananas (*genus Musa*)

Bananas are the fruit of what is, in effect, a gigantic herb springing from an underground root. Different varieties fruit at different times and one is thus able to obtain them all the year round. The banana is extremely nutritious and delicious, whether one eats it cooked or raw. The use of the fruit goes back a long way in South-East Asia, its place of origin, from whence it spread to every tropical region in the world. But the first definite mention of it is in the Buddhist Pali canon (dating from the fifth century B.C. in its oral form); in China it is not mentioned before A.D. 200. This, however, may be no more than a result of the disregard for their southern possessions by the ruling classes in the north, where the banana does not grow. The first European to come across it was Alexander the Great during his invasion of Northern India (327 B.C.), after which it finds mention in Greek and Roman authors. But it was not grown in the Mediterranean area until the Arabs brought it about A.D. 650. It may also have been they who introduced the banana into some parts of Africa, but no one seems very sure. A legend among the peoples of Uganda relates that it was brought by a warrior from the north some time before A.D. 1000, although the first real record of its growing in East Africa mentions it in the region of Mombassa about 1300. A possible theory is that Indonesians brought it during their period of influence in East Africa and dominance in Madagascar. In the former area bananas are generally of far more importance as a food than in the countries of their origin, and are more intensively cultivated.

The Portuguese took the banana to the Canary Isles some time after 1402, and from there it was taken to America in 1516. It has flourished particularly in Brazil, which is the principal exporter to such South American countries as Uruguay and Argentina, where the climate is not right for native banana growing. It also flourishes in the so-called banana republics of Central America where the industry grew up very rapidly between 1870–80, expanding first with the growth of railways as a fast means of transport, and then with the invention of refrigeration, which helped preserve them even longer. N. W. Simmonds, whose *Bananas* is probably the most inclusive work on this subject, has stated that 'the history of the Central American trades is very largely the history of the United Fruit Company'. This organisation, whether rightly or wrongly, has earned a sinister reputation of having a large part in the governing of these states.

Bananas grow in downward hanging clusters, several to a stem. Their appearance has given the technical name 'a hand' to the bunch, and 'fingers' to the individual fruit. Once the fruits are cut from the stem no more will grow; the Chinese usually cut off the stem as well and use it for feeding their pigs. A Malay proverb, 'the banana does not bear twice' (i.e. 'Once bitten, twice shy'), refers to this non-productivity. Another Malay proverb, the equivalent of 'I wasn't born yesterday', is 'I'm no banana-eating boy'.

Most edible varieties of banana have developed from two wild species, *M. acuminata* and *M. balbisiana*, whose fruits are small and full of seeds but occasionally eaten by aboriginal tribes in the jungles of South-East

Asia. One generic name for these is *tilai*, and Malay names refer to them as *jungle, torch* and *thrusting (sorong) bananas*. There are now upwards of a hundred edible varieties of all shapes and colours, dividing into two very broad groups, *Emusas* and *Australimusas*. The latter grow almost exclusively in the Pacific and are often referred to as *Fe'i* bananas, a name to be found in several Polynesian languages. All of these have to be cooked before they are really palatable since they are very starchy and hardly as sweet as plantains, the normal cooking banana. Their yellow flesh and reddish-violet juice discolours brightly the urine of their users.

Emusas are sweeter, on the whole, less starchy, and have a worldwide tropical distribution. The following are among the principal types:

Sucrier, which is of most importance as a food in New Guinea. More are grown there than in the rest of the world collectively. Because it is thin-skinned and bulbous the Thais refer to it as *egg-banana*, while in Malaya it is called *crocodile's fingers*, in Burma *sparrow* or *children's banana*, and in the French West Indies *bird's* or *bird of Paradise fig*.

Dwarf Cavendish (M. nana or cavendishii), also known as *Canary* or *dwarf Chinese banana*, was introduced from southern China to the Canaries, Europe and the Caribbean in the nineteenth century. It is the most widely grown of all varieties and adapts well to cool climates. In New South Wales (Australia) an *extra-dwarf Cavendish* (the name refers to the size of the plant) is grown, but it is of not much more than local importance.

Giant Cavendish, or *Chinese banana*, is much less generally grown, but is popular in Australia.

Gros-Michel, also known as *Jamaican banana* and in Malaya as *Amboyna banana*, is a very popular variety of the Cavendish group, much grown in Africa and South-East Asia. It is the principal type grown in the Americas. Another variety, known in Malaya as *green banana*, is sometimes sold in its stead. This retains a greenish shade even when ripe but is sweet flavoured; the fruit is easily detached from the bunch.

Red and *green-red* varieties, whose flesh is pink and delicious, are not grown on a commercial scale because they do not keep well and it is therefore impossible to export them. In the West Indies they go by such names as *red donkey's fig, copperbolt* and *mankiller*; a Malayan name is *king-prawn banana*.

Mysore, of hardy growth in poor conditions, is grown mainly in Asia. The Thais esteem it enough to name it *milk-of-heaven*.

Lady's finger or *apple banana*, so called because of its white flesh, is widely distributed but not a particularly important crop. It has a sour-sweet taste.

Silk banana (M. sapientum) is similar to the above in the width of its distribution, its very white flesh and sweet-sharp taste, which has made it one of the most popular dessert bananas in the tropics. In the French

West Indies it goes by such names as *plum, apple* and *pineapple fig*. Indians use the leaves of this as a remedy for snake bites.

Bluggoe, Whitehouse or *apple plantain*, also known in the West Indies as *frog plantain* and *coolie banana*, the fruits of which are almost square in shape, starchy and have to be cooked. It is a principal source of food to many peoples in the Pacific, the Philippines, South India, East Africa and the West Indies. A common dish in Grenada, for instance, is this banana and jack-fruit (*see under* Breadfruit).

Plantain (M. paradisiaca), also known as *Adam's fig* because its huge leaves were popularly supposed to be those Adam used to cover his nakedness in Eden, is an important source of food in South India, East and Central Africa and tropical America. The fruits must be eaten cooked because of their starchy nature and only marginal sweetness. In Africa the raw fruit is regarded as a good marching ration because, as they say 'it lies heavily on the stomach'; 'two and a bellyache' is a local proverb expressing the same sentiment. Plantains (the word is also used to cover any cooking banana) come in various colours. *Green plantain* is known in the West Indies as *May, maiden, common, creole* or *Congo plantain*; in addition there is the *pink plantain* and the bright red *wine plantain*, as well as the blackish-brown *black* or *snake plantain* and the spotted *tiger* or *flea plantain*. A very large variety, the *giant plantain*, is known in the French West Indies as the *hundred pound* or *fourteen paw plantain*.

Horn plantain (M. corniculata) is a separate cooking type, known in the West Indies as *cowhorn* or *horse plantain*, and to the Thais as *elephant's tusk*. The smaller *dwarf horn plantain* is known in the West Indies as *pig banana*.

Only a fraction of the names given to different species in different tongues are recorded above. But also when it comes to collective names for bananas there is enormous diversity, especially in Asia and the Pacific. This is indicative partly of how long the banana has been indigenous to a certain area and, in the Pacific especially, of how isolated certain peoples are from each other, despite their nearness 'as the crow flies', cut off as they may be by the sea or, in New Guinea at least, by high mountains and jungle.

Some of these names are descriptive, like the Chinese *tsiu, chiu,* etc. whose ideogram apparently combines three elements meaning 'herb-better-fire', perhaps referring to the fruit's need of warmth for ripening; then there is the delightful Burmese *nget-pyaw*, meaning 'the birds told', a reference to the old story that men first began to eat the fruit after they saw the birds doing so. The variety of names in the Pacific area is truly bewildering, though some have a strange beauty about them, such as the New Caledonian *dupijing* and *foexac*. There are, however, similarities between some, like the range that seem to have in common the root 'ut' or 'us': **usi**

(New Guinea); *uch*, *ut* (Micronesia); *vudi* (Fijian for cooking bananas); *futo* (New Caledonia); *futi*, *fusi*, *wuti*, *huti* (western Polynesia).

Turning to the Sanskrit and Sanskrit-derived names in India, one begins to discover more families of names, and possible evidence of how bananas were spread. The Sanskrit adjective *piçanga* (yellowish-brown) seems to underlie the Malayan and Indonesian *pisang*. *Pala* (a general name for fruit) was that by which the Greeks knew the banana; it is combined in *suphala* (Sanskrit), *bale* (Kannada), *vazhapazham*, *valapalam* (Malayalam). *Vazhei*, *vazha*, are the words for banana in the related Tamil. One combination has produced the present-day *kela* and its variants in the related North Indian group of languages (Hindi, Urdu, Punjabi, Marathi, Gujarati, etc.) and may also be present in the Thai *klue*. Another name, the Sanskrit *kadali*, reappears in such dissimilar and widely-separated areas as Nepal and Ceylon perhaps because of Buddhist influence. Lastly, the Sanskrit *moca* would appear to have become the Arabic *mouz*, *maouz*, *moz*, *moaz* and *mazw*, but there is an alternative theory that it may derive from the South Arabian town of Moka. Variant forms of the name, doubtless through Arab influence, appear in several African languages: *moz* (Somali); *maso* and perhaps *ndizi* (Swahili); *mazu* (Wadigo).

Native African names for the banana also seem to run in families, the most widespread being those based on the root 'konde': *ikundu*, *mikonde*, *maonde*, *nginda* (East Africa); *akondre* (Malagasy); *ikondo*, *ekon*, *digonde*, *liko*, (Gaboon); *ogede* (Yoruba). Another East African series is based on the root 'toke' (*kitoke*, *matoke*, *madoge*) while in Gaboon there is a series based on 'toto' (*itoto*, *ditotu*, *litoto*, *atora*). On the Guinea Coast we find the root elements of our own word, *banema*, *banama*, perhaps related to the Congolese *bonano* (a tree). That this is really where our word came from is perhaps reinforced by some Spanish Americans referring to the fruit as *guineo*.

The name *banana*, and its variants (the Portugese *bananeira*, Dutch *banaan*, etc.) is that by which the fruit is generally known in Europe now. Two other names generally used in South America are of obscure origin: *camburi*, a name also applied to a banana liqueur is one, *platano* the other; the latter, from which our word *plantain* is derived, is often used, as in Mexico for example, to cover any banana whether it is generally eaten cooked or raw. The Latin generic name, *musa* was coined by the eighteenth-century Swedish botanist Linnaeus, who did so much to systematise the science of botany. He drew it from the Arabic names already mentioned.

Lastly one must mention the name *fig* so often applied to various bananas in the West Indies, especially in those islands once held by the French. Behind this name also lies an Arabic origin, since the tree is mentioned in the Koran as 'The Tree of Paradise', and a later tradition grew up that Adam used banana leaves to cover himself upon discovering he was naked, as has already been mentioned. Since the leaves are called fig-leaves in the Bible the banana came to be known as *Adam's fig* (*fico d'Adamo* in Italian, *pomme* or *figue d'Adam* and *figue du Paradis* in French cf. the German *Adansapfel* and *Paradiesfeige*). It is interesting to note that a similar confusion of appelation exists here as well, for, whereas the French, German and Italian names cover any banana, the English is restricted to the plantain alone.

Of the edible bananas of the world, about half are eaten raw and ripe, and the other half cooked. Those cooked may be those which are only edible this way, immature fruits, or even those which are normally eaten raw; such sweet varieties are eaten fried in the West Indian dish 'rice creole', or with spaghetti ('spaghetti Carmen').

In many places the cooking banana is a staple food, especially in parts of East Africa where, beside frying, it may be boiled or steamed, baked or roasted. In Uganda the fruit is occasionally sliced up, dried in the sun, and the resulting 'chips' boiled into a porridge. A most nutritious beer is also made from the fruit in these regions. The Ugandan method is to trample them in a canoe, pour off the juice through a filter of grass and wash the remaining pulp, recombine it with the juice and sorghum flour in a clean canoe, and leave it covered with leaves to ferment for twelve hours; after this the mixture must be filtered once again and drunk quickly. In the Caribbean a whiskey, a wine, and other fermented drinks are occasionally made.

In some areas other parts of the banana plant are eaten. In South-East Asia the shoots and buds of wild bananas, and of the bluggoe variety, are eaten, although several changes of water are required to reduce their bitterness; in China the young flower-heads are pickled. The flowers also appear in Gaboon cooking, and these and the immature fruits are added to Malayan curries. The terminal buds of the tree are used in much the same way as the palm cabbage (*see under* Palms) in India, while in the Pacific the fleshy turnip-shaped root of M. sapientum var. oleracea yields a starchy food rather like yam. Those of the *bihai* (Heliconia bihai), a relation also known as *wild* or *false plantain*, are eaten in tropical America. Banana flour, and a fibrous, otherwise inedible, banana from South-East Asia (*tiparot*, from the Thai *teparod*, angel's food) are used in the making of sweetmeats. Other uses for the plant include the production of fibre, for which certain varieties are specially grown (e.g. M. textilis), and the gathering of its leaves for a variety of tasks, including wrapping, thatch, and umbrellas.

There are also a variety of uses to which the *ensete* or *Abyssinian banana* (formerly M. ensete, now genus Ensete) are put. This plant, whose fruits are inedible, originates from Ethiopia but has spread all over tropical Africa into Asia and the Pacific. It was brought to Europe in the nineteenth century and is to be found growing in warm areas. The species most used,

E. ventricosum, is to be found growing outside most African villages but gets most attention from the Ethiopians themselves. In the north the fruit sheaths are eaten and taste like soggy wheaten bread. In the south a starchy flour is extracted from the leaves and eaten with cabbage; it is the principal food of the Sidamo, who cultivate the tree for this purpose. Many Africans roast or grill the edible terminal bud, and the flower-buds are eaten in South-East Asia. It is also put to a variety of medico-magical uses; in Uganda, for instance, the juice is said to ensure the flow of milk in nursing mothers.

Beans

1. Broad beans (*Vicia faba*, or *Faba vulgaris*)
The origins of the broad bean are obscure, although it may have been developed from the French vetch Vicia narbonensis (*see under* Peas). Protein accounts for 25 per cent of their content, and they are as nutritious as wheat. They have been grown since prehistoric times and have often been used to eke out cereal crops. In Europe, remains have been discovered in Bronze-Age settlements in Switzerland and Italy, in Egyptian tombs and on the site of Troy. The beans that the prophet Ezekiel had to eat while acting out the future seige of Jerusalem were these (see Ezekiel 4:9). Cakes made from beans are offered to the various gods and goddesses in Homer's *Iliad* (IX. 589). By the first century A.D. they had been taken to China.

However, there were some curious beliefs about these beans. Herodotus (*History*, II. xxxvii) says that the Egyptian priests regarded them with horror as unclean. Pythagoras, who imported many Egyptian elements into his religion, similarly despised them. A tenet of his doctrine of metempsychosis is that souls may transmigrate into beans after death. This may have some connection with the fact that bean-feasts traditionally ended funerals, and that they figured in rites to rid households of the evil-effects occasioned by the nocturnal visits of *lemurs*, the wandering souls of the wicked. The pontifex of the official Roman religion was forbidden either to eat or mention them because they were a funeral plant and consequently inauspicious. And yet they were popular enough with the lay-folk, to whom beans were distributed by candidates for public office at election times. The politicians were not simply currying favour, since the beans were used as counters in the voting. A Latin proverb, 'abstineto a fabis' (abstain from beans), passed into English. No one is sure now whether this was an injunction to refrain from bribery, or from involvement in civil affairs, a continuing of the Pythagorean and priestly prejudice, or a warning against dabbling in the supernatural, since beans have been connected not only with ghosts and death but also with supernatural spirits and witches. Scottish witches, it was once believed, rode the air not on a broom-stick but on a bean-stalk. Ghosts were rid of by spitting beans at them, and roasted beans were buried to prevent toothache and small-pox. Among the Celts a *beano* or bean-feast used to be held regularly in honour of the fairies.

As a vegetable the broad bean retained its popularity in Europe not only because it could be dried and saved for eating later ('hunger makes hard beans sweet' runs a very old proverb) but because for many centuries it was the only bean readily available in Europe. So important was it, together with other pulses, that from the early Middle Ages there was a death sentence for theft from open fields of beans, peas and lentils.

The broad bean is sometimes classified into subspecies according to cultivars and their uses in various countries. Thus subsp. eu-faba var. minor is the *beck*, *tick*, or *pigeon bean*, mostly used for animal forage, like the *horse bean* (var. equina) specifically fed to horses. The broad bean proper, also known as *Windsor* or *straight bean* is var. major. Indian varieties, generally dried and eaten as a pulse, are classified as (subsp. paucijuga). There is generally a great consonance in the various names for this plant. The Latin *faba* develops into the French *fève*; in other linguistic groups it is *babe* (Basque), *babo* (Old Prussian), *bobu* (Old Slavic), *ffa* (Welsh), and *fav*, *fao*, in other Celtic languages. The Old English *bēan* derives from the same root as German *bohne* and Dutch *boon*. All are thought to be ultimately connected with the verb 'to eat' (Greek φαγειν and Sanskrit *bhag*).

Leicestershire men were traditionally fond of these beans, as several sayings testify:

> *Shake a Leicestershire yeoman by the collar*
> *And you shall hear the beans rattle in his belly.*
>
> *Shake a Leicestershire woman by the petticoat*
> *And the beans will rattle in her throat.*

The 'Leicestershire bean-belly' was proverbial. The vegetable was supposed to give rise to much gastric wind:

> *Every pease has its vease* [*fart*]
> *But every bean fifteen*

and yet the idiom 'full of beans' indicates a state of high spirits and well-being. The man who 'knows how many beans make five'[1] is full of sagacity

---

[1] This refers to a riddle, the answer to which is 'Two beans, a bean and a half, a bean and half a bean' or 'two beans and a bean and a bean and a half and half a bean.

1. BEANS

(*From left to right*) butter bean, soya, pea bean, Mexican black, runner bean, jack bean
(*From top to bottom*) soya, pea bean, runner bean, butter bean, Mexican black

and much worldly wisdom. Certainly he would not be the sort of person 'to sell a bean and buy a pea' or to go looking for a shilling and find a halfpenny. 'Three blue beans in a blue bladder' was once a tongue twister before the young lady who sold sea shells by the sea-shore was discovered.

Much country lore surrounds the planting and growing. 'A crooked man should sow beans, a wud (mad) man pease', for the one needs to be sown thickly, the other thinly. A pessimistic view is sometimes taken of their future progress:

> *Sow four beans in a row*
> *One for cowscot and one for crow,*
> *One to rot and one to grow.*

On the other hand

> *Sow beans in the mud*
> *And they'll grow like [a] wood.*

After this the people of Somerset pray for bad weather since 'dunder do gally [frighten] the beans' and they shoot up after a storm. The state of the moon also seems to have much to do with its future:

> *Plant the bean when the moon is light,*
> *Plant the potato when the moon is dark.*

> *Set garlic and beans on St Edmund the king,* **(20 Nov.).**
> *The moon in the wane, thereon hangeth a thing.*

> *Sow peas and beans in the wane of the moon,*
> *Who soweth them sooner he soweth too soon.*

2. Carobs and locusts (*Ceratonia siliqua*)

Carob beans are the seeds and pods of an evergreen Mediterranean tree. It was probably the husks of these that the prodical son in Jesus' parable (Luke 15:16) envied the pigs when he was down and out. He could have done much worse, since they provide an extremely palatable sweet pulp which Arabs sometimes use in place of sugar, and which has been used in Spain for making chocolate. The tree appears in Ancient Egyptian monumental sculptures, and it was probably the Arabs who brought it to Europe. However, it will grow no further north than the orange tree, and dislikes either too hot or too humid a climate. The seeds at one time served as the original 'carat' weight of the goldsmiths. From them a yellow dye may be extracted.

There is a tradition among Eastern Christians that carobs served as food for John the Baptist when he lived in the wilderness. This might explain the name of *locust* given to the tree, since the Authorised Version of the

Bible asserts that locusts were what the prophet ate. The tradition must also lie behind the German name *Johannis brod-baum* (Baptist's bread tree). The Arabic name *kharub* entered Spanish as *algarrobo* and Italian as *currabo* and *carubio*. The French is *caroubier*. It is from such sources that the names *algaroba*, *karoub* and *carouba*, current in the southern U.S., derive.

The related *black-locust* (Robinia pseudacacia) grows in northeastern America. The poisonous properties of its raw seeds may be dispelled by cooking, and they are eaten by the Indians. The Latin generic name derives from Jean Robin, who introduced the tree into France in 1601, from whence it has spread widely. It is sometimes wrongly referred to as an acacia.

There are several allied *locust-beans* (genus Parkia), growing in tropical areas, whose edible fruits are sometimes eaten. Those of the *fernleaf nitta* (P. filicoidea), from the West African sudan region, furnish a dry yellow powdery pulp and black seeds rich in protein. The tree has been taken to the West Indies, where the fruit serves as a basis for several dishes. The seeds of the *two-ball locust bean* (P. biglobosa) and *African locust bean* (P. africana) are parched, fermented in water, and ground to produce a flour used in condiments and as an aphrodisiac. A highly nutritious cheese-like substance is also made from this. Roasted, the seeds are used as a coffee substitute, and the large pods are eaten for their pulp. In Malaya the pods of P. speciosa, which smell and taste of garlic, are used as a food-flavouring.

3. Soya bean *(Glycine* or *Soja max)*

The soya bean or *white gram* is first mentioned in China about 2800 B.C. but had been cultivated for a considerable time before that. It is also an old staple of Japan. Missionaries sent it to Europe in the eighteenth century, and intensive cultivation grew up in the U.S. during the nineteenth. In Europe and South America, however, it is grown in only limited quantities. A Tyrolean variety was once used for making a coffee substitute which, though inferior in taste, had the right aroma.

There are about thirty varieties of soya bean, many of them of different colouring: white, black, brown, yellow, greenish-grey with white spots, various and variegated reds. Their high protein content enables many in the East to live healthily on a largely meatless diet, and there are many ways of using them, including as a flour. In China it was only the Tartars who used animal milk; in all cases the native Chinese equivalent was soya beans crushed in water, a substitute just as nourishing as the real thing. The beans require soaking and cooking in a chalkless water; if this is not obtainable naturally (from tap, well or pond), then rain-water or distilled water must be used.

Soy sauce is obtained from the beans, and in Japan they are used in the making of a cheese known as *to-fu* or *daizu*. In fact it was a preparation of these beans salted in oil, known to the Japanese as *sho-yu*, which the Dutch took to be their actual name, and which they transcribed as *soya*. Lastly the golden-yellow crunchy bean-sprouts of China are germinated from this species (cf. pea-sprouts, *under* Peas).

4. Dolichos beans (*genus Dolichos*)

European varieties of this bean were cultivated in classical times, and several species are still eaten in the East. The chief of the European types, also found in North Africa and as far as India, is the *lubia* (D. lubia). The name derives from the Greek λοβος (a lobe or pod); in Modern Greek the bean is known as λουβιον. Versions of the name appear in other languages: *lubia* (Arabic, Berber); *alubia* (Spain); *loba* (Hindi). In fact the name almost seems to act in as wide-ranging a manner as our word bean (*see under* Broad bean); the Hindi *lobia* applies to D. sinensis, and *lablab* to D. lablab.

The lablab, also known as *Egyptian*, *hyacinth* or *wall bean*, is often found separately classed as Lablab niger in botanical works. It is very commonly used in the tropics of Africa and Asia and has been cultivated for about 3000 years for its young pods and seeds, and for the young leaves, which may be eaten either raw or cooked. The ripe seeds are dried and used as a split pulse (*dhal*) in India, where they are very popular. They are also allowed to sprout, and are then soaked in water, shelled, boiled and smashed into a paste which is then fried. Still other names for lablab include *India*, *bonavist* and *seim bean*. There are a few varieties which include var. lablab or typicus, grown in India as a garden crop, mainly for the pods, and *Australian pea* (var. lignosus), which has a strong unpleasant smell and is grown in Asia mainly for its seeds.

Other tropical species include the *manioc* or *yam bean* (D. bulbosus) and D. bracteatus, whose young pods and seeds are boiled while they are still tender by the poorer classes. The *Madras* or *horse gram* (D. biflorus), known as *kulthi*, is widely grown for animal forage in southern India, where the poor also gather the tender young pods as food. The seeds may also be parched and eaten after boiling or frying. In Burma they are boiled, pounded with salt and then fermented to produce a condiment similar to soy sauce. Lastly the Japanese *kudzu* (D. hirsutus or japonicus, also Pueraria thunbergiana), which is also grown in China, is used in the making of textiles, and its red-violet leaves are dried and used as a pot-herb. The seeds, which resemble the yellow-podded beans which the French call *haricots jaunes*, may also be eaten, and a starch is extracted from the roots.

5. Jack beans (*genus Canavalia*)

Jack beans are of American origin. The commonest, *maljoe* or *horse bean* (C. ensiformis) and *sword bean* (C. gladiate), known as *overlook* in Jamaica, now grow wild in the tropics of Asia and Africa and are cultivated as vegetables in Australia. They were used in Mexico as early as 3000 B.C. The young flat pods, beans and leaves are used extensively in Africa and to a lesser extent elsewhere. The mature seeds are somewhat poisonous and require boiling in salted water, which must then be changed, after which they taste rather like broad beans. In East Africa there is a variety known as *gotani bean* (C. ensiformis nana). The *climbing jack bean* (C. plagiosperma) is little different except that it is of South rather than Central American origin and was being used in Peru in 2500 B.C. The *ground jack bean* (C. obtusifolia), known as the *seaside bean* in the West Indies, also grows in Australia, where the cooked pods and seeds are eaten by the aborigines.

6. The kidney bean and family (*genus Phaseolus*)
Since its introduction from America in the sixteenth century this bean, of which there are now about 500 varieties, has, together with its travelling companion, the potato, caused an entire revolution in eating habits in both Europe and Asia, and has been the cause of a growing lack of interest in whole ranges of vegetables. Because the plants spread so rapidly after their introduction from the New World, many nineteenth-century botanists had grave doubts upon the subject of their ultimate point of origin, and even suggested that they might have been developed from other plants, such as the genus Dolichos. But there is absolutely no proof of this and the facts of history seem to support the theory of this plant's American origin. Columbus remarked upon them very soon after he arrived in the Caribbean, and they were later found to be growing from one end of America to the other. Archaeologists are constantly digging up remains of them during investigations of old sites.

Best known to the English is the *common kidney bean* (P. vulgaris), so called because of the shape of the seeds. The stringy pods give them the name of *string beans*, a name often used in the U.S. But these pods are edible if they are not left too long on the vines, although in some varieties, such as the French *haricot jaune*, only the seeds are eaten. The *French* or *haricot bean* (*haricot vert*) is a tenderer, more tasty variety than the English. A dwarf variety is known in France as *flageolet* (the name is a corruption of the Latin *phaseolus* strengthened by an imagined likeness to the flute-like musical instrument), of which there are several sub-varieties: the mushroom-flavoured *Mexican black*; the floury brown *Dutch haricot*; and the *Chinese yellow*, which has yellow seeds inside a green pod.

In Latin America and tropical Africa the beans are used as a dried pulse and furnish a large part of the protein food of the local peoples. It is also these whole dried beans which are cooked with tomato sauce and canned as baked beans. The leaves of the plant are used as a vegetable in some parts of the tropics. It was introduced to South from Central America and remains of it have been found in Peru dating from 2200 years ago. In Mexico the Aztec kings used to receive 5000 tons of the beans in yearly tribute. The plant was brought to Europe in the sixteenth century, and also taken to Africa by the Spanish and Portuguese. The first record of it in England was in 1594. In spite of its recent introduction, however, there are two proverbs concerning it. 'A good year for kidney beans, a good year for hops' we are told. And, as with some other plants, one plans when to plant by watching the elm leaves:

> *When elm leaves are as big as a shilling*
> *Plant kidney-beans if to plant you're willing;*
> *When elm leaves are as big as a penny*
> *You* must *plant kidney-beans if you mean to have any.*

Allied to this plant, and similarly of Central American origin, though differing in the characteristics of both its flowers and fruits, is the *runner bean* (P. coccineus or multiflorus). Wild beans were eaten in Mexico about 7000–5000 B.C. and were eventually cultivated; and they are still mostly grown in America. In Central America the beans are eaten both green and dried, and the large fleshy tubers are boiled as a vegetable. The most popular variety is named *scarlet runner* after its flowers. The Czar variety has white seeds and broad pods which are sold in the U.S. under the names *butter nut* or *snail beans*. The *pea bean* is a fine-flavoured variety divided into such types as Du Pape, Red Douglas and Robin's Egg (the seeds go bright red as they mature). The Blue Coco (named after its flowers) has pods of a slight purple shade whose colour disappears on cooking. Other varieties include the *Painted Lady*, with scarlet and white flowers; *Giraffe*, with large nearly stringless pods; and *Blue Lake*, with stringless round pods.

The *scimitar-podded kidney bean* (P. lunatus), also known as *Hibbert*, *sieva* or *sugar bean*, grows wild in tropical America and is also cultivated; it has been in use there since prehistoric times. The vine has developed into a mutated bush, and its fruits are much used in the U.S., Africa and Asia. Dried and shelled the beans are of various colours but contain traces of hydrocyanic acid which must be dissipated by boiling in changes of water. The shelled green beans are also used, and occasionally the young pods and leaves. A variety known as *Lima bean* (var. limensis), also *Tonga, Burma, Rangoon, Java* and *Madagascar bean*, bears large white seeds which are dried, canned and marketed under the names *wax* or *butter beans*. Remains of these in Peru date from 7000–5000 B.C., but do not appear in Mexico until 500–300 B.C.

The plant and its variety were spread along the ancient Indian trade-routes: the Hopi or northern, from Mexico to the southern U.S.; the Inca or southern, from Central America to Peru; and the Caribbean or West Indian, over the Gulf of Mexico to the Guyanas and Brazil. The most poisonous variety was taken by the Spaniards from the Caribbean to the Philippines, from whence it spread to Asia. Brazilian plants were taken to Africa by slave-traders, and the Lima bean was taken early to Madagascar. It is now the main pulse crop in the wet forest regions of tropical Africa and widely grown in Burma. Similar to it, but cooked like a dwarf bean (*flageolet*), is the Asian *Bavia* or *Tonkin bean* (P. tunkiniensis).

The *Texas* or *tepary bean* (P. acutifolius) grows wild and is much cultivated in Mexico and the adjoining states of the U.S. Its popularity stems from its being both fast-growing and drought-resistant. In Mexico it was being used about 3000 B.C. Although it has been taken to Africa it is fast being replaced in cultivation by the common kidney-bean. Other more or less wild varieties are favoured by the American Indians still. The seeds of the *bean vine* (P. polystachys), alternatively known as the *wild* or *thicket bean*, are dried and cooked in the south and east of the U.S., where they are highly prized. The Louisiana Indians ate the roots of P. diversifolius boiled or mashed. Those of the tropical P. adenanthus are also eaten.

Many species have achieved great popularity in Asia. These include the *adzuki* (P. angularis), which is used in China and Japan as a pulse or ground into meal; and the *rice bean* (P. calcaratus), which is often served as a pulse with, or as an alternative to, rice in India, Burma, Malaysia, China and Fiji. The young pods and leaves are also used as a vegetable. The *aconite-leaved kidney bean* (P. aconitifolius), known by its Sikh name *moth* (*mout* in Hindi), and as *dew bean* or *Turkish gram*, grows wild in India and has spread to China and Ceylon. The green pods are used as a vegetable, and the ripe seeds serve as a pulse, whole or split.

Green or *golden gram*, also known as *mung* (P. aureus), is very important in India, esteemed for its tiny seeds, which become rather sticky on cooking, but are accounted both wholesome and nourishing. These are dried and boiled whole or split, or else parched and ground into flour. The green pods are also eaten as a vegetable. In China, where it is known as *lou-teou* (little pea), it is added to green noodles and used for bean-sprouts, a use to which it is also put in the U.S.A. Similarly known as *mung* or *urd*, P. mungo goes under the name of *wooly pyrol* in the West Indies. This is also highly prized in India, especially by high caste Hindus. The green pods are eaten, or the seeds whole or split, or parched and ground into a flour used for making spiced dough-balls, porridge or bread. Its variety, *black gram* or *mash* (var. radiatus), is especially popular, and its use has spread to the Arab countries. The seedlings, sprouts and leaves may also be eaten.

It is probably one of these Asian varieties of which the Hindi poet S. N. Pant is writing in his 'Fruits of the Earth':

Childhood. I planted pennies in the yard and dreamed
Penny trees would grow. I heard the air sweet
With the silvery ringing of the clustered crop
And strutted round like a big fat millionaire.
All fantasy! Not a single sprout came up;
Not one tree appeared on the barren ground:
Swallowed in dust, my dreams blighted.
On hands and knees I scratched for a sign of growth;
Stared into darkness. What a fool I was!
I gathered the fruit I had sown. I had watered coin.

Fifty years have passed. And passed like a gust of wind.
Seasons came – I hardly noticed them:
Summers blazed, swinging rains poured,
Autumn smiled and winter followed shivering,
The trees stood naked, later, the forest, green.
Again in the sky dark salve-like clouds,
Thick with healing elements. It rained.
Out of simple curiosity I ventured out
To the courtyard corner, and bending down,
Pressed rows of beds into soaked soil
And planted bean-seeds. Then covered them.
The hem of the earth's sari was tied with jewels.

Soon I forgot this simple incident –
No one could think it worth remembering.
Walking at evening in the yard, a few days later,
There they stood, suddenly before me,
A multitude of new arrivals, each
With a tiny green umbrella on its head:
Like young birds who had just cracked their shells,
Already fully fledged, trying the sky.
Hypnotized, I stared and stared wide-eyed –
This miniature army, dwarves arrayed in rows,
Just sprung from seeds I'd sown not yesterday,
Erect and proud, shook their feet as if to march.

I've spent days now, watching them grow.
Gradually the space around puts on light leaves
Then thickens to canopies of velvet green;
The tendrils rise winding and swinging on the frame
And spray out, fountains of fresh springs.
A stunning sight – the growth of a generation:
Star-like sprays of flowers scattered yet grafted
To the dark green undulating branches;
White foam on waves; a luminous new-moon sky;
Pearls in hair; a flower-patterned blouse.

At harvest time, millions of pods broke out:
Some were stringy, and some were fat – all sweet,
Long fingers, swords, or emerald necklaces.
If I say they developed like the moon
Or an evening sky growing into clusters of stars
It will not seem that I exaggerate.

All winter we ate them cooked for lunch and dinner.
Next-door neighbours, close friends, mere acquaintances,
Relatives, people we hardly knew,
Some who didn't even ask, all shared
The bountiful supply the earth produced.

She yields abundantly to her dear children;
I had not understood in planting pennies
The laws of her love. She bestows gems.
Storehouse of virtue and all-embracing love,
She serves on terms of truest equality;
She will only bear seeds of her own kind.
Then even her dust may yield crops of gold
And the directions burst with the joy of her works.

7. Others

The *Goa, Manilla* or *princess bean* (Psophocarpus tetragonolobus) is probably of Asia origin, and has since been taken to Africa and the West Indies. It has been known to the Melanesians for a very long time and is extensively cultivated in New Guinea, mainly for its immature pods which are eaten like string beans. The pea-sized seeds are rich in protein and require roasting when they are ripe; in Java they are eaten parched. The young leaves, shoots and flowers may also be eaten as a vegetable; the tuberous roots are somewhat similar to the yam bean or a sweet potato, and are eaten raw or cooked, especially in Burma.

Members of the Asian *velvet bean* family (genus Stizolobium) are much cultivated for animal forage, the beans and young pods of which are edible, although considered inferior to many others. Chief among them is the far-travelled *Lyon, Florida, Mauritius* or *Chinese velvet bean* (S. niveum). A relative growing also in the Pacific and Australia, and known in New South Wales as *cowitch* (Mucuna gigantea), is used by the aborigines, who eat the seeds after considerable preparation. The name derives from the Hindi *kawanch*, a name, together with its variants *cowage* and *cowhage*, applied to S. pruritum, whose hairy pods cause intolerable itching to any who touch them. It is no relation to New Zealand cowage (*see under* Spinach).

The Indian *guar* (Cyamopsis psorialiodes) is named *cluster bean* in reference to the way its hairy pods grow clustered together. It is cultivated for food and must be gathered young since the pods grow fibrous upon maturing.

The Somali *yebb nut* is eaten in the manner of a bean. It is in reality the fruit of a bean-sized one-seeded pod belonging to a shrubby desert tree, Cordeauxia edulis.

The tropical *tree beans* (genus Bauhinia) of Africa and Asia serve for food in various ways. The seeds of B. reticulata are eaten by the natives of Africa, and the pods of B. esculenta are an important food to those of

South Africa. The leaves of the *Malabar tree bean* (B. malabarica) are used for flavouring in Asia, and those of the *mountain ebony* or *Buddhist tree bean* (B. variegata), also known as *variegated St Thomas' tree*, serve as cigarette covers in India. Elsewhere in South-East Asia the leaves and pods are eaten as a vegetable, and the flowers are pickled.

The *tamarind* (Tamarindus indicus) is a widespread tropical plant, probably of East African origin. It was introduced early into India and taken to the north Australian coast (Arnhem Land) by Malaysian voyagers as a food plant. Its Arabic name, which means Indian date, seems to indicate that it entered medieval commerce from India. Its pods are hard externally but filled with a juicy acid pulp from which a pleasantly sour drink is made. Both the spongy insides and the bean-like seeds may be eaten after cooking; in Egypt the latter are made into a kind of bread. The leaves are added to curries in India.

Other plants also go by the name tamarind, notably the *sweet tamarind* (genus Inga) of tropical America, whose fruits serve as very popular desserts, and the *curl-brush bean* (genus Pithecolobium), alternatively known as *ape's ear-ring*. The pulpy seed-coverings of the *guamadil* or *Manilla tamarind* (P. dulce) are eaten by the poor in tropical America. The green pods of the *ebony curl-brush bean* (P. flexicaule), which grows from Mexico into Texas, may be cooked and the seeds roasted. The leaves, fruits and flowers of P. labatum are eaten as a vegetable in South-East Asia.

The pods of some of the *acacias* (genus Acacia), of which there are about 600 species, the larger number in Australia (many of which have not yet been particularly well-researched because of their rarity and inaccessibility), are eaten for food. A popular Australian name for the shrubs is *wattle*, their springy branches served the early-settlers for the making of 'wattle and daub' huts. Other species produce gums, tanning materials, timber, snuff and scent. The ripening pods of the *coast sallow wattle* (A. longifolia var. sophorae), known as *boobialla* in the Tasmanian aboriginal language, are roasted and eaten in Tasmania. In Western Australia they eat the seeds of the *food wattle* (A. cibaria), and in South Australia those of the *native willow* or *lightwood* (A. salicina), also called by such native names as *cooba* and *weeping myall*. *Myall* is a word of aboriginal origin also used to denote a wild native who has not come in contact with white civilisation.

Elsewhere the seeds of the *umbrella acacia* (A. oswaldii) are eaten. Those of the *scrub wattle* (A. dictyophleba or stipuligera) are ground for food by the aborigines of Central Australia. The seeds of A. rivalis are treated in the same way in South Australia, where the natives also gather and eat the tree's sweet gummy exudation. The seeds of the *mulga* (A. aneura) are ground for bread, and in Queensland the roots of the *waneu* (A. bidwillii or pallida) are baked and eaten.

The Mexican Indians eat the pods of the *prairie acacia* (A. augustissima), which grows from the southern U.S. into Central America; they also make a flour from those of the *cat's claw* (A. greggi). Elsewhere in Central America the flowers and buds of one of the related *coral beans* (Erythrina rubrinervia) are eaten like string beans, and the young leaves are added to soups. The bright red seeds of the long-podded E. edulis are cooked and eaten in Colombia. Other species are poisonous, however, and the numbing *curare* poison is obtained from them. What must be a very beautiful Indian species is said to have been stolen from Paradise by Krishna.

Beets and chards *(genus Beta)*

Edible beetroots (Beta vulgaris) are related to the sugar-beets (B. vulg. saccharifera), grown for the making of sugar, the extraction of which began in the eighteenth century; and to *mangel-wurzles* (B. vulg. macrorrhiza), grown as animal forage. The curious name of the latter is derived from the German *mangold* (beet) + *wurzel* (root). The beet was known in classical times but its development is comparatively modern. Anglo-Saxons used the root as an ingredient in a bone-salve, and in an emetic; the juice of the pounded root was recommended for all festering wounds and infectious bites.

The edible roots of B. vulgaris rapacea are of two colours, of which the *red* or *Roman* variety (so called because it came first from Italy) is the more recent. Because the English have pioneered this it is the better known in Britain, but on the continent the yellow-rooted variety was once much preferred; it tastes sweeter and is not so suitable for pickling in vinegar. The beet was developed at Castelnaudary in the Languedoc at the beginning of the nineteenth century. Originally the root was longer; round varieties of root are, again, a later development. There is a German mention of the beetroot in the thirteenth century; from here it was introduced to Italy, there bettered and passed on to the rest of Europe. In Turkey it was very popular in the fifteenth and sixteenth centuries. An old recipe advises cutting up a cooked root and putting it in a casserole with butter, parsley, chopped green (spring) onion, a little garlic, a pinch of salt, pepper, and flour, this to be cooked for a quarter of an hour. In western England a beetroot or two is sometimes added to cider-apples in the pressing in order to give the cider a golden colour.

There is a white-rooted beet (B. vulg. cicla) which is cultivated for the sake of its leaves and stalks, from the appearance of which it gains the names *thick-leaved* or *seakale beet*. The tops may be used as a spinach substitute, the ribs like asparagus or seakale. It is useful in warmer weather when it is too hot for other greens. Other names include *Sicilian beet, white* or *wild beet, leaf* or *spinach beet,* and *perpetual spinach.* But perhaps the better known name is *chard,* deriving from the French *chardon* (a thistle); *ruby chard* is a variety with red leaves. There is also a newly-developed variety with red stems and midribs, and known as *rhubarb beet.* Greek authors of the fifth century mention chard, but it was generally considered to be as indigestible as cabbage, needing to be highly seasoned and therefore best fit for the poorer classes. In the Middle Ages, however, no meal was considered complete without a soup made from the leaves. Often it was cooked with sorrel to tone down the acidity of the latter, especially of the English varieties and their dock relations. There is also a Chinese mention of chard in the seventh century.

Seakale beet is not to be confused with its wild relation, the *sea-beet* (B. maritima), which grows on the shores of Europe and Asia. The leaves of this are very good when cooked like spinach. In Ireland it is especially common on Rock Island to the northwest of Great Arran. There was once a belief that this 'wild spinach' would cure sick sheep; any Arraner who had one would take it to the island and leave it grazing there until it recovered (or otherwise).

Borage and family

Borage (Borago officinalis) is a member of the forget-me-not family. The name may be rhymed with either 'courage' or 'porridge'. It derives from the Latin, the root of which, *borra* or *burra,* means rough hair or short wool (cf. Old French *bourrace*; *borragine, borrace* (Italian); *borraje* (Spanish); *borragon* (Portugese)). Originally a wild plant from Syria, it is now widely spread over Europe and North America. It may be used in salads and as a pot-herb and tastes rather like cucumber. Its flowers and leaves have been used in wine, supposedly to make the effect even more enlivening, and in other drinks, giving a cooling and envigorating effect. For this reason it has acquired the name of *cool-tankard.* Other names are *tale-wort* and *bee-bread* (since it is often grown near hives). Gerarde says of this herb that 'Pliny calleth it Euphrosinum because it maketh a man merrie and joyful: which thing also the old verse concerning it doth testifie – *"Ego borago gaudia semper ago"*. In English, "I borage bring always courage"'.

Another plant once cultivated for the same purposes as the above was the rough and hairy *bugloss* (Anchusa officinalis). The general name given to all of this genus is *alkanet* or *orchanet,* derived from the French *orcanette*; the name is applied to the red dye which is extracted from the roots. Bugloss derives its name from the Greek βου γλοσσος, meaning *ox-tongue*, a name which the plant also bears, as well as *langdebeef*, from the French *langue de boeuf*, the name by which this plant was once cried through the streets of Paris. The *prickly ox-tongue* (Helminthia echioides), also goes under the name of borage and was once cultivated as a pot-herb. A sixteenth-century work on dietary remarks that 'the roots of borage and bugloss soden tender . . . doth ingender good blood'.

Sea bugloss (Mertensia maritima) is a native of the northern temperate and arctic zones, growing among stones or oaks like dried turf near the ground, although in some cases it reaches out some fifteen inches. Alaskan Eskimos eat the rhizomes; the leaves, whose taste has given it the name *oyster plant*, would appear to have been eaten in northern Scotland and the Shetlands. Other names for it include *lightwort* and *sea gromwell* or *bluebell*.

Comfrey (Symphytum officinale) is a European relation now introduced into North America and recommended as an emergency food plant. Its leaves may be eaten like spinach and the young shoots make a rough asparagus substitute. They may be blanched by forcing them to grow through heaps of earth. The plant's taste is somewhat astringent, as the name *alum* attests. It has a hairy appearance, in common with most of the family, and has thus been christened *ass-ear*. Most of its popular names refer to its use as a poultice or bone-setter: *backwort*; *knit-back*; *bone-set*; *bruisewort* (deriving from its Anglo-Saxon name). *Consound* may be compared with the French *consoude*, deriving from a verb meaning to make whole. It is from its Latin equivalent, *confirma*, that the name comfrey derives. In Anglo-Saxon times the plant went by the alternative Latin name *galla adriatica* which finally became corrupted into the old names *galloc* or *yalluc*. The German *schwartz wurtz* is echoed in the English *black root*.

Breadfruits and breadnuts (*genus Artocarpus*)

The true *breadfruit* (Artocarpus altilis) comes originally from the Malaysian region. It has been used from prehistoric times and during that period was spread over the islands of the Pacific. In some areas, particularly Samoa, it is a staple food. Of the two main varieties, seeded and unseeded, it is the latter which is usually preferred; there are numerous subvarieties ripening all through the year. The unripe fruit is gathered while the pulp is still mealy and has to be cooked, boiled or baked. Alternatively it may be

2. LEAVES AND WHOLE PLANT

comfrey celandine cowslip violet

cut into slices and fried like a potato; and, very rarely, it is eaten raw. Polynesians often combine it with coconut or make flour from it.

The fruit is about six inches in diameter and round. Cook first saw it and considered it would be good for the Negro slaves in the West Indies. Bligh was sent to bring specimens back on the *Bounty*, but before he could do so the crew mutinied and cast him ashore on the Pitcairn Islands. Eventually Rodney succeeded in bringing some back in 1782, but the experiment was not successful since the slaves preferred their own foods, bananas in particular. From the West Indies the plant spread into South America. In the Caribbean the creole name for the seeded variety is *chataigne* (chestnut), a fair description of the edible seeds of other species, particularly the *jackfruit*.

The *jacktree* (A. integrifolia) comes originally from India and its fruits are also used widely in the South Pacific. They are commonly four foot in diameter and weigh up to 70 pounds. The whitish pulpy flesh may be eaten but it is not as palatable as that of the breadfruit. It is cooked, fried, and occasionally boiled in milk. But the seeds (known as *breadnuts*), of which there are about 200–300 in every fruit, are the most used, and are an important element in the diet of the poorer classes living in the Asian tropics. They may be roasted or boiled. In Indonesia the young flower clusters are cooked with syrup and agar-agar (seaweed gelatine).

Rodney also took examples of this tree back with him, and it too has spread into the American tropics. To the Portuguese it is known as *jaca*, derived from the Malayalam *chakka*; its earliest European spelling (about 1328) was *chaqui*. The *Brazilian jack* (A. brasiliensis) is treated very similarly, both seeds and flesh being eaten, and there are many more edible species in tropical Asia. Edible seeded species include the *lakoocha* (A. lakoocha) and the inferior *wild breadfruit* (A. elastica). Some have more or less palatable fruits, such as the Indonesian *dadak* (A. dadak), the size of chicken eggs and rather acid; the Malayan *marang* or *Johore jack* (A. odoratissima) and *monkey jack* (A. rigida), are sweeter species. Another Malayan wild tree known as *champedak* (A. champeden) bears small fruits whose pungent flesh is most politely described as giving off a strong stench. Nevertheless they are used in soups while immature, and the fresh pulp of ripe fruits is eaten and the seeds cooked. From the same region comes the *kemando* (A. kemando), the latex of which tastes rather like coconut-milk and is used in sauces, although too much of it results in stomach-ache.

The true *breadnut* is the fruit of the related *snakewood* or *Ramon breadnut tree* (Brosimum alicastrum) which grows in tropical America. Its edible fruits are rather like chestnuts; they may be roasted or boiled and are very nutritious. The seeds of the *Costa Rican breadnut* (B. costaricanum) are treated similarly. Another species known as the *cow tree* (B. utile) grows in Venezuela, the latex caught from the slit trunk of which may be drunk like milk. Very similar to the American breadnut is the *African breadnut* (Treculia africana) from West Africa, whose large fruits contain edible seeds which may be boiled or roasted.

Like the breadfruits in certain respects are the fruits of some of the screwpines (genus Pandanus), so called because of the twisting of the stem in some species, growing in the same area of tropical Asia, and in Africa. In Madagascar those of the *common screwpine* (P. utilis) and the *edible screwpine* (P. edulis) are eaten; in Indonesia those of P. connoides. Here too the buds of P. ceramicus are eaten with rice, a use to which those of P. polycephalus are put in Malaysia; the leaves are also edible, as are those of the Indonesian P. odorus; they are also added to bean-curds.

The seeds of P. spiralis, which is commonly to be found growing all over the north of Australia, were once a principal part of the aboriginal diet. The fruits were baked in hot ashes, soaked in water and baked again until the seeds could be extracted from the broken fruit. The *fragrant screwpine* (P. tectorius or odoratissimus), also known as *keora* or *pandang oil-plant*, and *Moreton Bay* or *Nicobar breadfruit tree*, grows all over the Pacific and tropical Asia. The yellow pulp of its fruits and the white leaf-bases are eaten in India during times of scarcity. In Ceylon and elsewhere the leaves are used for making ornamental baskets, as are those of the common screwpine in Madagascar. The seeds are also eaten after careful preparation. The young shoots and leaves of various other species are eaten in the tropical Asian area. In Indonesia the fleshy young flowers of the related Freycinettia funicularis are eaten boiled as a vegetable, as are those of the *kie-kie* (F. banksii) in New Zealand, where they are known as *tawhara* to the Maoris. They and children also eat the fruit-spikes.

Cabbages and other coleworts

1. Brassicas *(genus Brassica)*
The crucifer family, to which the cabbage belongs, is characterised by hot or spicy smells and flavours, and cabbage seems to be one of the blandest in a family which includes cresses and mustards. Various wild-growing mustard plants – for instance England's *charlock* (B. sinapistrum or Sinapis arvensis) – would appear to be the cabbage's remotest ancestors. The next step in the cabbage's chain of development was into what is now known as *wild* or *sea cabbage* (B. oleracea), which grows wild in the British Isles and the European mainland. Other varieties are to be found in the Balearic Isles, Greece and Sardinia. It was out of these that most other

3. FRUIT

breadfruit and leaf olive pandan
akee okra

varieties of the cabbage developed since it began to be cultivated in prehistoric times.

The earliest variety, B. ol. capitata, appears to be that with a small round head, commonly referred to as a heart or an apple. The Romans are supposed to have introduced its cultivation into England; the more leafy varieties dealt with later are a British development. The vegetable was highly regarded by the Romans. In his *De Re Rustica* Cato advises eating it raw and pickled before a meal to counteract the effects of alcohol, and after it as an aid to digestion. But I have come across a fifteenth-century English poem in Latin which makes the counterclaim that it causes indigestion. Medicinally the Romans regarded it as a cure for melancholy and used it in poultices for sores and tumours. This may well be the origin of the Anglo-Saxon use of it as a poultice for 'black blains'. A fourteenth-century cookery book suggests what would seem horrifying today, that it is best and softest after the frost has got in it. Maybe this explains the first line of a Parisian street-cry:

Choux gelez, les bons choux gelez!
Ilz sont plus tendres que rosées.
Ilz ont cru parmi les poirées,
Et n'ont jamais été greslez.
    (*Les 107 cris de Paris*, A. Truquet, Paris, 1545).

Cabbage is the birthday plant for 19 September, symbolising gain and profit; it is the emblem of the self-willed and also, paradoxically, of the 'vegetating'. An old Greek proverb concerning it states that 'coleworts (or cabbage) twice sodden are death'. It is thus commented upon by a sixteenth-century author:

> Now this crambe[1] was in olde tyme much used in feastes and bankettes, but if it were twyse sod it was so lothed and abhorred that the Grekes made a proverb on it. For as often as they wolde signifi a thyng agayne and agayne repeted, not without tedyousness a grevaunce, they sayde forthwith in theyr langage: Crambe twyse served is death.

Spring greens, either hearted or leafy, are gradually superceding the tougher British coleworts, and the hearted summer and autumn varieties seem somehow to last the whole of the rest of the year. One can also buy *red cabbage* (B. ol. cap. rubra), a variety which could have had an early origin, although it is only mentioned in comparatively late documents. It is quite a favourite in the Netherlands; Belgian late-Impressionists and Expressionists seem to have gone through a phase of painting the fields in which they grow.

In winter one can buy *sauerkraut* 'fresh' from the barrel, if one knows the right delicatessen. This is finely sliced cabbage covered with salt and

---
[1] The Greek word for cabbage.

packed tightly, even trampled, into an earthern or wooden container and left to ferment for a few weeks, after which it is covered and kept for further use. It is advisable to wash at least some of the brine out of it before cooking.

The derivation of European names for the cabbage seem to stem from four heads. First the Greek καυλος (a stem), reappearing in Latin as *caulis* (meaning both a stem and a cabbage). In Old French this becomes *chol* and *col*, finally softened to *chou*. *Col* is also the Spanish. In German it is *kohl*, in Dutch *kool*, the same Germanic root which gives the English *kale, cole, collet* and *collard* (*cauliflower* looks back to the Latin). The second name-group centres on the word for a head, *cap* in Celtic and *caput* in Latin (these linger on in the word 'cape', meaning a promontory). Old Norman French uses the word *cabouche* for a head (cf. the Italian *cabuccio*) which is linked with the late Latin *cabus* (once used for cabbage by the French). Thus we get our word *cabbage*, and the Italians *cavolo*. The third group derives from the Latin name *brassica*. The Welsh is *bresych*, and other Celtic roots are *bresic* and *brassic*. This reappears in the Spanish *berza*, Portugese *verza*, and Roumanian *varza*. Lastly there is the Latin for seakale, *crambe*, and its Greek equivalent κραμβαι. This travels to or from the East, where we find the Persian *karamb, karam*, and *kalam*, the last being also the Kurdish, and *gaghamb* in Armenian. One might mention in passing the word *kraut*, which is really the German for a plant or herb, but is used for the cabbage in the south (elsewhere it is *weisskraut*).

The leafy coleworts are variously subdivided. The hearting variety whose terminal leaves are loosely folded on one another includes *Savoy greens* (B. ol. bullata major) which is a separate cultivation from the wild cabbage introduced to England from Savoy in the seventeenth century. Previously it had been widely grown in France, Holland and Italy. Since this is hardy it is very suitable for growing in the winter months, its season lasting from September to April. Other varieties are the ornamental, which is nevertheless good to eat, and the perennial, bushy and green or purple in colour.

The Scotch or curly variety, another hardy plant which can resist frost and snow, and is therefore not good for growing in warmer climates, is one that has the similar claim made of it that it improves after being touched by the frost. This includes *kale, borecole* or *curly greens* (B. ol. acephala), the original *colewort* of Middle English, later corrupted to *collard*. The name borecole may be compared with the Dutch *boerenkool*, a variety of kale grown in Holland and Germany, and there known as *peasant's cabbage*. The plant's rounded shape is referred to in the name *bow-kale* and *marrow cabbage*. Other names include *coolstock, greenwort, chitling* and *ragged Jack*. In England this is now grown largely as a forage crop for sheep, but it is still used as a vegetable, especially in Scotland and the southern U.S. It was once so important to the Scottish that not a meal passed without kale soup being present. Thus soup in general, and the pot in which it was cooked, became known as *kail*.

Burns, in his poem 'Halloween', records how the plant may be used for amatory divination. He notes that

'the first ceremony of Halloween is pulling each *stock*, or plant, of kail. The young people must go out hand in hand, with eyes shut, and pull the first they meet with: its being big or little, straight or crooked, is prophetic of the size and shape of the grand object of all their spells – the husband or wife. If any *yird*, or earth, stick to the root, that is *tocher*, or fortune; and the taste of the *custoc*, that is the heart of the stem, is indicative of the natural temper and disposition. Lastly, the stems or, to give them their ordinary apellation, the *runts*, are placed somewhere above the head of the door; and the Christian names of the people whom chance brings into the house are, according to the priority of placing the runts, the names in question'.

A sprouting variety of winter and early spring is considered especially delicious. The roots are accounted of fine taste and the young shoots may be cooked like asparagus. Examples of this are the *Egyptian*, *Jerusalem* or *asparagus kale*, *Buda* or *Deleware kale* and *Russian kale*, the shoots of which are highly prized. Lastly there is the tree-cabbage variety, of which *Jersey cabbage* is the crowning example. The leaves must be stripped from its thick stalk, which often grows up to twelve feet high and is said to be used for walking-sticks in the Channel Islands. This variety is commonly used for stock-feed now.

Two of the stranger developments of the wild cabbage are the *cauliflower* (B. ol. botrytis cauliflora) and the *Brussels sprout* (B. ol. bullata gemmifera). 'Cauliflower', growls Mark Twain, 'is but a cabbage with a college education'. Possibly it was known to the Romans, but we cannot be sure whether they have only broccoli in mind when they speak of a flowering variety of cabbage. But it is definitely mentioned by a Spanish Arab of the twelfth century, who says it comes from Syria and is known among them as *quonnabit*. It was from the Levant that the cauliflower was introduced into modern Europe, spreading through Crete and Malta, both Italian trading outposts in the sixteenth century, to the Italian mainland, and from there into Europe. A French name for it at the time was *chou de Chypre*. It was slow to take hold in other European countries; there is mention of it in England in 1597, and again in 1629, but it was rare. Eventually, however, the English variety came to be acknowledged as the best, and it was English seed that was sought after. On the other hand the cauliflower was taken to America very early, and was flourishing widely in Haiti as early as 1565. The French name is *chou-fleur*, and it is thought that the English derives from the Latin *caulis* plus the root of the word for flower, *flos-floris*. An intriguing alternative is that it is a combination of the

Picard diminutive for a cabbage, *caulet*, and *fleur*. *Broccoli* (B. ol. botrytis), whose Italian name means 'little sprouts', is a variety of cauliflower, sometimes known as *winter cauliflower*, but is without a head; the tops and stalks are eaten. There are several types, including the *English* ultimately of continental origin; *Roscoff*, of French origin; the American *Calabrese*; and the French *D'Angers*, having early, mid-season and late varieties. *Protector*, of Italian origin, is another winter varity. Apart from these there is *Italian, asparagus* or *green-sprouting broccoli* (B. ol. italica), a tall branching variety whose edible parts are the tops of the thick main central flower-stalks and all of the secondary stalks with unopened flower-buds.

The *Brussels sprout* (B. ol. gemmifera) positively sinks under the disdain and horror felt for it by members of the British Commonwealth, and the mystery which surrounds its beginning. The Romans might have known about it, but again they may have been talking about broccoli. Some hold that it is an early development from the small headed Milan cabbage; the Belgians like to think that Julius Caesar introduced it to their area, but for this there is absolutely no evidence. There is a vague mention of *spruyts* in some market rules of 1213, and in later documents, but there is no certain reference in authorative works until much later. No one seems to be sure when the plant was brought from Belgium to England either, or who taught the English to cook it to the consistency of a soggy orange. As well as the sprouts which grow along the tough central stalk, the cabbage-like leaves on the top of the plant are also edible, and may be bought in shops under the name of *sprout-tops*, but generally only in areas where sprouts are grown. Similarly edible are the tops of such rooted brassicas as kohl-rabi and turnip (q.v.).

Portugese or *Braganza cabbage* (B. ol. costata), also known as *couve tronchuda*, was introduced into France in 1612 but did not reach England until 1821. It is a leafy variety with an undeveloped head whose thick white ribs may be eaten like seakale, and the rest like ordinary cabbage. Two other species go under the name of *Chinese cabbage*. *Pe-tsai*, meaning white vegetable (B. pekinensis), is sometimes mis-named *celery cabbage* because it looks like a Roman lettuce. It is used throughout the East and is mentioned in Chinese works as early as the fifth century, although works on agriculture only record it in the fifteenth. There are various methods of conserving it for the winter. Either it may be pickled in vinegar or, having been half cooked in water vapour, it is exposed to the cold northern winds and then passed through mustard water and diced. The other, *Baak-choy* (B. chinensis), is also called *Chinese mustard*, and is noted for its lack of smell when cooking.

2. Seakale (*genus Crambe*)

Seakale (C. maritima) is quite distinct from the cabbage family, although

4. LEAVES

Brussels sprouts
watercress mustard seakale
rocket Chinese cabbage

*For my wife Ann
my sister Jayne,
and for Janet and Susan
my sisters-in-law*

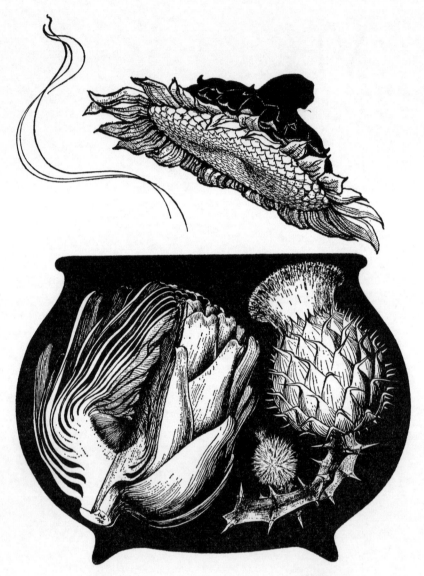

globe artichoke

sunflower
Scottish thistle

THE VEGETABLE BOOK

it may possibly have some connection with the cresses. It grows wild abundantly on the West European seaboard, especially in the British Isles. Its leaf-tips may be used as a pot-herb or in sauce. At one time it was used as a winter substitute for cabbage by fishermen's wives. It may have been this that Pliny was referring to in his *Natural History* where he says 'There is another kind of cabbage which has also some merit. It is called *halmyrides* because it only grows on the coast. It always stays green and is taken as a provision for long sea-journeys'. Seakale was first cultivated as a food in England in the seventeenth century, when it was known by its old name of *sea-colewort*; cultivation spread to Ireland after it had been brought to perfection in the eighteenth century. So that it should become white and tender it needs protection from light and air.

Another species, *Tartar-bread* (C. tartarica), grows from Hungary right over the sandy soils of the steppes. The Hungarian name for it is *Tatar-kenyer*, and the Bohemian is *Hieronymus-würzel* (St Jerome's wort). An explanation of the first name is that it was the Tartars who first taught the Magyars its use. In some districts the root is eaten either raw or cooked; the young leaves, before flowers have developed, may also be cooked like cauliflower. The *dog's cabbage* (Thelygonum cynocrambe) of the Mediterranean region is actually an isolated type. The young shoots are sometimes eaten as a vegetable.

3. Rockets

The *true rocket* (Eruca sativa), once known in England as *white pepper*, is characterised by its distinctive spicy flavour. It is a close relative of the mustards and grows wild in southern Europe. The Italians and Egyptians often use its leaves, anciently supposed to have an aphrodisiac effect, in salads. A classical Latin poem mentions them as 'foul provocatives of lust'. The name 'rocket' derives from the French *roquette*, a diminutive form of the Latin *eruca* (cf. the Italian *ruccetta*, and Provençal *roqueto*).

Aleppo rocket (Erucaria aleppica) grows at the eastern end of the Mediterranean and in Arabia. Its fleshy leaves and young shoots may be boiled or used as a salad. *Mustard* or *false rocket* (Bunias erucago) grows in the same area and serves as a salad or spinach substitute. In Russia the *hill mustard* (B. orientalis) is used in much the same way. *Sea rocket* (Cakile edentula) is a North American relative normally to be found growing on beaches. Its leaves serve as a salad and pot-herb, and the fleshy roots may be dried, ground and mixed with flour for bread in times of scarcity. *Sea stock* (C. maritima, formerly Bunias cakile), a common maritime plant, also goes by the same name. Its foot-long succulent branches resemble samphire (q.v.). The plant was spoken of by old writers as a good salad herb but, they cautioned, 'if rocket be eaten alone it causeth headache and heateth too much, therefore it must be eaten always with lettuce or purslane'.

4. Kerguelen cabbage (*Pringlea antiscorbuta*)
Kerguelen cabbage is one of the strangest and rarest in the family group. It is an edible plant closely resembling a leafy cabbage and has strongly antiscorbutic properties. One of the few flowering plants in the whole region, it is confined entirely to the Antarctic islands of Kerguelen (under French administration) and Heard (Australian).

Cacti

The best known of the cacti is probably the *prickly pear* (genus Opuntia) whose sweet fruits, know as *bastard* or *Indian figs*, are highly esteemed by many. They constitute the chief fruit crop of Sicily, largely because the cactus flourishes on extremely poor soils and requires little tending. The plant is originally from America, where it grows from British Columbia to Chile along the west coast, and grows as far east as Massachusetts. The Indians cultivated them before the coming of the Spaniards, who took it back with them to become naturalised in the Mediterranean area. From there it has been introduced into South Africa, Asia and Australia. It was taken to the latter continent because it was thought it might be useful both as cattle-food and as a water conserver. Its spread was so phenomenal in Queensland that it became a serious pest and only timely thinking prevented it from becoming worse. Cochineal grubs were known to feed on and destroy this cactus and were therefore imported. Apparently they found Australia very congenial too and rapidly fed upon the entire cactus population; now the population is under control.

The family is generally divided into two, cylindrical varieties being named *cholas*, a pest on the range with very little practical value. The name prickly pear is generally reserved for the flat-stemmed type which can serve as a cattle feed. It is first necessary to brush or burn off the spines. Cattle will very readily eat the cactus without this precaution and consequently get their intestines pierced. The young joints of several species serve as food for American Indians, chiefly in the southwestern states. The stems of O. engelmanii are fried, and the stems and fruits of O. clavata roasted. One of the most popular in the family is the *tuna* or *mission prickly pear* (O. megacantha) whose fruits are extremely sought after. Its joints are cooked

5. CACTI

prickly pear   sahuaro   beavertail

as a vegetable and from them a juice is extracted and used in poultices to reduce inflamation. It is also boiled with tallow to make candles. The joints, buds and flowers of the *beaver-tailed prickly pear* (O. basilaris) are prepared as a spring food by steaming them in a pit in the ground by the Indians of California and New Mexico.

In Mexico the fruit-pulp and seeds of the *cardon* (Pachycereus pringlei) are made into flour and used for *tomales*, a doughy bread-cake accompanying a meal, rather like the Indian *chapati*. The Spanish name means thistle. The *torch thistle* (Echinocereus pectinatus) of South and Central America and the West Indies is another cactus cultivated chiefly for the sake of its delicious gooseberry-like fruits, but the fleshy part of its stem may also be eaten after the spines have been removed. The true *West Indian* or *Barbados gooseberry* is the cactaceous *lemon vine* (Pereskia aculeata) also growing in Mexico. The leaves of this serve as a pot-herb, those of the Columbia *bleo* (P. bleo) as a vegetable.

Yet another cactus whose fruits are savoured is the *nopal* (genus Nopalea); however the Mexican Indians eat the joints of the *nopal chamacuero* or *Panama nopal* (N. dejecta). Lastly there is the *sahuaro* or *giant cactus* (Carnegiea gigantea) which serves as the emblem of Arizona, and from the woody stems of which Indians made their lances. The seeds are used to make a butter-like paste and *pinole* meal.

Carrot (*Daucus carota sativa*)

Our garden vegetable has developed through cultivation from the acrid and unpleasant *wild carrot* (D. carota). The Latin comes from the Greek name, which is thought to derive from the verb δαιω, to burn. The wild carrot goes under a variety of names, several of which were eventually transferred to our vegetable. *Dauke* is coined from the Latin. *Field-more*, and the northern Scottish *mirrot*, derive from the Old English *mora* and Scandinavian *morot*, meaning a root[1]. Several names refer to the plumed leaves: *bee's* or *crow's nest*; *bird's nest* (c.f. the German name *vogelnest*); *fiddle*; *kecks-head* and *hill-trot kecks*; *rantipole* (from *poll*, meaning a head). It is also known as *mug* and *yellow root*. The name 'carrot' also comes via the Greek and Latin and is said to be connected with the Irish *curran*, meaning any long root; one may compare with this the French dialect word for horseradish, *cran*.

The Greeks and Romans made little use of carrots, principally because no good crop can be obtained in the warm south, where they tend to

[1] The German for a carrot is *möhre*.

become bitter and revert towards their wild state. But Pliny mentions them as very popular in Gaul in the first century, from whence they were exported to Italy under the name *pastinaca gallica*. Much confusion arises from the word *pastinaca*, which also covers the parsnip and some other roots. The legacy still remains in some of the French dialect names for the carrot: *pastenade, pastenague, patenaille*. In spite of the carrot's bad record in warm climates we find the Arabs growing them in Spain and saying they were used throughout the Islamic world. Indians imported them from Persia and they reached China in the time of the Mongol Yüan dynasty (1280–1368). In Europe the Dutch have always been forward in cultivating the carrot and some of the best varieties come from Holland. Some say it was the Flemings who introduced it to England in the time of Elizabeth I, although it may have been that all they brought were their own more edible varieties. In Anglo-Saxon medicine we find the carrot as an ingredient in a drink against the devil and insanity, although this might have been the wild carrot.

On the whole the carrot is not very nourishing, although the redder it is the more vitamin A it contains; but for a long time it was the lighter coloured varieties which were most favoured. Old recipes recommend roasting them and then mixing them with oil, vinegar, spices and warm wine; other suggest frying. The colours of the carrot turn the range from near-white through yellow and orange to near-red. There are small succulent varieties and enormous 'cow carrots', so called because the largest varieties are fed to cattle, like the thin *Altrigham carrot*, which is half a yard long. In Australia the species D. brachiatus is used as sheep-fodder, while the wild *south-western carrot* (D. pussilus) of America, alternatively known as *rattlesnake weed*, is eaten raw or cooked by the Indians of the southwestern U.S.

Celandines and family (*Ranunculaceae*)

Most of the crowfoot family, to which the buttercup belongs, are extremely acrid and more or less poisonous. However, one or two have their uses, as do several relations. Alaskan Eskimos eat the roots of the *Arctic crowfoot* (R. pallasii), but the chief of the edible-rooted species is none other than the *lesser celandine* (R. ficaria). Wordsworth wrote three poems about the flower, of which two are entitled 'To the Small Celandine' and classed as *Poems of the Fancy*. Their general tenor is that the plant is a harbinger of spring and has gone unnoticed too long. The first of them was written on 30 April, 1802, and in it we are informed that:

> ... *the thrifty cottager,*
> *Who stirs little out of doors,*
> *Joys to spy thee near her home;*
> *Spring is coming, Thou art come!*

One imagines that the thrifty cottager had more to her delight in viewing the plant than that spring was round the corner. Lenten months were lean, and an assurance of greens soon to come must have been welcome. And to begin with there was the small celandine itself, whose foliage may be eaten like spinach, or bleached first.[1] In addition its tubers serve as a sort of potato substitute. Curiously enough the plant is assigned to birthdays falling on 22 July, thus obscuring its significance as the flower of joy to come. It used to be carried on the person, together with a mole's heart, as a talisman for success against enemies and law-suits.

Wordsworth's second poem was written three days later. In it he seems to have repented of emphasising the flower's obscurity and humbleness, and makes up for it with a mixture of folk-lore, observation and autobiography:

> *Pleasures newly found are sweet*
> *When they lie about our feet:*
> *February last, my heart*
> *First at sight of thee was glad;*
> *All unheard of as thou art,*
> *Thou must needs, I think, have had,*
> *Celandine! and long ago,*
> *Praise of which I nothing know ...*
>
> *Soon as gentle breezes bring*
> *News of winter's vanishing,*
> *And the children build their bowers,*
> *Sticking 'kerchief-plots of mould*
> *All about with full-blown flowers,*
> *Thick as sheep in shepherd's fold!*
> *With the proudest thou art there,*
> *Mantling in the tiny square ...*
>
> *Blithe of heart, from week to week*
> *Thou dost play at hide-and-seek;*
> *While the patient primrose sits*
> *Like a beggar in the cold,*
> *Thou, a flower of wiser wits,*
> *Slip'st into thy sheltering hold;*
> *Liveliest of the vernal train*
> *When ye all are out again.*[2]

---

[1] Perhaps the German name, *pfennigsalat* (penny-salad), indicates this.

[2] This stanza was transferred by the author from the first poem in the 1845 edition of his works.

82 / *Celandines and family*

Drawn by what peculiar spell,
By what charm of sight or smell,
Does the dim-eyed curious Bee,
Labouring for her waxen cells,
Fondly settle upon Thee,
Prized above all buds and bells
Opening daily at thy side,
By the season multiplied?

One might hazard a guess at the problem that puzzled Wordsworth in the stanza above. The bee, far from being dim-eyed, is equipped with vision sensitive to ultraviolet light; obviously the celandine must reflect this. The conduct of the flower in rough weather, mentioned in the preceding stanza, gives rise to yet another poem which Wordsworth assigns to *Poems referring to the Period of Old Age*, seemingly paradoxical until one has read it:

There is a flower, the lesser celandine,
That shrinks, like many more, from cold and rain;
And, the first moment that the sun may shine,
Bright as the sun himself, 'tis out again.

When hailstones have been falling, swarm on swarm,
Or blasts the green fields and the trees distrest,
Oft have I seen it muffled up from harm,
In close self-shelter, like a thing at rest.

But lately, one rough day, this flower I passed
And recognised it, though an altered form,
Now standing forth an offering to the blast
And buffeted at will by rain and storm.

I stopped and said with inly muttered voice,
It doth not love the shower, nor seek the cold:
This neither is its courage nor its choice,
But its necessity in being old.

The sunshine may not cheer it, nor the dew;
It cannot help itself in its decay;
Stiff in its members, withered, changed of hue.
And in my spleen I smiled that it was grey.

The celandine's name derives from the Latin *chelidonium*, an adjective formed from the word for a swallow; an old writer explains the name as 'swallow-herbe, bycause (as Plinie writeth) it was first found out by swallowes, and hath healed the eyes and restored the sight to their young ones'. The small celandine is named for its likeness to the great celandine (genus Chelidonium) or, as the same author explains, 'bycause that it beginneth to floure at the coming of the swallowes, and withereth at their return'. Since it belongs to the same family as buttercups it inevitably shares several names with them, including such references to the yellow flowers as *king* or *gold cup*; but names like *gilding cup, golden guineas* and *bright eye* seem to be peculiar to it. Other names refer to its roots, as in the official *fig-rooted buttercup* and the more popular *figwort*. The name *pilewort* (cf. the French *herbe aux hémorrhoides*) is explained as deriving from the Latin *pila* (a small ball) in reference to its small tuber; however that may be, it was once considered by country people to be good for piles.

The *marsh marigold* (Caltha palustris), also a member of the buttercup family, as such names as *great butterflower* and *big, water, horse* or *bull buttercup* indicate, has also been called *brave celandine*. The early leaves and stems may be eaten like spinach, and the flower-buds preserved in salted vinegar and used as capers. Thus it has been used in Europe and by the early settlers in America; the Red Indians have also used it, and the Ainus eat the roots.

The name marsh marigold derives from the Old English *mersc mear gealla*, best translated as *marsh mare (horse) blob*. Part of the name persists in the already-mentioned *horse buttercup*, and there are others incorporating the last element: *May, Molly, yellow, water, mire* or *horse blob*. A French name, *bouton d'or* (gold button), has influenced such English names as *boots*; *Billy, Bobby* or *soldier's buttons*; *meadow bout* or *bright*. Other references to the golden flowers appear in such names as *fire o' gold* and *water gowan, gowlan, golland* or *goggles*. It is also there in the unlikely name *Crazy Betty*, since 'crazy' (a corruption of *Christ's eye*) is applied to many bright orange or yellow flowers. *Claut* is a regional name surviving in the old Wiltshire simile 'yellow as a claut'. The rounded shape of the flowers has suggested *brave bassinets* (a small basin or skull-cap) as a name. Since the flower is usually to be found growing near water and in marshy places it has been called *drunkard*; from this grew up an old country belief that anyone gathering the flowers would become a drunkard. Also stemming from this name comes the alternative *publican*; buttercups and marsh marigolds growing together used to be referred to as *publicans and sinners*. Another French name, *jaulnette*, lingers on in the old Scottish *jonette* and probably lies behind such alternatives as *John Georges* and *Johnny cranes*. The 'crane' element is one that the plant also shares with buttercups in general; it is also known as *crow* or *cow crane*, and as *bull flower*.

To North Americans the plant is usually known as *cowslip* (cow's lip, by the way, not cow-slip), presumably because there are, or were, no true cowslips in the continent. The *elkslip* or *white-flowered marsh marigold* (C. leptosepala), which is to be found in the mountain chain from Alaska to New Mexico, is accounted an excellent pot-herb.

Many species of the *anemone family* (genus Anemone) are highly poisonous and must be approached with caution. The best-known European representative, the *wood anemone* (A. nemorosa), causes such animals as chance to eat it to die in convulsions and passing blood. Of the flowers in general Turner comments in his *Herbal* (1551) that 'the Anemone hath the name in Greeke of wynde because the floure never openeth but when the wynde bloweth . . . it may be called wynde-flour'. However, the primitive Ainus of northern Japan (Hokkaido) and Sakhalin gather large quantities of A. flaccida in the spring, drying the leaves and stems and storing them for use in the winter. Alaskan Eskimos use the leaves of the *narcissus* or *zephyr anemone* (A. narcissiflora) as cress, eating them soured or in oil. The related *wild rue* or *rue anemone* (Anemonella thalictoides) has edible starchy roots, referred to as *wild potatoes* in the mountainous districts of Pennsylvania, where they are still occasionally eaten by the local population.

The young sprouts of another relation, the vine-like *biting clematis* (Clematis vitalba), of European origin, are eaten in some parts. The wood of this plant used to be crumbled and smoked by young boys, from which practice it has gained such names as *smoke wood*, *tombacca* and *boy's bacca*. The tendrils have been used for binding up other plants in gardens, and thus named *bindwith*; other names for them include *bull-bine*, *love bind*, *belly wind*, *hedge vine*, *hag rope*, *crocodile* and *Robin Hood's fetter*. In autumn the white flowers develop into silky plumes in which the seeds are contained; these have suggested such names as *grey beards*, *old man's woozard*, *silver bush* (in Jersey), *beggar brushes* (cf. the French *herbe aux mendiants*) and *lady's* or *Virgin's bower* (also with French parallels). The plant is also known as *Devil's cut*, *honey stick* and *maiden's honesty*. A close relative very similar to the above is the *columbine meadow rue* (Thalictrum aquilegifolium), which grows in the mountainous districts of Europe and northern Asia. To gardeners it is known as *plumed columbine* and elsewhere as *feathered* or *tufted columbine* and *Spanish tuft*. Its slightly purgative roots are eaten raw or roasted by the Ainus.

A last relative is the *peony* (genus Paeonia), known to the Chinese as the queen of flowers (*mu-tan*), an emblem of March and the beginning of spring, symbolising wealth, honour, love and feminine loveliness. In the West, however, it is the birthday flower of 21 June and stands for anger, indignation, bashfulness, lowliness and shame. It is also known as king of flowers (*lo-yang*) to the Chinese, and is a symbol of longevity associated in their art with the phoenix, pheasant and peacock. A peony and a crowing cock signifies honours and success. To the Japanese it is the herald of summer, symbolic of brightness, virility and high rank.

Some of these qualities are the subject of Jon Silkin's poem concerning the flower:

It has a group of flowers.
Its buds shut, they exude
A moisture, a gum, expressed
From the sepals' metallic pressures.
Its colour shows between shields,
Cramped where the long neck
Swells into the head. Then they open.
They do it gradually,
Stammer at first. It is a confidence
Permits this; push aside
The shield, spray outwards,
Mount in height and colour
Upon the stem.
They claim the attention, up there,
The focus of all else. Not aloof at all;
Brilliantly intimate,
They make the whites of others
A shrunk milk. They must draw
To them, the male ardours,
Enthusiasms; are predatory
In seeking this. Obliterate the garden
In flikerless ease, gouging out
The reluctant desires . . .
By nature, a devourer. Cannot give.
Gives nothing.
In winter shrinks to a few sticks.
Its reversion, bunches of hollowness.
Pithless. Insenate, as before.

One species much grown as an ornamental flower is the *white flowered* or *Chinese peony* (P. albiflora), a native of Siberia, whose roots are used for food by Tartars and Mongolians. The hot seeds of the *common garden peony* (P. officinalis) were once ground into a spice in Europe. Langland mentions this as used, either in ale or with accompanying food, in his poem *Piers Plowman* (B. v: 311-13). Various dialect names for the plant play upon its name: *pianet, piny, posy, pie-nanny, nappie* (in Yorkshire) and *piano rose* (in Ireland). It also goes by the curious name of *sheep-shearing rose* 'from the rough joke of filling the folds of its petals with pungent snuff or pepper at sheep-shearing feasts, in order to enjoy the torments of those who innocently smell it'.

Celery and substitutes (*Apium graveolens*)

Celery is an umbelliferous plant with white edible stems. It is a relative of parsley, and of *wild celery* which was once cultivated and used as a condiment. Captain Cook's sailors are said to have used it as an antiscorbutic.

Celery and substitutes

The wild form of celery is also known as *marsh parsley*, and it was from cultivating this that our modern vegetable was developed. In classical times the celeries and the parsleys were both called by the same name so that one can never be quite sure which is meant. The classical and Egyptian funerary plant is often thought of as parsley, but it is most likely that a form of celery was used. There was a proverb of the time used of a man gravely ill: 'He only wants celery [parsley]'.

In the Middle Ages these plants were used as medicines, principally as laxatives and diuretics, but also for breaking up gall-stones and soothing swellings, and against jaundice oedema and wild beast bites. The vegetable variety, A. grav. var. dulce, was developed in Italy and, having lost its characteristic bitterness and foul smell, was eaten either raw or cooked from the sixteenth century onwards. There are now innumerable varieties, many of them force-grown, the result of which is that the celery season tends to begin and end somewhat earlier than formerly. *Australian* or *New Zealand celery* (A. australe) is a separate species used similarly in those countries.

Most of our names for celery have derived from old names for parsley. The word itself has come to us via Italian dialect and French from the Latin *selinum*. Latin *apium* becomes *ache* in French, and as *small-ache* in English develops into *smallage*. A Germanic root gives us *march*, which was *merce* in Old English and *merk* in Old Saxon; one may compare these with Danish *mærke* and Swedish *mörki*.

Knob or *turnip-rooted celery* (var. rapaceum), probably best known as *celeriac*, whose shoots are inedible, is probably older than our present celery, but for a long time was less popular. This too is of Mediterranean origin. The root is generally cooked, but may be eaten raw, as it was often formerly. It is very like *turnip-rooted parsley* (Petroselinum sativum var. tuberosun), which had long and ancient usage in Holland, Germany and Poland as a winter vegetable. Its leaves may be used as a condiment like those of the ordinary parsley (q.v.). In the sixteenth to seventeenth centuries its use gradually spread into the rest of Europe. The alternative names of *Dutch* or *Hamburg parsley* indicates its places of origin.

CELERY SUBSTITUTES

1. Alexander parsley *(Smyrnium olustratum)*
There was a time, before celery was widely cultivated, that *Alexander parsley* was used in every way that celery is used now. In a document of Charlemagne's time it is mentioned as an ordinary vegetable. Its English name, together with the alternative *Macedonian parsley*, refer to its supposed origin in Greece. A common shortened version of the name is

Alexanders, which is often corrupted into various forms, such as the Scottish *Elshinder*. Other names are *megweed* and *stanmarch*, which has developed from the Old English name for it, *stan-merce* (stone-celery). In the Isle of Wight it was frequently called *wild celery*, which rightly belongs to the wild version of celery proper. It is also known as *horse-parsley* and abounds in England where, indeed, it was once widely cultivated until about the end of the seventeenth century, when its popularity waned before the newer vegetable.

In classical times both the roots and the stem were used, eaten raw or, in the case of the roots, boiled with oil and vinegar. They are still reputed to be very good if kept all winter in a cool place away from the air, in order to make them more tender. In Anglo-Saxon medicine the stem was used in the Holy Salve, and the root was used in a bone salve and a medicine against typhus.

2. Lovage (*Levisticum officinale*)
This is another umbelliferous plant, a relative of the preceding (which was known as *black lovage*). It is of Mediterranean origin, and grows wild in that region. Its name derives ultimately from the Latin *ligusticus* (Ligurian), the name of the Italian shore opposite Genoa. The Old French was *levesche* or *luvesche*, which accounts for the English name *love-ache*, meaning love-parsley. It was formerly cultivated for the same purpose as Alexanders and was once widely used in England. Lovage leaves may be infused and used to provide a very popular cordial bearing the same name as the plant. In Anglo-Saxon magic it was an ingredient in a diuretic, in a spell against elf-disease (jaundice), in the green salve, and in a charm against typhoid fever. A relative, *Scottish lovage* (Ligusticum scoticum), is eaten both raw and boiled by Shetlanders, but the taste may come as rather a shock to those who are not used to it. This plant is also known as *sea-parsley*, and in Skye by the Gaelic name of *siunas*. The related *Canadian lovage* (L. canadense) is eaten by Alaskan Eskimos.

Chervils

There seem to be an indecent number of plants that go under this name, although a certain amount of confusion is understandable in the case of umbellifers. They are distinct from chervells, like honeysuckles, which derive their name from the French *chèvre-feuille* (goat leaf). Chervil, however, comes from a Greek word meaning 'leaf of rejoicing' or 'cheer-leaf'. The Latin was *chaerophyllum* or *cerifolium*, and the Old English *cærfille*; the Latin also underlies several of the European names: *cerfeuil* (French); *cerefoglio* (Italian); *kervell* (Dutch); *korffol* (German). Gerarde, a herbalist at the end of the sixteenth century, seems to confirm the original meaning with his praise: 'It is good for old people – it rejoiceth and comforteth the heart and increaseth their strength'. This may hold good in relation to the plants discussed here, but it does not with the root chervil (*see under* Parsnip), whose leaves are poisonous.

Salad chervil (Anthriscus cerefolium) is of East European origin and is to be found growing wild in southeast Russia and Persia, where it was introduced by the Greeks during the time of their domination over those areas. It was brought to England by the Romans. The leaves and stems are much used, especially in France, for seasonings, salads, soups and as a pot-herb. Its aromatic relative, *wild* or *mock chervil* (A. sylvestris), was also once used in England and Holland. Several animals are fond of it, especially cows and rabbits, as appears from the following names: *Common* or *smooth cow-parsley*; *cow chervil, mumble* or *weed*; *coney* or *hare parsley*; *rabbit meat* or *food*. The parsley connection is further underlined by the name *ass parsley*, which is probably a variation on *ache-parsley* (i.e. parsley-parsley), together with *sweet ash* (*ache*), *beaked hedge-parsley, dog's* or *pig's parsley, gipsy's parsley, wild* or *wood beaked parsley*.

At the same time a suspicion of the plant because of its likeness to hemlock gives it the cries of agony, *da* or *ha-ho* and *hi-how*, which are variations on hemlock's *hech-how*. Other sinister names which it sometimes shares with hemlock and its associates are *hemlock* itself, *humlock, adder's* or *devil's meat, bad* or *naughty man's oatmeal, rat's bane, scab flower, scabby hands* and *stepmother's-blossom*. An occasional link with other plants occurs in the names *wild carraway* or *ciceley*. Other names refer mostly to its provenance and the appearance of its flowers: *orchard weed; ghost key; gipsy curtains* or *umbrella*; *Queen Anne's, old lady's* or *my lady's lace*. In Cambridge during Elizabethan times the stems were called *cashes* or *caxes*, really a variation on *kex*; but under this name they were used as bobbins in weaving for winding yarn, a use to which others in the family were put elsewhere.

Sweet Ciceley (Myrrhis odorata) is another umbelliferous plant very like chervil which may originally have come from Asia. The name is from the Greek σεσσελι; it is also called *great* or *sweet chervil, sweet ciss, angelica, bracken fern* or *humlock*; in the north it may still be known as plain *sweets*. All refer to its sugary taste. It nows grows wild in the British Isles and is very common in Europe, especially in mountainous districts. In England it was very popular during the sixteenth and seventeenth centuries and was widely used on the continent, especially in France and Germany, still later. The leaves served as a pot-herb or were used in salads, and the roots were eaten either raw or boiled. The aromatic seeds were highly regarded, and in the north of England were once used for polishing furniture. The plant is still used as a stimulant for various disorders and sometimes for rubbing on hives, since bees are supposed to be fond of it. It also goes under the name of *British myrrh*, and in Lancashire it is called the *Roman plant*. A sixteenth-century writer gives the explanation for this: 'They do commonly call al such straunge herbs as be unknowen of the common people Romish or Romayne herbes, altho the same be brought from Norweigh, which is a country far distant from Roome'.

In North America various members of the *sweet-rooted ciceley* family (genus Osmorrhiza) are used as pickles or condiments, as for instance *sweet anise* (O. occidentalis). The roots of O. claytoni are eaten by Wisconsin Indians who wish to gain weight. The *honeworts* (genus Cryptotaenia) are another family of North American origin, also to be found in the Far East. In America the young leaves of the *American honewort* (C. canadensis)[1] are used in salads and soups. The whole plant serves as a vegetable, and the roots are also eaten. In Japan, where the plant is known as *mitsuba*, the latter are fried. The plant is there cultivated and the young aromatic shoots serve as a spinach or chicory substitute.

In northwestern America there are yet other edible aromatic umbelliferous relatives, such as the *ribweeds* (genus Aulospermum). The leaves of the *purple ribweed* (A. purpureum) are used for seasoning and as a potherb by the Navajos; other tribes use the leaves of A. longipes. The roots of the *harbinger of spring* (Erigenia bulbosa) are eaten in the Mid-West, where it blooms very early in the year in the moist lowlands.

The young plants of the European *small hartwort* (Tordylium apulum), growing in the Mediterranean area, are eaten as a vegetable in Greece while it is used as a condiment in Italy, from which connection it gains the name *Roman pimpernell*. In both Greece and Asia Minor the young plants of the *great shepherd's needle* (Scandix grandiflora) are much esteemed as a salad. The young tops and stems of the related *Venus, lady's* or *shepherd's comb* (S. pecten veneris), a native of Europe and Asia, and introduced into North America, may also be eaten. The curious name is explained by Gerarde: 'After the flowers, come uppe long seedes very like unto packneedles, orderlie set one by another like the great teeth of a combe'. From this likeness the plant receives several other alternative names: *needle* or *wild chervil*; *devil's darning needle* or *elshins* (i.e. alexanders, *see under* Celery); *Adam's, old woman's, beggar's* or *tailors needle*; *clock, crake, crow, poke, powk,* or *pound needle*.

[1] The original honewort is the English *corn parsley* (Petroselinum segetum).

Chicory and endive (*genus Cichorium*)

There is a certain confusion of names when it comes to dealing with this vegetable. The French *chicorée frisée* refers to the slim tattered leaf of what we normally call endive. The more bitter Roman-lettuce-shaped chicory proper (C. intybus) is called *chicorée sauvage*. Indeed, the uncultivated variety may be found growing wild under the name *blue succory*, named after its flower, which has been described as looking like a blue dandelion. It was once known as *wild* or *bitter lettuce* and also went under the names *hard ewes*, *monk's beard* and *strip for strip*. The first edible varieties were specially developed in Syria and Egypt, although even the leaves of the wild variety are just about passable if picked young enough. In the late nineteenth century special cultivation of a large-rooted variety used as a coffee substitute developed rapidly. The roots may also be used in salads. In France a blanched cultivated variety goes under the name of *barbe de capuchin*. But the most successful blanched variety, *Witloof* or *large-rooted Brussels chicory*, was discovered by accident about 1850 during an experimental attempt to grow white-leaved winter vegetables on a mushroom bed. Although the method was kept secret, news of it gradually spread along the gardeners' grapevine. It is a forced growth, taking about thirteen days and making a passable winter vegetable.

Witloof is the Flemish for 'white-leaf', an old name for wild chicory. The name chicory comes from the Greek κιχορεια, borrowed originally from the Coptic. The *intybus* of the Latin generic name comes via the Greek ἐντυβον, from the Syrian word for a flute (*ambubaia*) in reference to the thin hollow shape of the vegetable. In Byzantine Greek the word had become spelt ἐνδιβον, and the *beta* (β) was pronounced as a 'v'; thus it entered late Latin as *endiva*, which gives us the name of the *endive* (C. endiva).

The endive is of doubtful origin, some saying it was developed in the Mediterranean, others that it came originally from India. Although it was cultivated by the Greeks and Romans, the men of the Middle Ages do not seem to have distinguished it from succory until about the twelfth century, and even then it was not generally eaten as a salad vegetable until the sixteenth. At about this time it began to be cultivated in England. Of its varieties the narrow curly-leaved var. crispa is eaten in salads while the broad-leaved *Batavian endive* (var. latifolia) may be cooked.

Chrysanthemums (*genus Chrysanthemum*)

These flowers are members of the daisy family and closely allied to the sunflower; the majority of them are of Eastern origin and associated with the East. It is the national flower of Japan. There its petals dipped in saki are said to confer grace, health and longevity – the latter because it is a bringer of blessings from the fountain of youth in which resides Kiku-Jido, the chrysanthemum-boy, a sort of Japanese Peter Pan. For the Chinese the mid-autumn flower typifies joviality, the life of ease after retirement from public office; and also one whose beauty shows late. A chrysanthemum and oriole signifies the wish that the receiver's house be happy; a red one means that it is. A yellow flower means dejection and slighted love, a white one truth. About the latter there is a poem by the Japanese poet Oshikochi no Mitsune (active at the end of the ninth century):

> *I would have liked*
> *To pick one, groped for it,*
> *But white frost had*
> *Covered and hidden from me*
> *The white chrysanthemum.*

In the West the flower is the emblem of the solar wheel and governs the astrological house of Sagittarius. As birthday plant for 24 December it signifies abundance, wealth, regal beauty and cheerfulness in adversity.

The best known European species is the *ox-eye daisy* (C. leucanthemum), also growing in temperate Asia and introduced into North America. The young leaves are sometimes eaten as a salad and used as a pot-herb. It has been recommended as an emergency food-plant and is a home remedy for catarrh. In Britain the plant is also known as *great, bull, butter, midsummer, moon, thunder, poor-land, field, London* or *Devil's daisy*; *large dicky daisy*; *daisy goldins*; *dog flower* (or *daisy*); *horse penny* (or *daisy*). Regional names include *Dutch Morgan, white bothen* or *bozzum, espibawn* in Ulster, *gandervraw* and *cow's eyes* in Cornwall, and *catenaroes* in the North.

The leaves of another wild flower, the *corn, field* or *wild marigold* (C. segetum), were once used as a pot-herb. Otherwise it was regarded as a very troublesome weed, and in several North European countries there were laws passed against those who permitted it to grow on their land. Most popular names for the plant refer to its brilliant golden flowers: *Yellowby*; *yellow oxeye* or *cornflower*; *golden daisy, cornflower* or *sunflower*; *lady's roses*; *moon. Bigold* meant false gold; *boodle* or *buddle* came from the Dutch for a purse (*buidel*), in reference to another name for the flower, *goldins* (from the Dutch *gulden*, gold florins). Other names include *daisy* or *hoary thistle*, *harvest flower* and *St John's bloom*.

Several other species are used in various ways in China and Japan. The *garland* or *Sicilian chrysanthemum*, alternatively known as *crown daisy* (C. coroniarum), is of Mediterranean origin and widely grown as an ornamental flower. In the East it serves as a vegetable and the young seedlings as a pot-herb. The flower-heads of the *Japanese* or *mother chrysanthemum* (C. indicum) are preserved in vinegar, those of the *florist's* or *common garden chrysanthemum* (C. sinense) boiled as a vegetable or dried for use as a pot-herb; the leaves are also eaten. In Indo-China the young aromatic leaves of the related *ox-eye* (Buphthalum oleraceum) are used as a flavouring.

Costmary (C. pyrethrum or Tanacetum balsamita) is of oriental origin, now naturalised in southern Europe, and grown in English gardens since the sixteenth century. Its leaves may be blanched as a salad or used as a pot-herb. These are spicy and mint-like and have given the plant its names *balsam mint* and *mint geranium*; they were once added to ale as a flavouring, for which reason the plant is also known as *ale cost*. In Medieval times it was associated with the Virgin Mary, a fact attested by its present name and the older *herb Mary* (cf. the French *herbe Ste Marie* and the German *Marienmintz*). *Cost*, deriving from the Latin through the Greek, was transferred to it from another plant whose thick aromatic root was known as *kustha* in Sanskrit.[1]

The bitter-tasting juice of another relative, the *tansy* (Tanacetum vulgare), is poisonous if taken in large quantities but was once used in cooking omelettes and puddings at Easter-tide. These were themselves known as 'tansies' and eaten in memory of the bitter herbs accompanying the Jewish Paschal lamb. The plant was considered important enough for the early settlers to take with them to America. Its narcotic qualities made it useful for relieving the pain of ulcers, bruises and rheumatic complaints, and it is still used in stomachic drugs. At one time the oil was used as a poor man's substitute for nutmeg and cinnamon, and both the leaves and flowers had their uses. It figured in the flavouring of cheeses, from which it gains the obscure name of *Joynson's remedy-cheese*. Several names attest to its taste, such as *bitter-buttons* and *ginger plant* or *wood*. Others include *English cost*, *French weed*, *king's plant*, *mushroom* (referring to the round flower-heads, also known as buttons), *golden* or *batchelor's buttons*, *scented* or *parsley fern*.

[1] This was the *cane-reed spiral-flag* (Costus speciosus), used medicinally as a diuretic and eaten by the Asian poor; it has been taken to Brazil, where it is used as a condiment under the name *wild ginger*.

Clovers

The best known of the clovers, all of which are members of the pea family, is *herb-trefoil* (genus Trifolium), so called because it has three leaves growing together on one stalk. It was first introduced to England from Flanders in the seventeenth century and used for soil improvement when the rotation of crops was becoming accepted agricultural practice. It is also much relished by cattle, of course; so much so that the phrase 'to be in clover' is used of one living in great comfort or luxury. The trifoliate leaf is taken as being symbolical of the Trinity and therefore a good protection against the evil eye. A four-leaved clover, perhaps because it is so rare, is emblematic of good fortune, but also symbolises the cross. In the language of flowers clover speaks of promising and a white clover carries the message 'I think of you'.

It seems to be the Indians of America who make most use of the clovers in this and other families, either by using the seeds for bread or eating the plants raw or cooked. The seeds of the *Aztec clover* (T. amabile) are mixed with those of white maize and others to make the *chucan* meal of the Peruvian Indians. The leaves and stems of *foothill clover* (T. ciliatum) are eaten by the Indians of California. Elsewhere the following among others are eaten: *pinpoint clover* (T. gracilientum); *small-headed clover* (T. microcephalum); *tomcat clover* (T. tridentatum). The roots of the *slender white prairie clover* (Petalostemon oligophyllum) are eaten by the Indians of New Mexico. Elsewhere in the U.S. those of the *bush clover* (genus Hedysarum) are said to be eaten.

In Eurasia the Kalmuks eat the roots of various *sweet clovers* (genus Melilotus), among which are included the South Russian *Ruthenian sweet clover* (M. ruthenica) and *yellow sweet clover* or *common melilot* (M. officinalis). The latter, also known as *hart's, king's* or *plaster clover* and *wild laburnum*, serves to flavour cheeses elsewhere, as does M. altissima, a constituent of Swiss green cheese. This too is mixed with snuff. In the Mediterranean region and in East Africa M. elegans is mixed with butter.

Another clover used in Swiss green cheese is *blue* or *sweet trefoil* (Trigonella coerulia). This belongs to the same family as *fenugreek* (T. foenum-graecum) whose name, a corruption from the Latin, means 'Greek hay'. Its grain-like fruits are ground to produce a dye, a condiment and a curry-spice used in India, Egypt, and elsewhere in the Mediterranean region. The leaves may also be cooked and eaten. *Sweet fenugreek* (T. suavissima), an Australian relative alternatively known as *Australian shamrock*, and to the aborigines as *calombo*, grows in dried-up water courses and on lands subject to flooding. An early explorer spoke highly of its delicious flavour and the fact that it retained its green colour when boiled. Other species serve as forage crops.

Colocasia and family

1. Taro and dasheen (*genus Colocasia*)
Taro (C. macrorrhiza), the principal member of this family, grows wild easily in the dampness of the tropics; but it is cultivated principally in the Pacific islands and is particularly important in the Hawaian group. Captain Cook was the first European to discover it in the Sandwich Islands, although it is mentioned in Chinese writings about 100 B.C. and is also grown now as a food in Japan, Africa, the West Indies and South and Central America. It is mostly for the root that *taro* is cultivated, but the leaves and stem may be eaten cooked, and thus they lose their acridity. There are some species which are grown only for their green parts. The name *taro* is that used by the Maoris and other peoples of the Pacific; another form of it is *tallo*, which may be compared with *dalo* (Fiji), *tallus*, *tallas*, *tales* (Malaysia), and *kandalla* (Ceylon). In South Carolina, apparently, it is known as *tanga*. *Eddo* is the name it bears in Ghana, from which has come the alternative *eddy-root*

Kenko, the fourteenth-century Japanese essayist, records the very strange passion for this vegetable of a Buddhist monk:

There was an Archdeacon named Joshin, venerated for his wisdom, at the Shinjo-in monastry, who was extremely fond of *imogashira*,[1] which he ate in large quantities. Even when he was preaching he had a bowl piled high with them by his chair and would eat them while expounding his text. Whenever he was ill he used to shut himself up for one or two weeks, giving out that he was undergoing treatment. Having chosen himself the very best of this vegetable he indulged himself even more than usual and thus cured himself of all his maladies. He would not share them with others but always kept them for himself. Once he was extremely poor, but his teacher left him on his deathbed 200 *kwans*[2] of copper money and his hermitage besides. This he sold for a further 100 *kwans*, making 30,000 coins in all, and used the whole as capital for buying taro. He deposited the money with someone living in Kyoto, and buying them in 10 *kwan* loads at a time, so indulged himself that he spent the whole sum on this and no other item. That a poor man so suddenly

[1] Literally 'potato heads', taro.
[2] Thousand-coin rolls strung together.

6. SWOLLEN LEAF-BASES

       celeriac   dasheen
garlic   kohl-rabi
   Florence fennel   onion   leek

enriched should dispose of his money this way must argue, so men said, that he was of a truly exceptional piety. (*Tsurezuregusa*, section 60)

Another species, C. antiquorum, was noted by Europeans as growing in Egypt during the sixteenth century, where it was known as *khoulkas*, from which the Latin name is derived. The Arabs introduced it into Portugal, where it is called *alcoleaz*. This plant was the one connected with the naming of the yam (q.v.). In some parts of southern Italy it is called *aro di Egitto* (Egyptian arum).

Dasheen (C. esculenta) is a native of the East Indies which has also spread round the world; both the leaves and the tubers may be eaten. Other names for it are *elephant's ear*, *Egyptian ginger*, *Chinese potato*, *Tanyah tuber* and *poi-plant* (from the name of its pounded meal). *West Indian kale* is a name applied to its leaves.

2. Spoonflowers (*genus Xanthosoma*)

Members of this family grow chiefly in the tropical areas of South and Central America and the Caribbean. The use of both their leaves and roots goes back to before the European conquest. Although the roots do not grow very abundantly they are accounted among the best in the region; the rather tasteless leaves are boiled as a vegetable. Chief among these plants is the *arrow-leaved spoonflower* (X. sagittaefolium), known by such names as *tannias* and *yautia malanga* (or *yellow yautia*). Although the leaves are eaten, it is grown chiefly for the large central root and surrounding nut-sized tubercules. *Primrose malanga* (X. violaceum), also known as *oto* and *Indian kale*, now spread to the tropics of Africa and Asia, is also cultivated for its tubers, which are cooked and eaten like potatoes. The young leaves are chopped and boiled like spinach. *Dark-leaved malanga* (X. atrovirens), on the other hand, is cultivated mainly for the leaves, as the alternative names *Caribbean cabbage* and *West Indian kale* demonstrate. The corms are sometimes eaten, however. Much the same is true of *caracu* (X. caracu) of Central America, the West Indies and Puerto Rico, whose young unfolding leaves also serve as a pot-herb. But the tropical South American *belembe* (X. brasiliense) is only grown for its spinach-like leaves.

Among related plants, both the leaves and tubers of Caladium striatipes are eaten in some parts of Brazil; the roots only of the *meadow cabbage* (Symplocarpus or Spathyema foetida). The thick root-stocks of this plant send up grotesquely swollen purple-brown leaves enclosing many flower-clusters. They serve as an emergency food for the Indians of northeastern America; these and the fruits are most often reduced to flour and made into bread. The plant also rejoices in such elegant and enticing names as *skunk* or *pole-cat weed* and *swamp* or *skunk cabbage*.

The fruits of the Madagascan *viha* (Typhonodorum lindleyana) are occasionally eaten after much boiling. This is another strange-looking plant with a long white spathe, the whole plant sometimes growing over four metres high. The young leaves of the tropical Asian Lasia aculeata, and the peeled leaf-stalks, are used as a pot-herb in several countries. In Java they accompany rice, and in Ceylon they are added to curries.

3. Spoonlilies (*Alocasia macrorrhiza*)
Large rooted alocasia is also known as *roasting coco*, and by the Australian aboriginal name of *cunjevoi*, as well as by the family name of *spoonlily*. It is to be found growing wild in India, Ceylon and the Pacific area, including Australia, where it is the commonest species in its family. It has also been cultivated over a long period for the sake of its large fleshy rhizomes which extend for several feet over the surface of the earth. It is to be found growing in dense masses on river banks and in the shady moist forest patches of the tropics. The roots must first be cooked in order to dispel their bitterness. The Australian aboriginal method was to scrape the roots, divide them and place them under hot ashes for half an hour, after which they were pounded and all watery pieces rejected. This process could be repeated several times. The juice of the plant has been used as an antidote for the vicious sting of the Australian nettle tree (*see under* Nettles). Other native names include the interrelated *apé* of Tahiti, *kappe* in the Friendly Isles, and *habara* in Ceylon.

The corms of another species A. odora, growing from eastern India to China, are occasionally eaten. *Indo-Malaysian alocasia* (A. indica) is a native of South-East Asia. In Bengal its stems are eaten after a certain amount of special preparation and boiling. The nearly related genus Typhonium grows widely over eastern Australia; its tubers are eaten after roasting, or are pounded into flour by the aborigines. Other relatives include Cyrtosperma merkusii, whose large starchy root-stocks are eaten in the Philippines, and Anchomanes difformis, a native of tropical West Africa whose tubers are eaten in times of scarcity.

Corn salad and valerian (*Valerianella olitoria*)

This plant grows all over northern Europe, North America, Asia Minor and the Caucasus. Peasants used to gather it as a winter and Lenten substitute for lettuce, as they did the dandelion and wild species of the lettuce. In the sixteenth or seventeenth century it began to be cultivated, and was traditionally popular accompanying beetroot in the eighteenth. Nowadays its use is mostly confined to Germany and France. English names, however, testify to its having once been used: *lettuce-valerian, white pot-herb,* and in Cornwall *cornel-salette.* It is also known as *lamb's lettuce,* owing to a belief that it is particularly sought after by sheep; the name *milk-grass* comes about because flowering patches of it look like splashes of milk. Two stranger names are *fetticus* and *lob-lollie.*

Italian corn salad (V. eriocarpa), cultivated in the Mediterranean area, is used in much the same way as a salad plant. Another substitute from the same area is *African valerian* or *horn of plenty* (Fedia cornucopiae). The *spur valerian* (Centranthus macrosiphon), of Spanish origin, is somewhat inferior. The true valerian (genus Valeriana) is not used very much, although to the Arabians and medieval Europeans it was an aphrodisiac, the Greeks used it to prevent bleeding, and the Romans as incense. In fact the old spikenard perfume was made from its root. Animals are very fond of it. A piece of the root was used as a bait to catch rats, and cats are apparently in the habit of tearing the plant to pieces and rolling in it, and of digging up and eating the roots. The North American Indians used the root of the *edible valerian* (V. edulis) both as a boiled vegetable and for making bread.

Cowslip, oxlip and primrose (*genus Primula*)

Nowadays the *cowslip* (P. veris) is best known for the wine made from its flowers, once known as *culverkey wine* in Kent. These flowers used to be hung from maypoles; young girls used to pelt each other with them, or else tie them up into a *tosty ball* and play a game of catch known as tisty-tosty. Their likeness to a bunch of keys has given them such names as *key flowers, lady's keys, marriage keys, bunch of keys* and *keys of heaven.*[1] These last are the badge of St Peter (and therefore of his representative, the Pope) after whom the plant is also called *herb-Peter* and *St Peters-wort.* They are therefore thought of as unlocking and opening spring, and there is a belief they will lead one to buried treasure.

The cowslip is the birthday flower for 22 September and represents comeliness, pensiveness, rusticity and winning grace. In the language of flowers it signifies 'You are my divinity'. Shakespeare (*Cymbeline*, II. ii) makes a perceptive comment on its appearance when describing a mole on a lady's breast,

> *Cinque-spotted like the crimson drops*
> *I' the bottom of a cowslip.*

[1] Cf. similar German names such as *schüsselblume, St Petersschlüssel, himmelsschlüssel, himmelsschlöschen.*

100 / Cowlip, oxslip and primrose

Both the flowers and the leaves may be eaten in salads; and in Essex they are cooked inside a suet pudding known as cowslip pudding. Gerarde comments on the plant's medicinal virtues: 'They are named *Arthreticae* and *Herbae Paralysis*, for they are thought to be good against the paines of the joints and sinewes ... a conserve made with the flowers prevaileth woonderfully against the palsie'. His learned names lingered on in the country in such corrupted forms as *artetyke* and *palsywort* or *passwort*.

The name 'cowslip' is of doubtful derivation, although it is most likely to be a form of 'cow's lip'. Other names include *cowflop*, *cow-stropple* (or, *stripling*), and *cowsmouth* (Scotland). Another country name for it, *paigle* is also of dubious origin. It is used for the most part in the Eastern counties of England, but 'yellow as a paigle' is a popular simile in Kent and Suffolk, and the Kentish name *horse buckles* is reckoned to contain a corrupted form of the name. Other names include *horse knot, longlegs, lady's candlestick, racconals, galligaskins, drelip, crewel* (in the West), and *fairy cups, basins, flowers* or *bells*.

The true *oxlip* (P. elatior) is a rare plant in England, found only in the Eastern counties, Its name is elsewhere used for a large cowslip or variety of primrose. Alternative names includes *summerlocks* and *great cowslip*. The flowers may be used in salads. So too may those of the *primrose* (P. vulgaris) which is, in England, the prime representative of spring. One of its alternative names is *Spring flower*, and the name 'primrose' has developed from the Middle English *primerole* (also corrupted later to *primet*), derived through the French *primeverole* from the Italian *primaverola*, a dimunitive of *fior di prima vera*, the flower of early spring.

The flower is supposed to govern the astrological sign of Aquarius. In Japan it is known as *tsuki-miso* (the grass that looks at the moon), and thus, as the sunflower follows the sun like a lover, this is regarded as the friend of the moon. In Celtic areas it is thought of as a fairy plant, but in England it typifies wantonness. So, as the birthday flower for 7 May, it symbolises inconstancy and a lover's doubts, although in the language of flowers it is supposed to say 'Believe me'. As a flower of spring it also stands for early youth, innocence, and sadness. This last point Herrick makes in his poem 'To Primroses, filled with morning dew':

> *Why do you weep, sweet babes? Can tears*
>   *Speak grief in you*
>   *Who were but born*
>   *Just as the modest morn*
>   *Teem'd her refreshing dew?*
>
> *Alas! you have not known that shower*
>   *That mars a flower,*
>   *Nor felt the unkind*
>   *Breath of a blasting wind;*

Nor are ye worn with years,
  Or warp'd as we,
Who think it strange to see
Such pretty flowers, like to orphans young,
Speaking by tears before ye have a tongue.

The primrose was reckoned to be the Tory Prime Minister, Disraeli's, favourite flower. He was therefore known as 'the primrose sphynx', and his birthday (19 April) is known as 'Primrose Day'. The Conservative Primrose League was founded in his memory. Nevertheless, 'to walk the primrose path' (a phrase which seems to have originated with Shakespeare) is to go the road of pleasure without regard to virtue. John Clare likened the flower's colour to sulphur and brimstone, sufficient warning of what is considered to lie at the end of that path.

The flower looks nothing like a rose, of course, but the corruption of its original name has brought about such alternatives as *butter* or *golden rose*, *early* or *first rose*, *Lent* or *Easter rose*, and *rose-without-a-thorn*. The curious Irish name of *buckie-faulie* includes a local name for the dog-rose (*buckie*). Other names include *golden star*; *butter and eggs*; *jackets and petticoats*; *boys and girls* (a variety with short and long pistils); *darling-of-April*; *May spink* (in Scotland, where the season is later, of course); *plum-rocks* (Scotland); *simmeren* (Yorkshire). Several garden varieties go by the names *beef and greens*, *lady's frills*, *King-Charles-in-the-oak*, *Jack-in-the-green*, and *Jack-in-the-box*.

Cresses

Cress is an imprecise name generally covering a range of plants eaten as a sharp salad, sometimes as a condiment, and used as a pot-herb, the majority of which belong to the mustard family. At one time the name was thought to derive from the Latin *crescere* (to grow), but more recent opinion inclines to the Greek χρεσαιν (to crawl) as the word of origin, the majority of cresses being creeping plants. In Old English metathesis of the 'r' changed the word to *kerse*, forms of which persisted late in country dialects. Several are to be found growing in water or damp sites, and thus the name has been extended to plants in other families, sharp tasting and used similarly, found in the same situations. Finally there are a number of plants that go by the name of cress, more specifically pig's or swine's cress (cf. members of genera Coronopus, Lapsana, Anthemis and others), which are not eaten at all, or only in the greatest emergencies, as far as I have been able to discover. A selection of the more commonly used species follows,

1. Rock cress (genus *Arabis*)

The plants of this family are, as the name suggests, usually to be found on stony ground and growing against walls. They make quite attractive rockery plants, falling over the stones in a mass of flowers. The leaves are eaten in some areas of the U.S.; this might argue a prior European usage, and doubtless they are eaten elsewhere. Other general names for the family include *molewort* and *wall* or *tower cress*. *White Allison* (A. alpina), one of the commonest species, shares many names in common with *sweet Allison* (Alyssum maritimum), a related member of the mustard family which might also have served as a cress. Most of these names refer to the white flowers; *white rocket*, *snow-in-harvest* (or *summer*), *snow-on-the-mountain*, *snowdrift*, *Lady's cushion*, *bishop's wig*, *dusty husband*, *March-and-May*.

2. Winter cress (genus *Barbarea*)

These cresses are extremely resistant to cold and were once in high repute as a winter salad and pot-herb in the temperate regions of Europe and the U.S. The commonest species is *yellow* or *winter rocket* (B. vulgaris), also known as *wound rocket* since it was once used as a vulnerary. Its season has also endowed it with such names as *wintergreen cress* and *winter hedgeweed*. The Latin name probably derives from its maturing on St Barbara's Day (4 December), and it is thus known as *herb Barbara* and *barber's* or *St Barbara's cress*. Other names include *land cress* (distinguishing the plant from water cress), *wild cabbage* and *cassabully*. The most commonly cultivated in this family, however, is *yellow*, *bank* or *early winter cress* (B. praecox), also known as *American*, *Normandy* or *Belleisle cress*, from whose seeds oil is also obtained.

3. Shepherd's purse (*Capsella bursa-pastoris*)

The leaves of this plant serve both as a vegetable and salad, and the seeds, of which birds are very fond, have been used in bread-making. It is an antiscorbutic crucifer very closely related to *penny cress* (genus Thlaspi). The name is suggested by its heart-shaped capsule, also referred to as *shepherd's pouch*, *bag* or *scrip*, *lady's purse*, *witch's pouch*, *case-weed* (from the French *caisse*, a cash box) and *clappedepouch* (a bit of mongrel French, best expressed in the alternative name *rattle-pouch*). *Pick-purse* is either a play on names or an aggrieved realisation that the plant takes up good farming land. *Mother's heart* or *pick-your-mother's-heart-out* stems from a common children's game. The seed case is held out to a companion, cracking open when he takes hold of it, whereupon the triumphant cry of 'You've broken your mother's heart' or 'picked your mother's heart out' is raised. In the Swiss version the unfortunate is told that he has stolen a purse of gold from his mother and father. Other names include *cocowort*, *toywort*, *ward seed*, and *pepper-and-salt* (in reference to its taste).

4. Bitter cress (*genus Cardamine*)

Members of this family resemble water cress and are often used as a substitute. That most often used is *hairy bitter cress* or *cardamine* (C. hirsuta) from Northeastern America, also known as *Pennsylvania cress*. Its seed-pods also explode to the touch and it has thus been named *spiky-flower* and *touch-me-not*. The *Chilean bitter cress* (C. nasturtioides) is also used as a salad, while both the leaves and young root-shoots of the *Yeso bitter cress* (C. yesoensis) are eaten by the Ainu in Japan. The commonest species, however, is the *cuckoo flower* (C. pratensis), so called because of its appearance in spring, a native of Europe and North America recommended as an emergency food-plant. This is also known as *less* or *wild water cress* and *meadow cress*. Several names refer to the time of its appearance: *Whitsuntide gilliflower* and *Pentecost eye*, corrupted to *pinksten eye* and thus to *pigeon's eye*, *piggesnie* or *pig's eye*. It is connected with the Virgin, whose day falls on 25 March, and thus called *Our Lady's flock, smock, cloak, mantle, glove, shoe* and *milksile*. The birds and beasts of the season are celebrated in *lamb's lakins* (i.e. playthings) and *cuckoo* (or *cuckoo bird's*) *bread, buttons, buds, flower, shoes* and *stockings*. Its fondness for damp places is alluded to in such names as *swamp's companion* and *bog spinks* (a corruption of bog's pink). Other names include *bread-and-milk, laylock, Lucy Locket, smick-smock* and the Scottish *carsons*.

5. Spider herb (*genus Gynandropsis*)

This antiscorbutic cruciferous family is to be found growing in the tropical regions of Africa, Asia and Central America. In the latter area the leaves of the *showy spider herb* (G. speciosa) are eaten as a vegetable, while in Nigeria and India those of the aromatic *five-leaved spider herb* (G. pentaphylla) serve as a spinach substitute, pot-herb or food flavouring.

6. Pepperwort (*genus Lepidium*)

Pepperworts are named after their hot-tasting roots and can be found all over the world. The cultivated species is known as *garden cress* (L. sativum) or alternatively as *town cress*, a name deriving from the Old English *tun* (a garden or clearing), or *garth cress* in some areas of Northern England, the name deriving this time from an Old Norse word for yard. In Ireland it was sold under the name *tongue grass*, a name it shared with the chickweeds. The plant was supposed to give spirit to the fool and Romans used to taunt the cowardly and lazy with the saying 'Eat some cress'. It is this which is eaten in the combination known as mustard-and-cress. The pepperwort is generally oversown with white or black mustard after about three days, and the taste of the similar plants (both belong to the same family) are supposed to complement one another. The roots are occasionally used as a condiment. Those of the *mace* (L. mayenii) reach turnip-like proportions and are eaten as a vegetable by the Indians of the upper Andes in South America.

Among the various wild species one might note the North American *wild peppergrass* or *Virginian cress* (L. virginicum), and *pepperweed* or *hoary cress* (L. draba), whose seeds are used in sauces instead of pepper, the plant also being cooked as a vegetable in Asia Minor. The seeds of the *desert pepperwort* (L. fremontii) serve the Indians of Arizona (U.S.) as a flavouring and for food. *New Zealand cress* (L. oleraceum) was used as a pot-herb by the early settlers on South Island (N.Z.), and *round cress* (L. rotundum) is cooked as a vegetable by the aborigines of South Australia. *Field* or *wild cress* (L. campestre), also known as *poor man's pepper, churl's mustard* and *cow cress*, is the commonest European species. Old alternative names such as *mithridate* or *treacle mustard* record the fact that it was once used as an antidote to poisons and the bites of animals.

7. Loosestrife (*genus Lysimachia*)

Loosestrife is a member of the speedwell family, many of which are to be found in watery situations and were once counted as cresses. It is a country belief that the plant acted as a peace-maker between quarrelsome cattle, and for this reason a spray was tied to the yokes of such beasts before setting out on a long journey. In the East L. clethroides and L. fortuneii serve as condiments and L. obovata as a pot-herb.

8. Watercress (*genus Nasturtium*)

Watercress grows wild in running water and flooded places all over Europe, America and the East. It is high in iron content and an antiscorbutic. It may be cooked like spinach, but then, together with its extremely biting taste, it loses most of its good qualities. On the other hand, this way of treating it would certainly kill the liver-fluke it is suspected of carrying and passing on both to cattle and humans. The Persians regarded watercress as a valuable food for children; Romans and Anglo-Saxons advised it for falling hair and scurf. It grew so commonly in its wild state that there was no need for large-scale culture until the nineteenth century. Its cultivation was pioneered by Nicholas Mesmer of Germany in the sixteenth century and taken up in England in the seventeenth. It is now cultivated very widely in the various river valleys of Hertfordshire and the surrounding region, especially in chalky areas. It grows so rapidly that it may be cut every ten to fifteen days. The beds, however, are invaded by all sorts of weeds which are sometimes included with the cresses when they are sold. Occasionally the seeds are ground and added to mustard.

There is some confusion over the Latin generic name since the plant generally known as *nasturtium* is completely distinct (genus Tropaeolum; *see below*). Botanists have still not made up their minds which alternative name they prefer; both *Radicula* and *Roripa* have been suggested. On the other hand this chopping and changing is mischievous from a semantic point of view, cancelling out the amusing fact that the name derives from *nasus* (nose) and *torquere* (to twist), with the idea that the plant's hotness entails much screwing up of the nose and sneezing. The Old Latin name still survives in the Spanish *mastuerzo*, and *nasitor* in the Languedoc.

The cultivated species, N. aquaticum or officinale, also goes under a variety of interesting names, including *eker* (of Old English origin) and *billers* or *beller*, an equally old name still surviving in parts of Ireland. It was once known as *bilure* in England and *beler* in Cornwall, possibly deriving from the Latin *berula* or Welsh *berw* (to glow). Other names include *rib*, *black* or *brown cress* (cf. the German *brunnenkresse*), *teng* (sting) *tongue*, *long-tails* and *well grass*. A wild variety known as *marsh watercress* (N. palustre), *creeping* or *yellow cress*, is still gathered for salads in some parts of Europe. This also goes by such names as *annual radish*, *water rocket* and *bellragges* (another variation upon *beller*). In Asia the poor gather the *wild Indian watercress* (N. indicum).

9. Water naiad (*Najas major*)
This is a submersed plant very much like watercress, eaten with salt as an appetiser in Hawaii, and to be found on sale in the local markets. In Eocene times it grew in large quantities all over the world.

10. Brook weed (*Samolus valerandi*)
This plant, also known as *round* or *water pimpernel*, is another member of the speedwell family, commonly found growing in swampy ground. It is widespread, has antiscorbutic properties and is recommended as an emergency food plant. The young leaves serve as a salad or may be cooked like spinach.

11. Stonecrop (*genus Sedum*)
Many species of stonecrop have been cultivated for their hot bitter leaves, which may be used in soups and salads, as a pot-herb and a condiment. They were especially popular in the seventeenth century and are still used in Europe and North America. Among these was the *white stonecrop* (S. album) which was once used medicinally, like so many others in the family, particularly for treating worms and other stomach diseases. The name *wormgrass* is still used for this and one or two other species. It is also called *prick-madam*, a corruption of the French *trippe* or *trique madame* more usually applied to the equally edible (S. reflexum), which also goes under the names *stone-hot*, *dwarf house-leek*, *sow's ear*, *love-in-a-chain*, and *Indian fog* in Ireland.

Orpine (S. telephium), also known by such corruptions of the name as *alpine* and *orphan* or *harping Johnny*, is another edible species used medicinally and as a vulnerary. The name is supposed to allude to the gold-coloured pigment known as orpiment (Latin *auripigmentum*), which would be very appropriate to others bearing gold-coloured flowers in the stonecrop family but fails to fit this plant with its red blossoms. Other names include *midsummer men* and *Solomon's puzzles*. Another, *live-long*, refers to the plant's great tenacity; it used to be placed in various locations about the house as an ornament and was supposed to last several months with only a sprinkling of water once in a week. Such vitality led to another custom. Betrothed girls would pick two flowers on Midsummer Eve and observe how long they flourished, thus estimating the fidelity of their lovers. From this practice arose the alternative name, *live-long love-long*.

Biting stonecrop (S. acre) had only a medicinal use, serving among other things as an emetic for drunkards, which is the reason behind one of its more appealing alternative names, *welcome-home-husband-though-never-so-drunk*. Among other edible species may be included *St Vincent's rocks stonecrop* (S. rupestre), alternatively known as *jealousy* and *link-moss*; and *rosy-flowered stonecrop* (S. roseum), also known as *rosefoot*, and in Wales as *Snowdon rose*. Alaskan Eskimos eat the leaves of the latter soured or in oil.

12. Swine's cress (*genus Senebiera*)
Members of this family, also known as *wart-cress* because its juices were once used to remove warts, is another antiscorbutic crucifer. One species, S. lepidoides, grows in the Algerian Sahara region and is often eaten by the Tuareg, while *crowscress* (S. pinnatifida) serves as a condiment or salad in some parts of Europe.

13. Spot flower (*genus Spilanthes*)
Para cress (S. oleracea) is a very hot plant of Central American origin used as a condiment and considered good for the digestion, and as an antiscorbutic. In Madagascar it goes under the name *anamalahobe*. Other species, such as the *alphabet plant* (S. acmella), are considered good for toothache and are used in the manufacture of dentifrices. Its leaves may be eaten boiled as a vegetable or in salads.

14. Nasturtium (*genus Tropaeolum*)
The nasturtium is of South American origin. The *tall nasturtium*, popularly called *stortioner* (T. majus), and known in America as *Indian*, *Mexican* and *Peruvian cress*, together with the *dwarf* or *bush nasturtium* (T. minus), were brought to Europe in 1684. Their green parts taste rather like watercress and may be used in salads, while their fruits serve as capers. Other Andean species which may be used in the same way include the *yellow rock* or

wreath nasturtium (T. polyphyllum) and T. sessilifolium. In Brazil and Paraguay the leaves of the *five-finger nasturtium* (T. pentaphyllum) are used as a pot-herb. The root of the *Peruvian nasturtium* (T. tuberosum), known as *capucine*, is a staple food for some people in the Andean region. No larger than an egg, conical with swellings, it is a red-striped yellow in colour and very bitter tasting, although some of this is lost with cooking. It has been grown in France but was never very popular there. In Peru the roots go under the name of *ysaño* and *taiachas*. Other edible-rooted species include the *Patagonian nasturtium* (T. patagonicum), and T. edule. of Peru and Chile, only eaten by the Indians in time of scarcity.

15. Speedwell (*genus Veronica*)

Members of this family are also known as *cancerwort*, possible because they were once used medicinally to alleviate growths. The *great water speedwell*, *faverell* or *water pimpernell* (V. anagallis) is to be found throughout the temperate zone. Its leaves are eaten as a salad in Japan. *Brookline* (V. beccabunga) is also widespread and has antiscorbutic properties. The young shoots and stems may be eaten. Popular names for the plant include *pig's*, *horse* or *fool's cress*, *becky leaves*, *water purple* (in Northern England and Scotland) and *well ink* (in Scotland and Ireland). The name *beccabunga*, by which it also passes, is derived from the Flemish *beckpungen* (mouth-smart) in allusion to the pungency of its leaves.

Cucumbers and melons

The ordinary cucumber, Cucumis sativus, a member of the gourd family and possibly of Indian origin, has a very ancient history among the early civilisations. It has been found in Egyptian tombs of the twelfth dynasty and was among the vegetables the Jews sighed for after leaving Egypt. When they had settled in Palestine the Jews began growing cucumbers for themselves; they are mentioned in the prophecy of Isaiah: 'And the daughter of Zion is left as a cottage in a vineyard, as a lodge in a garden of cucumbers, as a besieged city'. (1:8). The explanation of this image is that the cultivator used to build a little cabin of branches in the middle of his field and stay there until harvest, driving off jackals and other night-prowlers; the abandoned cabin was then left to fall down. In classical times cucumbers enjoyed a very high reputation although they had been in use among the peoples of East Europe since time immemorial. The Emperor Tiberius is reputed to have eaten them every day of the year. By the second century A.D. they had been taken to China.

During the Middle Ages cucumbers were known, for we find them among a list of the plants growing in Charlemagne's garden. But they do not seem to have been grown in England until the fifteenth century. Nevertheless, the northern countries are now the principal growers in spite of the cucumber's need of heat to ripen. Among the many varieties are some small ones, such as the *apple* or *lemon cucumber*, also called *crystal apple*, and the *petite pepina*. Eastern varieties such as the Japanese *climbing cucumber* and the Himalyan *Sikkim cucumber* are considered some of the best.

The cucumber may be eaten raw as a salad vegetable, but it is very pleasant, and some might find it less indigestible, when lightly cooked, as it is often served in the East. In Indonesia and Malaya the young leaves and stems are eaten raw as a salad or cooked like spinach in a little water. The seed kernels, which yield a cooking oil, are also eaten occasionally.

The Latin *cucumis* becomes *cocombre* in Medieval French, and it is from this that we derive our name. It is tenuously linked with various Eastern forms of the word: *soukasa* (Sanskrit); σικυος (Greek); *kischuim* (Hebrew); *kissa* (Arabic). The *gherkin*, for which immature garden cucumbers are occasionally substituted pickled in France and Britain, derives its name from various Germanic names for the cucumber, although the old Indo-European root is present in ἀγγουρια in modern Greek and *agurka* in Romany. Compare these with *agurk* (Danish), *gurke* (German), and the Dutch *augurkje* (from which the English is said to derive).

The true *gherkin* (Cucumis anguria) is a prickly West Indian species covered in warts that may originally have been taken there from Africa. It is also to be found growing in Brazil. In the West Indies it is known as *little wild cucumber*, or in the creole-speaking areas as *maroon concombre* (the adjective signifies a plant that has become wild). In the U.S. it is often called *Jerusalem cucumber*. The fruits may be eaten cooked and are also used for pickling.

The Zanzibari *mandera cucumber* (C. sacleuxii) is also gathered young and used for pickling. Other members of the family include C. metuliferus, cultivated in tropical Africa and used in salads, and the Australian C. trigonus, which is eaten by the aborigines. *Cucumber root* (Medeola virginica), on the other hand, is actually the tuberous rhizome of a member of the lily family growing in northeastern America. Indians taught the early settlers its use as a pickle; it is also sometimes eaten in salads.

There are various other plants that go under the name of cucumber or bear a resemblance to one. The South American *climbing cucumber* (Cyclanthera edulis) is of Colombian origin and is used by Peruvians in soups. It is related to C. pedata, whose small fruits, like gherkins, are used in pickles; and to the *exploding cucumber* (C. explodens), so called because its ripe fruits burst at a touch. The *camias* or *cucumber tree* (Averrhoa bilimbi) belongs to the sorrel family and bears acid-pulped green fruits which are widely cultivated for food in South America and are equally esteemed in Ceylon, where it goes under the name *bilimbi* (Tamil) and *bilin* (Singhalese); the Malay is *bilimbing*. The tropical African *naras* (Acanthosicyos horrida) is a prickly shrub related to the cucumber bearing pleasantly acid fruits about the size of an orange which are eaten fresh or preserved. The Hottentots eat the seeds, which are known as *butter-pits* in South Africa. Lastly there is the tropical American *casabanana* or *melocoton* (Sicana odorifera), the last name being originally applied to a peach grafted with a quince by the Spaniards. The cucumber-like fruit is eaten when unripe. When ripe the fruit is brownish red and has a pleasant smell; it is then used in preserves.

Melons (C. melo) belong to the same family as the cucumber and are of African origin; the fruits are mostly used for dessert, although some may be used as a vegetable. These are grown chiefly in India, China and Japan, and are elongated like cucumbers. The Japanese *conomon* is an example. The seeds are edible and a cooking oil is expressed from them. The *melon tree* (Carica papaya), of Central American origin, bears fruit known as *pawpaw* or *papaya* (from the Carib *ababai*) which are also used principally as a dessert. However, the unripe fruits may be cooked as a marrow substitute, and the young leaves like spinach. This plant was first taken to the Philippines by the Spanish, from whence it spread through the tropics to India, and was taken from there to East Africa.

The *water melon* (Citrullus lanatus) is a native of Africa which was cultivated anciently in the Mediterranean area. It reached India in prehistoric times but was not taken to China until the tenth or eleventh centuries. It was also taken to America and is now cultivated in the U.S. The sweet juicy pulp of the ripe fruits, which may weigh up to 75 pounds, is eaten fresh and serves as an alternative to water in parched areas. It may be bitter sometimes and the Africans generally test it by striking the fruit with an axe and tasting the juice. A variety grown in India (var. fistulosus) has small round fruits the size of turnips which are cooked as a vegetable and pickled. From the seeds a cooking oil can be obtained; these are sometimes ground and baked into bread or, especially in southern China, parched and chewed. In Africa the young fruits are often eaten cooked and, under the name *egusi*, are very popular in soups in Nigeria.

Finally the *bitter melon* or *balsam pear* (Momordica charantia), also known as *carilla fruit* (from the Singhalese *karawila*), is of African and Asian origin, but now introduced into North and South America. It is a warty ovoid fruit which turns from green to orange. A small variety (var. abreviata) is the one most used for cookery in China, Peru, and in other localities. The fruit is usually gathered before maturity while it is still firm. It is so bitter that it is advisable to boil it lightly first and then throw away the water before continuing. Another edible relative is the *balsam* or *marvellous apple* (M. balsamina), also known as *apple of Jerusalem*. A seventeenth-century writer mentions 'the balm apple whose oyle doth close up wounds like balm'. In India they eat the fruits of M. dioica, which goes under the same names. The young shoots, leaves and roots are eaten also.

Dandelion *(Taraxacum officinale)*

The dandelion grows all over Europe, Central Asia, and North America. It is related to chicory, with which the Romans confounded it. Although it has been eaten and otherwise used since time immemorial, it is only since the end of the nineteenth century that it has been cultivated in such places as market gardens. In England it has lost its popularity, but it is still widely used on the continent, either as a salad vegetable or cooked. The leaves taste very bitter, and should be cooked in at least two changes of water; when blanched they loose their bitterness. As a matter of fact they are extremely nourishing, containing more iron than spinach, four times more vitamin C than lettuce, and vitamin A. For this reason the juice is considered good for the teeth and gums. The roots especially have been used medicinally, as a diuretic, a stimulant, and for liver disorders. Because of their iron content the cooked leaves are very good for dermatitis and have been claimed to cure skin diseases when doctors have given up.

John Silkin has described the plant in a poem:

> *Slugs nestle where the stem*
> *Broken, bleeds milk.*
> *The flower is eyeless: the sight is compelled*
> *By small, coarse, sharp petals,*
> *Like metal shreds. Formed,*
> *They puncture, irregularly perforate*
> *Their yellow, brutal glare.*
> *And certainly want to*
> *Devour the earth. With an ample movement*
> *They are a foot high, as you look.*
> *And coming back, they take hold*
> *On pert domestic strains.*
> *Others' lives are theirs. Between them*
> *And domesticity,*
> *Grass. They infest its weak land;*
> *Fatten, hide slugs, infestate.*
> *They look like plates; more closely*
> *Like the first tryings, the machines, of nature*
> *Riveted into her, successful.*

Dandelions / 111

The roots of the dandelion, the tenaciousness of which is referred to in the last line above, are roasted and used in a bitter dusty coffee available from health-food stores. Needless to say, its resemblance to real coffee is extremely tenuous. From the leaves a kind of beer is made, and from the crushed flower-heads a light golden wine. The making of the latter is described in Ray Bradbury's haunting novel *Dandelion Wine*: it contains and keeps for winter, he says, the warmth of summer. In fact the flower's glowing colour, attested to by such names as *burning fire, sunflower of Spring*, and *golden sun* or *lady*, has made it an emblem of the sun, and in astrology it is governed by the house of Leo (in which the sun is said to dwell).

Most people know that the name is derived from the French *dent-de-lion*, the meaning of which is retained in the name *lion's teeth*. Some of the original spellings in the dialect variants are close to the French: *dentelion, dantdalyon*, and the country name *dandy-go-lion*. Very similar, too, is the Welsh *dant-y-llew* preserved in a thirteenth-century document. There are further parallels with the French: the diuretic properties of the roots is referred to in the name *pissenlit*, in English *piss-a-bed* with variations on it with the words dirt, mess, pee, shit and wet, together with the names *pissy-mudder* and *pissimire*. Again, *tete-de-moine* (cf. the German *mönchskopf*) reappears in *monk's head* and *priest's crown*, referring to the tonsure-like flower head after the seeds are dispersed; the same idea lies behind *swine-snout*.

By far the majority of names concern themselves with the feathery seed-head and the practice of blowing it either to tell the time or one's fortune, or even telling the weather, since the seeds are supposed to float off when it is going to rain on a windless day: *Bessy clock, blowball; blower; childrens'* or *schoolboy's clock; fairy clock; farmer's, old man's* or *shepherd's clock; clock flower* or *posy; clocks and watches; down balls; fluffy-puffy; fortune teller; one-two-three; tell-time; time-tables.* or *flower; weather-glass; white clock; what's-o'clock; twelve, one* or *four-o'clock*. The last name might also have something to do with the belief that the dandelion closes early, recorded also in the name *lay-a-bed*. Another children's practice of picking the stalk and blowing through it gives the name *bum-pipe*; the bitter milky fluid that flows from the stalk suggests *devil's milk-pail, milky disle, milk gowan* and *witches gowan* (a gowan was, at one time, a gallon measure of liquids).

The bitter taste of the plant's parts is recorded in the name *bitterwort*, which goes back to Anglo-Saxon times, and its medicinal use is referred to in *heart-fever grass*. Several names describe the tattered leaves, or compare them with the smooth hairy flower: *barbed arrows and fishhooks; combs and hairpins; canker* (cf. the Dutch *cankerblæmen*). Other names describe its tenaciousness and give a harsh judgement of its food value:

conquermoors; *grampher*; *livelong*; *male*; *crow-parsnip*; *dog's grass* or *posy*; *rabbit meat*; *grigglesticks*; *Irish daisy*; *stink Davie* (Scotland). Some of these names it shares, wholly or in part, with various of the thistles and groundsels, with which it would seem to be associated in the minds of the countryman. For instance, children used also to blow at the fluffy tops of thistles-heads in their games. One of the rhymes they sang while doing this explains many of the dandelion's names:

> *Marian, Marian, what's the time of day?*
> *One o'clock, two o'clock, it's time we were away.*

In certain instances each puff at the head counted as three hours.

Yet another tradition that 'as the lover blows off the downy heads they carry his thoughts to the absent one' may lie behind one more of its many names, *wishes*. Alternatively it can be blown to the accompaniment of 'she loves me, she loves me not', although it is easy to cheat here. A better oracle is obtained if one follows the tradition that one is loved passionately if all the seeds blow off at one breath; with reservations and some unfaithfulness if a few are left; and if many, with indifference or not at all. As the birthday flower of 27 September it is emblematic of bitterness, coquetry and grief, which is not a particularly fortunate association.

There are, in fact, one or two plants, in the same family and bearing a similarity to thistles or groundsel, members of the genus Senecio, which are also eaten. The *American fireweed* or *burnweed* is now classed as Erechthites hierarcifolia and considered good for piles (*pilewort* is an alternative name); this is occasionally used as a vegetable in North America. *Kaempfer's golden-ray* (S. or Ligularia kaemferi) grows in Japan, where the flower-head is boiled or salted; in the same area the Ainus eat the young leaves of S. palmatus. The *leafy-stemmed false dandelion* (Pyrrhopappus catolianus) is a North American plant whose roots are eaten in autumn by the Kiowa Indians. *Cat's ear* or *gosmore* (Hypochoeris maculata), a plant growing in Europe and Eurasia, is considered an emergency food plant. Its leaves look like those of the dandelion and may be used in salads.

Egg plant (*Solanum melongena*)

The name derives from the edible fruits of this plant, a smooth shiny-skinned deep purple fruit, not always ovoid and very much larger than an egg. It is a relative of the tomato, the potato and the deadly nightshade, the best known to the English of the many solanaceous plants. In Asia and southern Europe it is very popular, and is becoming increasingly so in

CONTENTS

Acknowledgements   *page* 13
Introduction   15

1. Leaves, stems, roots and fruits

Akee	*page* 25	Lentils	*page* 123
Amaranths	25	Lettuces	124
Angelica	28	Lily, iris and crocus	126
Artichokes	29	Mallows	145
Asparagus and substitutes	33	Milkweeds	148
Avens	38	Mustards	149
Avocado pear	39	Nettles	151
Bamboos and canes	39	Olives	154
Bananas	41	Onions and family	155
Beans	47	Palms	163
Beets and chards	60	Parsnip and related roots	168
Borage and family	61	Peas and vetches	171
Breadfruits and breadnuts	62	Pignuts and carraways	178
Cabbages and other coleworts	66	Poppy	179
Cacti	76	Potato	180
Carrot	79	Potato-beans and other substitutes	189
Celandines and family	80	Purslanes	191
Celery and substitutes	85	Radish	193
Chervils	87	Rampion	195
Chicory and endive	90	Reeds and rushes	195
Chrysanthemums	91	Salsify	201
Clovers	93	Samphires	202
Colocasia and family	94	Scorzonera and burdock	203
Corn salad and valerian	98	Seaweeds	204
Cowslip, oxlip and primrose	99	Skirret	215
Cresses	101	Sorrels and docks	216
Cucumbers and melons	107	Spinach, orach and blite	222
Dandelion	110	Sweet potato and other creepers	230
Egg plant	112	Thistles	234
Evening primrose	113	Tomatoes	237
Ferns	114	Turnips and other worts	240
Gourds, marrows and squashes	117		

Contents

Violets	244	Water plants	251
Water lilies	245	Yams	255

2. Miscellaneous nuts, grains and pastas

Acorns and beechnuts	page 261	Lichens	page 285
Almonds and other fruit kernels	263	Maize	287
Arrowroot, salep and orchids	264	Millet, sorghum and other grasses	289
Arums	268	Oats	291
Barley	270	Pasta	295
Brazil nut and relations	273	Pine nuts	296
Buckwheat	274	Queensland nut	297
Cashew	275	Rice	298
Cassava	276	Rye	300
Chestnuts	279	Sago palms and cycads	301
Groundnuts	282	Walnut	305
Hazel nuts	282	Wheat	306

3. Various herbs, spices and condiments

Balm	315	Marigold	333
Basil	315	Marjoram and oregano	334
Bay	316	Mints	335
Burnet	318	Nutmeg and mace	339
Capers	319	Parsley	340
Capsicums	320	Pepper	341
Cardamoms	321	Plantain	345
Cinnamon and cassia	322	Rosemary	348
Cloves	323	Rue	349
Coriander	324	Sage	350
Curry flavourings and spices	324	Savory	352
Dill	327	Sesame	352
Fennel	328	Tarragon and family	353
Ginger	329	Thyme	355
Horseradish and susumber	330		

Index of Latin names	359
Index of common names	373

Britain. The *egg fruit* (sometimes called *garden egg* in the West Indies) is probably of Indian origin. But it appears to have been the Arabs who first thought of eating it, and we find mention of it in the works of Avicenna in the fourth century. The first European mention is in the thirteenth century, when it was imported by the ships of the Italian trading states of Pisa, Genoa and Venice, coming from the East. At first, as with most of the other edible solanaceous plants, all of which were regarded as highly poisonous, it was distrusted and grown only as an ornamental plant. By the end of the fifteenth century people had begun to eat it, and it was taken with the Spanish and Portuguese to their American possessions, where it thrived excellently. The fruit comes in several shapes and sizes, and there is a yellow-white variety as well as the purple. In the Malagasy region they eat a greeny-yellow species called *grosse anguine* (S. macrocarpum) which is inferior and rather bitter tasting.

There are several other names for the egg fruit; it is not known by that name even by all people in England. *Guinea squash*, and *Jew's apple* are two of the most colourful. But the Bengali *brinjal*, and the French *aubergine*, ultimately connected etymologically, are equally popular names in England. The Sanskrit name was *vatingana*, and the Persian *badin-gan*. Arab variations were *badinjan* and *badinjal*, the latter of which developed into Bengali *brinjal*. It must have been members of the Indian nation arriving in the West Indies who were ultimately responsible for the corruption of the name there to *Brown-jolly*. The Arab presence in the Iberian peninsula has left its impression in the Portugese name *beringela*, and the Spanish *alberengena*, to which the Arabic for 'the' (*al*) is prefixed. On the other side of the Pyrenees these are supposed to have been corrupted into the Provencal *meringeane* and the French *verangène, melongène* and *aubergine*. An alternative explanation for the last name (but one which I doubt holds much water) is that the word is a diminutive of an old French word for peach, *auberge* or *alberge*, with which one may compare the Spanish for apricot, *alberchigo*, or *alverchiga*.

Evening primrose *(Oenothera biennis)*

The evening primrose is a plant of American origin now naturalised in Europe, where it grows in waste places and along railway embankments, and is considered a troublesome weed. It is named for its primrose-like flower which opens only towards night, but is not a member of the primrose family itself. John Clare, the nineteenth-century peasant poet, had a particular affection for this plant:

> *When once the sun sinks in the West*
> *And dewdrops pearl the evening's breast,*
> *Almost as pale as moonbeams are*
> *Or its companionable star,*
> *The evening primrose opes anew*
> *Its delicate blossoms to the dew*
> *And shunning-hermit of the light*
> *Wastes its fair bloom upon the night;*
> *Who, blindfold to its fond caresses,*
> *Knows not the beauty he possesses.*
> *Thus it blooms on till night is by*
> *And day looks out with open eye;*
> *Abashed at the gaze it cannot shun*
> *It faints and withers, and is done.*

It is sometimes cultivated for the sake of its sweet-tasting roots, particularly in Germany; in England they go by the name *large rampion*. These are said to have a vinous scent when dried and were formerly eaten as an incentive to wine-drinking (hence the Greek-derived generic name, cf. the German *weinblume*). There is a story that the Romans, who knew the plant at first simply as *onagra* (ass-food, cf. the French *herbe aux ânes*) swiftly changed the name when they found out about its marvellous properties from the Greeks.

The young shoots of the evening primrose are also eaten, and in Chile the leaves of two other species, the *dandelion sundrop* (O. acaulis) and O. mollissima. In North America the young shoots and leaves of the related *rosemary-leaved willowherb* (Epilobium rosmarinifolium), which grows in Arctic and temperate regions, are used like asparagus and as a pot-herb.

Ferns

Ferns represent some of the oldest forms of vegetation in the world. Fern-fossils are frequently to be found in coal. They and their allies, giant tree-ferns and cycads (*see under* Sago), were once spread over the greater part of the earth. Little attention is paid to them now apart from by palaeobotanists. They are used very rarely even for such one-time purposes as animal fodder and roofing material. Several were once used medicinally, and one of the explanations of how the maidenhair fern came by its name is that young girls used its roots for dying their hair yellow. Nevertheless, several ferns are edible in one way or another, even if their users are now largely restricted to the more primitive or poor peoples of the world.

which apparently means 'eaten green'. 'Pumpkin' goes back to the Greek πεπων (Latin *pepo*), which becomes *poupon* in early French and, having been nasalised into *poumpon*, enters English as *pompion*, to which was added the diminutive '-kin' ending.

Another plant whose fruits may be used like a vegetable marrow is that known in the French West Indies as *christophine* (Sechium edule), called elsewhere in the area *cho-cho*, *choke* or *chaka plant*, and in Brazil as *chayote*. It is a creeper bearing pear-shaped warty fruits very popular both in the Americas and in warm climates in other continents. There are two main types, white or green fleshed, which are boiled before fully mature. The young leaves and tendrils are also edible, while in Mexico and elsewhere the tuberous roots are occasionally eaten.

The gourd-like fruit of the *Zanzibar oilvine* or *fluted pumpkin* (Telfairia pedata), which grows exclusively in Africa, has an edible seed known as *telfairia nut*. This may be used in confectionery, boiled or eaten raw, or roasted and made into a paste. Although the fruit itself is inedible the leaves and young shoots may be cooked. So may those of the *krobonko* (T. occidentalis), whose large flat seeds have long been relished by African natives and are now entering the international market under the name *oysternuts*.

Gourds, of course, have many uses, domestic and ornamental. There are inedible varieties looking like waxy warted apples, lemons and tangerines, all the fascinating produce of the same plant. During classical times and the Middle Ages a certain kind of edible gourd was known, although no one is quite sure which it was. It was probably one of the bottle gourds (genus Lagenaria). These have woody rinds which are used as bottles, cooking and eating implements, and a whole range of other useful objects. The inside pulp is bitter and carminative, even poisonous in some cases. One of these (L. vulgaris), however, may be eaten like a marrow when young. The young shoots and leaves are occasionally used as a pot-herb, and the seeds may be chewed or pressed for cooking-oil. It is especially popular in India. The plant seems to have been common to both Africa and America since remote times. It was probably borne over the sea from one to the other and thus propagated, since the fruit is capable of withstanding sea-water without coming to harm for up to 224 days. Remains have been found in Egyptian tombs dating from 3500–3000 B.C., while in Mexico it was being used about 7000–5500 B.C. and in Peru 4000–3000 B.C. Roman authors mention it, and it seems to have reached China not later than the first century. The Maoris took it with them to New Zealand in the twelfth century.

The *wax* or *white gourd* (Benincasa cerifera) is from China and India and was first noted by Europeans in 1827. It also grows wild in Java but is little cultivated outside Asia. It may be used as a superior cucumber or marrow and has long been consumed in the East. The seeds also are fried and eaten, and the young leaves and flower-buds used as a vegetable.

There are several edible varieties of the *snake gourd* (genus Trichosanthes), originally from India and now spread over the tropics. The *club* or *viper gourd* (T. anguina), known in South America as the *serpent cucumber*, has a long thin fruit which is eaten when young, either pickled or seasoned with butter and mixed herbs. The *Japanese snake gourd* (T. cucumeroides) grows as far west as India and in Western Australia. The very bitter T. ovigera grows in Indonesia, while a less bitter version (T. dioica) has yellow-orange fruit eaten cooked when ripe in India, where it is known by such names as *palbal* (Hindi) and *patol* (Bengali). The *redball snake gourd* (T. bracteata) is also eaten in the same way. In Australia the aborigines of Queensland eat the roots of T. subvelutina.

Another gourd of Indian origin, now growing in tropical Asia and West Africa, is the fruit of the *strainer vine* (genus Luffa). The *dish-cloth* or *towel gourd* (L. acutangula), also known by the African name of *singkwa*, and in Sierra Leone as *long okra*, is gathered when only semi-developed; cut up, boiled and prepared with butter and salt, it is said to taste as good as garden peas. Also growing in the same regions, similar to the above but with a rounder body, is the *vegetable sponge* or *washing gourd* (L. aegyptiaca), which does under such African names as *suakwa* or *soolyqua gourd*. This is still quite edible in a young developed state; the mature seeds may be roasted, and the leaves and flower-buds are eaten as well.

The reason for some of the curious English names is that when these gourds become fully mature they are fibrous, with a consistency somewhat like Shredded Wheat, and are then often used as the household implements represented in the names. Europeans are perhaps best acquainted with it in the form of our bathroom friend the back-scrubbing 'loofah', whose name derives from the Egyptian via the Latin, since it was Egyptian varieties of this which were first used for the purpose.

The *scarlet-fruited* or *Indian ivy gourd* (Coccinea indica) grows in tropical Asia, and in Africa along the Red Sea coast and into the Sudan. The fruits are eaten raw or cooked in Africa, while both the young fruits and shoots are eaten with rice in South-East Asia. In Central America the young fruits of other members of the gourd family are boiled as a vegetable. These include Elaterium ciliatum and the Costa Rican *tacaco* (Polakowskia tacaco).

Various members of the gourd family figure in the iconography and mythology of several races and traditions. To the Hebrews it is the symbol of resurrection through its connection with Jonah. He sheltered under a gourd-tree waiting for the destruction of Nineveh which never came. Justifiably angry that his prophecy had not come to pass after he had gone through so much, he was taught the meaning of mercy and compassion said that the eagle when he will upfly, in order that he may see the more brightly, will touch his eyes with the juice and wet them, and he through that obtains the greatest brightness'. The Latin name *lactuca* refers to the plant's milky, mildly soporific juice, which has a slightly relaxing narcotic effect. Most European names derive from the Latin: *lattuga* (Italian); *lachucha* (Spanish); *laittues* (the Old French, from which the English name is derived); *latuw* (Dutch); *lattich* (German); *laktuk* (Russian).

Where the lettuce was first cultivated is unclear. A form of it was grown in Egypt, from whence it was introduced to the Jews. Among the bitter herbs to be eaten at the feast of the Passover (Exodus 12:8) were probably members of the chicory family such as this. The Greeks and Romans also cultivated a lettuce. In ancient Rome it was eaten at the end of a feast to dispose the reveller for sleep; a little later it was eaten with radish at the beginning of a meal to stimulate appetite. In the Near East the lettuce was connected with the cult of Ishtar and Thammuz (whose classical equivalents were Venus and Adonis, the goddess of love and the beloved hunter killed in the chase). The dead Thammuz was supposed to have been laid on a bed of lettuce. When his death was celebrated several quick growing plants (with lettuce as a favourite) reared in pots, were borne in procession as a symbol of the transitory nature of life. The potted plants were known as 'gardens of Adonis'. In the Far East we find the lettuce first mentioned in China between A.D. 600 and 900.

There are now several hundred varieties of lettuce, of which the three major categories are crispy-leaved, round-hearted (like a cabbage), and cylindrical-hearted. There are some strange hybrids like the *Batavian lettuce*, a tough and bitter cross between endive and lettuce. *Celtuce* (var. agustana) looks like a a cross between celery and lettuce; also known as *asparagus* or *stem lettuce*, it is grown almost exclusively in China for the young fleshy stems, which are cooked; the outer leaves are coarse and unpalatable. A cylindrical-hearted variety known to the Romans as *Cappadocian lettuce* is now called *Roman lettuce* or *romaine* by many. This dates from the time when the popes moved from Rome to Avignon in the fourteenth century, bringing this variety with them and having it grown in the palace gardens. It was therefore known as *Avignon lettuce* as well. In England, however, it is called *Cos lettuce*, after the Greek island which was the birthplace of the famous doctor Hippocrates; it was brought to England in the time of Charles I. In the Pamir region and in China it is also grown and eaten raw or cooked.

Lily, iris and crocus

(*a*) LILY FAMILY

Some members of this extremely wide and varied family are treated elsewhere; they include asparagus and its cousin, the broomrape; and the members of the onion family, including garlic and leek. These have always been regarded as vegetables and for the pot, while the bulb-borne flowers of the lily and others are regarded more as decorative plants. Nevertheless there are some members of the family which have had a long history of being eaten; the majority of lilies proper are more or less palatable and have often been used in times of scarcity or by nomadic tribes. The best method for cooking them is to peel the bulbs scale by scale and wash carefully; then boil them slightly in salted water and afterwards sauté in butter or cook in a sauce. The following are such members of the family as I can discover that have been or still are used.

1. Asphodel (*genus Asphodelus*)

The greyish leaves and yellowy flowers of this plant suggested the palor of death in classical times, and it was therefore connected with the Underworld, where the dead were supposed to feed on them. They were planted on graves, and the flowers made up Persephone's crown as queen of Hades. The English daffodil probably bears a corrupted form of the name – the German name for an asphodel is *affodill*. In the Mediterranean region there are about seven species of the asphodel. In classical times they served as a food for the poorer Greeks and they figure largely in Roman cookery. One method of cooking was to roast the bulb over hot ashes and then add oil and salt. Pounded together with figs, it was considered a great delicacy. That chiefly used was the *yellow asphodel* (A. luteus), also known as *Jacob's rod* (cf. the French *bâton de Jacob*) and *king's spear*, originally from central Europe and Asia Minor. The *onion asphodel* (A. fistulosus) of the Mediterranean region and Arabia is still eaten boiled by the Bedouin tribes.

2. Blue Dicks (*Brodiaea capitata*)

The sweet bulbs of this plant, native to the western U.S., are eaten by the southwestern Indians.

3. Mariposa lily (*genus Calochortus*)

The plants go under several other names, including *butterfly lily*, *tulip* or *weed*, and *star* or *globe tulip*. The corms, many of which are eaten by through his sorrow for the tree, which had withered in the sun and thus deprived him of shade. His sojourn in the belly of the whale and consequent escape is taken as a type of resurrection. In the Christian tradition the gourd is the attribute of Christ, St James and the Archangel Raphael.

The peoples of Indo-China and Burma are supposed to have sprung from a pumpkin-like gourd. The first two people, Yatawm (Ta-hsek-khi) and Yatai (Ya-hsek-khi), were shaped like tadpoles and had a daughter who became an earth-goddess. But the first couple, brought into being by a gourd, eventually met their deaths by eating a gourd.

Lentils (*Ervum lens*, or *Lens culinaris esculenta*)

Lentils are small pea-like fruits which are used for cooking when dried. Two varieties are obtainable in England. The French, sold entire, usually ash-grey or yellow in colour; and the Egyptian, husked, split, reddish-yellow, small and rounder than the former, and more readily obtainable. In the East they are grown as far as India, where they are a principal form of nourishment, especially to those whose diet is vegetarian. They are rich in protein, which is essential to those who also abstain from eating eggs and cheese as well. Lentils are also grown in North Africa where, prepared with oil and garlic, they are very popular; in Europe they are grown as far north as Germany, and they are also cultivated in both North and South America.

The origin of lentils was probably in Asia Minor, from whence they spread to southern Europe and western Asia as men first began cultivation. They have always been a field rather than a garden crop, and no wild plant now exists. They have been in use since prehistoric times, and remains of them have been found in a Hungarian stone-age site and in Egyptian tombs of the twelfth dynasty, about 2400–2200 B.C. The Louvre is, or was, a proud possessor of three uncooked lentils from such a source; they were observed to be no different to those still sold in Egyptian markets and similar to the French domestic varieties. They are first mentioned in Egyptian writings of the nineteenth dynasty, when they were called *arshana*, possibly a corruption of a Semitic word still present: the Hebrew was *adash*, the Arabic is *adas* and the Berber *adès*.

Of their use by the Hebrews the Bible gives us several examples. It was a pottage of red lentils for which Esau sold his birthright to Jacob (Genesis 25:34). Together with parched corn and pulse, lentils were brought as a gift to David by his people when he was fleeing from a revolt raised against him by his son Absalom (2 Samuel 17:28); and it was pulse, amongst which lentils would figure, that Daniel asked to eat, rather than meat sent from the king's table, when he was a captive in Babylon (Daniel 1:12–15).

The use of lentils continued through classical times, and they continued to be grown in southern Europe. During the nineteenth century they began to be attacked by various insect pests and the French moved their production to Lorraine, where the colder weather killed off the insects but the lentils were unharmed. Shortly after this France lost the area to the Germans, who took up lentil production themselves and introduced it to other areas in the north with wide success.

In most countries the word for lentil is very similar. Latin *lens* was typically diminished to *lenticula* and thus became the French *lentille*, originally spelt *lentile* in English. Incidentally the country people shortened this to *tills* at one time, declaring that the 'Lent' part was superflous since it did not fruit at that time! The old Slavic word is *lesha* and the Old German *linsi*; between them they give rise to the following modern forms: *linse* (German); *linze* (Dutch); *lenszik* (Lithuanian); *lencse* (Hungarian); *leča* (Slovene); *leče* (Serbo-Croat). *Dhal* is the Indian word used in restaurants, where it is frequently served.

Lettuces (*Lactuca sativa*)

The lettuce is a relative of the chicory family that grows all over Europe, Mediterranean Africa and the Near East. It is related to the *poison lettuce* (L. virosa), which is also a common plant in Europe. This, like most members of its family, is narcotic, from which is gains its name *sleepwort*. It is also known as *strong-scented* or *medicinal lettuce*, *green endive* and *horse thistle*. Another relation is the blue- or violet- flowered *perennial lettuce* (L. perennis), which grows on stony and chalky ground, and is considerably removed from the yellow-flowered annual vegetable. It used to be eaten by French peasants in the same way as dandelion leaves (q.v.). Other wild varieties of edible lettuce include the *Canada wild lettuce* (L. canadensis) of North America, *least* or *willow lettuce* (L. saligna), and *oak-leaved lettuce* (L. quercina). There are also some oriental cultivated varieties: the Japanese *tsitsa* (L. tsitsa), and *Indian lettuce* (L. indica), grown mostly in China, Japan and South-East Asia, the leaves of which are usually cooked.

Our lettuce was probably cultivated from the *prickly, wood* or *wild lettuce* (L. scariola), originally a very bitter plant. It was this species that was once used for dimness of the eyes. An old author explains that 'It is

8. BULBS

Florentina

martagon lily
dog's tooth violet

tulip

Solomon's seal
wild tulip
lily

various Indian tribes in northwestern America, go under the name *Indian potato*. The most celebrated is the *sego lily* (C. nuttallii) of the central and southern U.S., whose roots were said to have been eaten by the Mormon settlers at Salt Lake City (Utah), for which reason the flower is the emblem of that state. The corms of this and its variety, the *golden sego lily* (var. aureum), are boiled or roasted by various Indian tribes, especially the Hopis and Navahos. The Indians of California eat those of the *pretty grass* (C. pulchellus), known locally as *wild tulip*, raw or roasted, and further north those of the *north-western pussy-ears* (C. elegans) are used. Finally the Cheyenne Indians eat the corms of the *sagebrush mariposa* (S. gunnisonii) either boiled or dried and pounded into porridge flour.

4. Camass root *(genus Camassia)*
The name for the plant derives from the Indian *quamash*. By them the bulbs are boiled, roasted or eaten raw. Taught by them the early settlers ate those of the *common camass* (C. quamash) of northwestern America. Also used was the *Atlantic camass* (C. esculenta or scilloides) of the eastern and southern states of the U.S. The plant also goes by such names as *wild hyacinth, bear's grass* and *indigo squill*.

5. Bead lily *(genus Clintonia)*
This plant is a member of the lily-of-the-valley family; of its six species, four are of American origin and two Asian. The best known in America are the *yellow bead lily* (C. borealis) and the *white* or *speckled bead lily* (C. umbellata). They are said to be eaten as a vegetable in eastern North America.

6. Palm lily *(genus Cordyline)*
Palm lilies grow in the Pacific area and are relatives of the dragon trees (*see under* Asparagus). Indeed, there is a record of a Maori who hollowed a storeroom with a door out of the trunk, and the tree still continued to live. This might be compared with the story of the Canary Island dragon tree. The trees are remarkably tenacious of life. A settler once used the stems of one as a chimney and kept a fire going in it for many months. After he abandoned his hut the burnt-through stems eventually sprang to life again. It is said that even a chip flying from an axe will put down rootlets if it lands in damp earth.

In some Pacific islands the prepared roots of the *ti-pore* or *common club-palm* (C. terminalis) serve the natives as an important part of their food, and an alcoholic drink is also prepared from it. The *ti-kouka* (C. australis) was christened *cabbage tree* by the early settlers in New Zealand who used the young and tender heads as a cabbage substitute. The Maoris use the heads of some palms (q.v.) in the same way.

7. Desert candle *(genus Eremurus)*

Two Central Asian species of this plant are used. The leaves of one (E. aucherianus) are eaten as a vegetable, and the young shoots of the other (E. spectabilis). These are often to be found sold in the markets of the Crimea, Caucasia and Kurdistan.

8. Fawn lily *(genus Erythronium)*

There are about fifteen species of this plant, all but one of which can be found in North America, where they are also known as *trout lilies*. The corms of several are said to have been eaten by the Indians, for instance those of the *lamb's tongues* or *Oregon fawnlily* (E. grandiflorum), which are used by the Indians of the Pacific coast. They must, however, be specially prepared and cooked; in a raw state they are used as emetics by country folk. The plant is also to be found in Japan, where the leaves are boiled as a vegetable and a flour is extracted from the corms and often used for making noodles. In Siberia and Mongolia the root extract is cooked with cow or reindeer milk.

In Europe, however, the only representative seems to be the *dogtooth fawnlily* (E. dens canis), best known as the *dog's tooth violet*, and also as *yellow adder's tongue*, one of the earliest of spring flowers, which grows into eastern Siberia. In Asia it is very popular and cultivated as a vegetable. A supply was once sent to the Tsarist court in Russia, and it was considered an aphrodisiac. In his novel, *The Zemgano Brothers*, Edmond de Goncourt describes how a circus caravan draws up in a field for the night and a search is instituted for this plant, which he says 'makes such excellent salad'.

9. Fritillary *(genus Fritillaria)*

The name of this plant is derived from the Latin *fritillus* (dice box) in reference to the squarish dark markings on the flower. Thus the English species is often referred to as *chequered daffodil, lily* or *tulip* (F. meleagris). This is to be found growing in some parts of southern England and very occasionally further north. Victoria Sackville-West gives a vivid description in the spring passage of *The Land*:

> *And then I came to a field where the springing grass*
> *Was dulled by the hanging cups of fritillaries,*
> *Sullen and foreign-looking, like Egyptian girls*
> *Camping among the furze, staining the waste*
> *With foreign colour, sulky, dark and quaint,*
> *Dangerous too, as a girl might sidle up,*
> *An Egyptian girl, with an ancient snaring spell,*
> *Throwing a net, soft round the limbs and heart,*
> *Captivity soft and abhorrent, a close-meshed net,*

–See the square web on the murrey flesh of the flower–
Holding her captive close with her bare brown arms,
Close to her little breast beneath the silk,
A gipsy Judith, witch of a ragged tent.

I doubt whether the bulbs are at all edible. Those of a garden species, the *crown imperial* (F. imperialis), contain a bitter substance which is poisonous if taken in large amounts, and also smell disagreeable; a passable flour can be extracted from these nevertheless. A very palatable flour is obtained from the bulb of the *Kamchatka fritillary* (F. camschatensis), once called *black lily*, and known as *sarana* to the local inhabitants; it is also to be found in the Kuril and Aleutian Islands. This grows wild in considerable numbers and is also cultivated in kitchen gardens and even on the thatched roofs of some cottages. It is gathered in autumn, dried in the sun and then stored until needed. The bulb is either roasted in embers and eaten like a potato, or baked in an oven and pounded into a flour used for bread, in soups and many other dishes. It has a pleasantly bitter taste and is very nourishing. The Ainus and Aleutians eat the bulbs with fat or rice.

Royle's fritillary (F. roylei) grows on the high plateaus of Tibet, where the bulbs are cooked in sugared water. It is very popular with birds and was once used in Chinese medicine under the name *pe-mou*. In north-western America the bulbs of the *yellow fritillary* (F. pudica) are eaten raw or boiled by the Indians, who also dry and store them for winter use.

10. Day lily (*genus Hemerocallis*)

The name of this plant stems from the belief either that it is only the flower of a day, or else that it is only a day flower, closing at night. The dried flowers of the *grass-leaf day lily* (H. minor) are used as a pot-herb in China, and in Japan to a certain extent. There the Ainus use the related *shiboshi* (Funkia ovata), otherwise known as *common blue plantain lily* or *Japanese day lily*, boiling the white parts of the leaf-stalk. The buds and flowers of the *lemon day lily* (H. flava) are added to soups in North America. A Japanese species (H. aurantiaca), bearing faintly fragrant yellow flowers, is known as 'forgetting grass' (*wasuregusa*) and symbolises coquetry and short lived affection; for this reason it appears often in poetry. The poets occasionally affect to confuse its spear-like leaves with those of an evergreen fern (Polypodium lineare, *see this family under* Ferns) conveniently known as 'herb of remembrance' (*shinobu*).

One day a man was walking along the corridor between the Koroden and the Seiryoden (two sections of the Imperial Palace in Kyoto). A hand thrust out a sprig of greenery from inside a high-ranking lady's apartment and he was asked 'Can forgetting grass be called herb of remembrance?' He took the plant (and the hint that his attentions to the lady had been remiss of late) and replied:

> *Altho the fields seem*
> *Overgrown with forgetting grass*
> *This is the herb of remembrance;*
> *And remembering,*
> *I look forward to the future.*[1]

11. Common lily (*genus Lilium*)

The white lily has always been a symbol of purity in Europe, the attribute of Jesus, the Virgin Mary, and the Archangel Gabriel. On the other hand, if one looks closely at the paintings in which they appear, one sometimes finds that the lily is of quite a different family – the calla or arum lily, in fact, a member of the arum family and relative of the cuckoo-pint. The compliment has been repaid by the Americans who have christened one of the lily's relatives a wake-robin, a name usually applied to cuckoo-pint-like arums!

The lily also symbolises a host of other things in the Christian West: bashfulness; beauty; celestial beatitude (heavenly bliss); chastity, purity and sinlessness; good works; grace and innocence; divine marriage and eternal love; queenliness; trust in God and the joyful mystery of the rosary. It is said to have sprung from the repentant tears of Eve as she left Eden and was the emblem of the Israelite tribe of Judah (to which Jesus belonged). Thus one sees how the flower has been linked with the three personages already mentioned. It is also the emblem of several saints: St Anthony of Padua, a Franciscan monk whose eloquence was such that an ass once knelt to the sacrament when he was preaching; St Catherine of Siena, the Dominican mystic; the Irish St Kenelm; St Clara; St Isabella; St Othila; St Nicholas of Tolentino. It also belongs to St Joseph the Worker, Jesus' father, whose staff was said to have blossomed with lilies as a sign that his wife was still a virgin. It is the emblem of Upper Egypt and the badge of Florence. A more unfortunate association is recorded by Reginald Arkhill in his collection of humorous poems upon gardens and gardenings:

> *She walked beside the lily-bed*
> *And, as she looked at them, she said:*
> *'They make me think of being dead'.*
>   ('Green Fingers Again')

In China the flower symbolises showiness and short-lived beauty, but it is believed to be favourable at childbirth and is worn by expectant mothers who want a son. It is also regarded as an antidote to sorrow and is known as 'forget-grief herb'. Painters both in the East and the West are fond of painting it, and it frequently appears in poetry. Especially in the later

---

[1] Section 100, Ise Monogatari, *trans.* Helen McCullough. Stanford University Press, 1968.

Middle Ages it was usually associated with roses in descriptions of the peerless complexion of the beloved. Lines like:

> With lilies white
> And roses bright
> Doth strive thy colour fair

reappear *ad nauseam*. But it need hardly be pointed out that there are other more cheerful colours than white to be found among lilies. Jesus assures us that 'even Solomon in all his glory was not arrayed like one of these'. Some hold that the particular 'Lily of the field' to which he was referring was the scarlet martagon lily (L. chalcedonium).

The field lily is the birthday flower for 13 November and symbolises humility; 13 January is the imperial lily and symbolises majesty. The yellow or golden lily belongs to 14 November and stands for playful beauty, coquetry and falsehood; in Christian art, however, it symbolises divine light. A lily and a dove stands for annunciation and may be compared with the similar Eastern white swan and lotus.

In the Old Testament the word for lily probably covers the iris and gladiolus as well. It figures suggestively in Solomon's 'Song of Songs'.

> Thy two breasts are like two young roes that are twins, which feed among the lilies;
>
> My beloved is mine and I am his: he feedeth among the lilies.

This sort of thing should lead the more thoughtful person to enquire just what it is that is so attractive about feeding in such surroundings. It may be the very satisfying smell of some species which attaches itself to the person, for in the same poem the beloved's lips are described as two lilies 'dropping with myrrh'. On the other hand, less romantic though it may sound, the object may have been to eat them. It seems to be standard Eastern practice. The following are a selection of edible species:

Gold-band or *gold-striped lily* (L. auratum), a hardy large-bulbed plant which grows wild in Japan. It is much cultivated for its flowers, which figure frequently in Japanese paintings. The bulb is eaten.

Bulbil or *orange lily* (L. bulbiferum); both the bulb and the flower are eaten in China.

Warty red Japanese or *Slimston lily* (L. callosum), one of the most cultivated for food in Japan.

Heart-leaved lily (L. cordifolium); in Japan the young leaves are used as a vegetable and a food starch is obtained from the bulbs. A variety known as the *turep* (var. glehni), which sounds like a Japanese attempt at 'tulip', or as *oba-ubayuri*, is eaten by the Ainu, but the bulb requires some preparation first.

Dahurian or *Siberian orange lily* (L. dauricum), also eaten by the Ainus, and possibly elsewhere. Its name comes from the ancient kingdom of Dahuria (Davuria or Dauria) which once existed about where the borders of Siberia, Mongolia and Manchuria meet. It is also to be found growing in Sakhalin, Japan and Korea. The plant is one of the parents of the *Thunberg lily* (L. elegans) whose flowers are considered a great delicacy in China and Japan.

Japanese lily (L. japonicum) has only a mediocre bulb. It grows wild in the mountains, where it is gathered by the poor and referred to as *sassayuri* (little bamboo-leaved lily).

Spear-leaved Japanese lily (L. lancifolium) is cultivated for its edible bulbs.

Easter or *common trumpet lily* (L. longiflorum) grows wild in the Riu Kiu islands to the south of Japan. The bulb is eaten.

Martagon or *turk's cap lily* (L. martagon) grows wild in Europe and Asia and is the most widespread of all the lilies, which has led some to conjecture that perhaps it is the oldest existing species in the world. The origin of the name has been much debated and is generally taken to have some reference to the turned-up petals, resembling a Turkish turban (cf. the tulip); these have also suggested such alternative names as *turn-cap* and *turn-again-gentleman*. The bulb is eaten in Eurasia and the Far East, either roasted as in Siberia, or dried and eaten with cow or reindeer milk, as among the Volga Cossacks and in northern Mongolia.

Medeola or *wheel lily* (L. medeoloides); the Ainus eat the bulbs with rice or millet.

Columbia or *Oregon lily* (L. parviflorum) grows in northwestern America; the bulbs are eaten by the Indians of British Columbia.

Whorl leaved American lily (L. philadelphicum), also known as *wood* or *wild orange-red lily*, is a North American species. The Indians of the Mid-West eat the bulbs like potatoes.

Pompon or *turban lily* (L. pomponium) is eaten in the Kamchatka region.

Sargent lily (L. sargentiae); the flowers are eaten in some parts of China.

Spotted or *showy lily* (L. speciosum) is cultivated in Japanese gardens for its flowers (a great favourite in the arts and crafts) and its bulbs.

Siberian lily (L. spectabile); the bulbs are eaten roasted.

Swamp lily (L. superbum), also known as *great American turk's cap lily*, is a species belonging to the eastern U.S., where the bulbs are used by Indians for thickening soups.

Tom-thumb or *coral lily* (L. tenuifolium) is exceptionally sought after for its bulbs in Siberia.

Tiger lily (L. tigridum) is symbolic of pride and wealth to the iconographers. In England it is known as *crumple lily* because its petals turn up at the end. In spite of its rather bitter bulbs it is cultivated as a food plant in China and Japan; they are either boiled as a vegetable or prepared as a sweet.

12. Snake's head lily (*genus Ophipogon*)
The name refers to the appearance of the flower. The tubers of a Japanese species known as *dwarf lily turf* (O. japonicus) are eaten in the East.

13. Solomon's seal (*genus Polygonatum*)
This is another near relation of the lily-of-the-valley. Its thick white rootstocks give rise to the name *white-root* or *wort* for the English P. multiflorum. The general name of Solomon's seal also applies to the roots which, so Dioscorides says, were good in ointments for sealing up wounds; their markings are thought to be reminiscent of the interlinked triangles in stellar form which were said to be used by the wise man as his seal. The very young shoots of the European *common Solomon's seal* (P. officinale) are eaten like asparagus and recommended as a food in times of emergency, while the rhizomes of the *Japanese Solomon's seal* (P. japonicum) are similarly prepared. Those of the *giant Solomon's seal* (P. giganteum), which grows in temperate Asia and in northwestern America, are used for an edible starch by the Japanese, while the Ainus eat them boiled or roasted. The Indians of northeastern America eat the starchy root-stocks of the *small Solomon's seal* (P. biflorum), which are also recommended in time of want. They may be used for bread, and the stems are sometimes eaten as a vegetable or pickled.

14. Cat or green briars (*genus Smilax*)
These plants are tough twining shrubs with prickly stems, the rhizomes of many of which are used to produce sarsaparilla, and some of which are used as aphrodisiacs. Most are of American origin. Dioscorides mentions that the shoots of the *prickly ivy* (S. aspera) were used like asparagus, as are those of many American species; they are also used in soups along with the young leaves, particularly those of the *American green briar* (S. rotundifolia). The pounded root-stocks of this and several others are used for bread or porridge by various Indian tribes, particularly the Seminoles. In South-East Asia the rhizomes of S. megacarpa are eaten, and in Japan they boil the young leaves of the *Japanese carrion flower* (S. heracea var. nipponicum).

15. American wood-lily (*genus Trillium*)
These plants, also known as *wake-robin* or *three-leaved nightshade*, are members of the asparagus family. Those with purple flowers have an unpleasant smell resembling rotting meat; this acts as a lure to the carrion fly upon which the plants are largely dependent for fertilisation. Several in

the family are used as a bitter-sweet vegetable which, however, must first be cooked. The roots, known as *beth* or *birth roots*, are used as an emetic in their raw state but are eaten cooked by North American Indians. Those plants principally used include *snow* or *great wood-lily* (T. grandiflorum), *ill-scented wake-robin* or *purple-flowered wood-lily* (T. erectum), *Indian balm* (T. pendulum), *mountain* or *rose wood-lily* (T. stylosum) and the beautiful *painted wood-lily* (T. undulatum). Species are also to be found in Asia.

16. Tulip (*genus Tulipa*)

'The tulip is the peacock among flowers: the one has no scent, the other no song; the one glories in its gown, the other in its train'. Thus an old French book on gardening of the showy Eurasian flower which, for all that there are wild species abundant in Italy and Greece, is not mentioned in Europe before 1554, when cultivated Turkish varieties were remarked upon by a visiting ambassador. From Turkey the plant was introduced into Europe, first to Austria and Germany, from whence it spread to Flanders in 1562, and Holland shortly afterwards. In about 1578 it was brought to England, but France does not seem to have taken it up until 1608, after which it was a great favourite among women of fashion, being worn in their low-cut dresses in much the same way as orchids were in the nineteenth century. By Charles I's time in England the flower had become more popular than the former favourites, the rose and daffodil.

At first bulbs were rare and costly. The sixteenth century professor of botany at the University of Leyden, Carolus Clusius, began cultivating the bulbs, but charged such extortionate prices that no one could afford to buy them. Eventually a group of gardening enthusiasts stole the best of his plants by night, after which Clusius lost heart. Prices did not seem to go down, however, and all over Europe the most fantastic speculations in bulbs were indulged in, especially in Holland, where the madness came to a climax during the years 1634–7. At this time the following goods were given in exchange for just one bulb: 2 loads of wheat; 4 loads of rye; 4 fat oxen; 8 fat pigs; 12 fat sheep; 2 hogsheads of wine; 4 barrels of 8-florin beer; 2 barrels of butter; 1000 lb of cheese; a bed; a suit of clothes; a silver beaker – the whole valued at 2500 florins. Another brought twice this amount, with a carriage and pair thrown in. Similarly, in France, a miller exchanged his mill for a bulb; another man exchanged a flourishing brewery valued at 30,000 francs. 'Brummagem millionaires' were made overnight. And several were broken when at last the amateurs grew tired of the whole business and started flooding the market with bulbs. At about this time Clusius' successor at Leyden, Evrard Forstius, could not pass a tulip without attacking it violently with his stick.

In Britain there was rather more moderation, and the enthusiasts were gently mocked, as by Addison in an essay in *The Tatler* in 1710:

> As I sat (sheltering from the rain) in the porch, I heard the voices of two or three persons, who seemed very earnest in discourse. My curiosity was raised when I heard the names of Alexander the Great and Artaxerxes; and as their talk seemed to run on ancient heroes, I concluded there could not be any secret in it; for which reason I thought I might very fairly listen to what they said. After several parallels between great men, which appeared to me altogether groundless and chimerical, I was surprised to hear one say, that he valued the Black Prince more than the Duke of Vendosme. How the Duke of Vendosme should become a rival of the Black Prince, I could not conceive; and was more startled when I heard a second affirm, with great vehemence, that if the Emperor of Germany was not going off, he should like him better than either of them. He added, that though the season was so changeable, the Duke of Marlborough was in blooming beauty. I was wondering to myself from whence they had received this odd intelligence: especially when I heard them mention the names of several other generals, as the Prince of Hesse and the King of Sweden, who, they said, were both running away. To which they added, what I entirely agreed with them in, that the Crown of France was very weak, but that the Marshall Villars still kept his colours. At last, one of them told the company, that if they should go along with him, he would show them a Chimney-Sweep and a Painted Lady in the same bed, which he was sure would very much please them.

The plant, of course, is frequently mentioned in English poetry. Browning has some memorable lines on a wild Italian variety in 'Up at a Villa – Down in the City':

> '*Mid the sharp, short emerald wheat, scarce risen three fingers well*
> *The wild tulip, at end of its tube, blows out its great red bell*
> *Like a thin clear bubble of blood, for children to pick and sell.*

One of the most recent poems concerning a tulip is Alan Jackson's 'Loss':

> *A tulip fell deid*
> *bi ma doorstep the day*
> *dark rid the colour o blood*
> *Wis the only yin come up this year*
> *A imagine it fell wi a thud.*

In the Middle East, where the flower seems to have first been cultivated shortly after 1500, it was very popular. The Persians made it the symbol of perfect lovers, and represented its origins as being much the same as those of love-lies-bleeding (*see under* Amaranths). The Turkish Osmanli house adopted it as their emblem, after which it became even more fashionable, so that the very style of tying a turban was modified in order to conform to the shapes of various blossoms. In fact it is from the colloquial Turkish *tülbend* (a turban) that the West derives the plant name. To the Turks and Persians it is *lalé*.

The tulip figured in several fantastic fêtes, one of which, held in 1726 by the vizier Ibrahim Pasha for the Turkish sultan Ahmed, was thus described by the French Ambassador:

> Care is taken to fill the gaps where the tulips have come up blind, by flowers taken from other gardens and placed in bottles. Beside every fourth flower is stood a candle, level with the bloom, and along the alleys are hung cages filled with all kinds of birds. The trellises are all decorated with an enormous quantity of flowers of every sort, placed in bottles and lit by an infinite number of glass lamps of different colours. These lamps are also hung on the green branches of shrubs which are specially transplanted for the fête from neighbouring woods and placed behind the trellises. The effect of all these varied colours, and of the lights which are reflected by countless mirrors, is said to be magnificent. The illuminations, and the noisy consort of Turkish musical instruments which accompanies them, continue nightly so long as the tulips remain in flower.

A more intimate kind of tulip show was later held for the ladies of the Sultan's harem in the Spring:

> A great amphitheatre of wooden stands was erected, fitted on both sides with shelves to support vases of cut flowers. Among the vases were placed lamps, glass globes filled with different coloured waters, and cages of canaries; and here and there the flowers were grouped into pyramids, towers or archways, or drawn into patterns upon the ground like carpets. At sunset he [Mahmud I] caused complete privacy to be enforced and the outer gates of the courtyard were closed. Then, as the fortress cannon fired a salute, the doors of the harem were flung open, and in the sudden magic light of a thousand sweet-smelling torches borne by eunuchs, the women would rush out on all sides, like bees settling on the flowers and stopping continually at the honey they found there. To the shrill cries of the eunuch gardeners and to the roar of the cannon were added the joyful ululations of the imperial concubines. Sometimes one of these women, jealous of the beauty of the blooms, half maddened by year-long confinement, would fall upon the flowers and tear them to pieces. But for the most part they had but one object to accomplish – to attract the attention of their royal master. When it grew late the Mistress of the Harem presented to the Grand Seigneur the girl who had most taken his fancy, and the handkerchief which he threw signified his wish to be alone with her. (Wilfrid Blunt, *Tulipomania*)

It seems to have been the Germans who first experimented with the tulip bulbs to establish how good they were as vegetables, although several Eurasian tribes could doubtless have taught them a thing or two. A Frankfurt apothecary preserved them in sugar and pronounced them better than orchid bulbs. One of Addison's characters bewails the fact that his maid had mistaken the bulbs for onions and cooked them, thus nearly ruining him. Now the Italians often use them in their cookery, and the Dutch resorted to them in the times of scarcity towards the end of the last war. These, of course, belonged to the *common* or *garden tulip* (T gesneriana) and its many varieties. The Kalmuks eat the bulb of the *wild* or *Florentine tulip* (T. sylvestris) despite the fact that it is considered an emetic elsewhere. Also eaten in the region of the Don is the *perfumed* or *Duc van Thol tulip* (T. suaveolens). The *Cilician tulip* (T. montana) grows in the highlands of Baluchistan and is eaten in the neighbouring areas of Iran, Pakistan and Afghanistan. From the bulbs of the Oriental *edible tulip* (T. edulis) the Japanese extract an edible starch; the leaves are also eaten as a vegetable.

17. Bellwort *(genus Uvularia)*
This is yet another member of the lily-of-the-valley family, of American origin. In northeastern America the stems are eaten like asparagus, especially those of the *little merrybells* (U. sessifolia), whose roots are edible cooked and considered good as an emergency food in times of want.

18. Yucca *(genus Yucca)*
This, the state flower of New Mexico, originates from the former Spanish possessions in the U.S.A. Its hard-pointed leaves have given it such names as *bear's grass, Spanish bayonet, needle and thread, Adam's needle*. The last is given it because the leaves were supposed to have been used by Adam in the garden of Eden to sew together his first suit of clothes. The leaves are also used by the Zuni Indians as whips in exorcism rites to drive out demons and misfortune. The flower sepals are sometimes eaten as a salad by Indians, especially those of *Whipple's yucca* (Y. whipplei), whose seeds are also ground and used as a porridge flour. The seeds of the *Torrey yucca* (Y. macrocarpa) and the *Joshua tree* (Y. arborescens) are similarly used. In California and district the green pods of the *Mohave yucca* (Y. mohavensis) are roasted amongst coals by the Indians, and the flowers are boiled as a vegetable. The buds of the *datil* (Y. baccata) are eaten by the Indians of the southwestern U.S. and of Mexico, while those of Central America use the buds of the *ozote* or *bulb-stem yucca* (Y. elephantipes). These are often put in soups and are to be found on sale in the markets.

The *baretta date yucca* (Samuela carnerosa) is a close Mexican relative whose flower-clusters are boiled or roasted by the Indians. Like many yuccas, it is a tree which also bears edible fruits. An Australian family of trees (genus Xanthorrhea) also goes by the name of 'yucca' (or yacca). Most of the sixteen species provide a resinous gum, after which the trees are also known as *grass-gum trees*. After bush fires the trunks retain the charred remains of the leaf-bases for years, for which reason they are often called *blackboys* in Western Australia. But this and the expression *blackfellah* (the pidgin form) refers to the aborigine, and it has been suggested that the name for the tree comes about because its charred and twisted shape suggests an aborigine standing still in the bush. The extremities of the young shoots and tender leaf-bases are eaten by the natives.

(b) AMARYLLIS FAMILY

The majority of plants in the amaryllis family are poisonous. Some however, mostly of American origin, are edible in various ways.

1. Agave aloe (*genus Agave*)

These plants are to be found mostly in the very warm parts of America, from the south to the north. The flowers are considered extremely beautiful and some species were brought to Europe in the sixteenth century, becoming naturalised in southern countries. Of the *American aloe* (A. americanum) it has been claimed mistakenly that it does not bloom until it is a century old. This might be so in cooler climates to which the plant is not well adapted; in hot desert areas it does not take nearly as long. From the leaves of various species a fibre is obtained (*sisal*), and many are used for making alcoholic drinks such as *mezcal, tequila* and the cider-like *pulque*. One such is the *desert aloe* (A. deserti), whose sweet and juicy leaf-bases are roasted in the Indian mescal pits for food. Here too are roasted the central parts of the *Wheeler sotol* (Dasylirion wheeleri), strictly a member of the lily family proper, in New Mexico and Arizona, The bulbs of the *Utah aloe* (A. utahensis) are also roasted for food by the Indians. The *tuberose* (Polyanthes tuberosa) is a distant relation, whose name refers to its tuberous root, not the affinities of its flower; it is originally from Central America and now widespread. The Javan Chinese use its white flowers in soups.

2. Herb lily (*genus Alstroemeria*)

These plants are to be found in South and Central America. That chiefly used is the *purple-streak herb lily* (A. ligtu), commonly to be found growing by streams in Peru and Chile and known locally as *liuto*. A kind of flour, referred to as *chuño de concepcion*, is extracted from the roots, excellent in soups and easily digestible but not very nourishing. The Chilean Indians treat in the same way the *purple-spot herb lily* (A. haemantha), the *dwarf purple-spot herb lily* (A. versicolor), and the *purple-petal herb lily* (A. revoluta). There are other colourings, but only the purple seems to be edible. The tubers of a West Indian species (A. edulis) are sold in the markets and eaten like potatoes. This might possibly be the same as Bomarea edulis, eaten in Dominica, since the genus is accounted by some as merely a subsection of the herb lily family, the starchy tubers of which are eaten by the Indians of various countries: B. ovata, also in Dominica; B. acutifolia in Central America; B. glaucescens in Ecquador; and B. salsilla in Chile.

3. Edible bulbs
In the eastern U.S. the bulbs of the *atamasco lily* (Zephyranthes atamasco) of the zephyr lily family are eaten by the Creek Indians in times of scarcity. Similarly in times of want, an arrowroot can be obtained from the bulbs of the Australian *Darling lily* (Crinum flaccidum), ordinarily used by the aborigines.

(c) IRIS FAMILY

1. Crocus, saffron and safflower (*genus Crocus*)
In some areas the corms of various croci are eaten by the local peasants. For instance, the chestnut-flavoured corms of the *cross-barred crocus* (C. cancellatus) and the *Damascus crocus* (C. damascenus or edulis), which grow from the Ionian islands to Syria, Armenia and Persia. *Sieber's crocus* (C. sieberi) grows in the Greek mountains and the corms are eaten by shepherds. They are said to taste like hazel nuts. Those of a grape hyacinth, the *tassel, tufted* or *fair-haired hyacinth* (Muscari comosum), are also eaten in this area in the spring. The plant is of middle and southern European origin. A sixteenth-century English work says the plant was known as *tuzzy-muzzy, purple tassel* and *purse tassel*, since 'the whole stalke with the flowers upon it doth somewhat resemble a long tassel purse, and thereupon divers gentlewomen have so named it'.

But the best known species of crocus is the *purple saffron crocus* (C. sativus), from the pungent dried orange stigmas of which saffron is made. The name of the spice, which is used as a dye as well as for flavouring and colouring food, comes to us via the old French *safran* from the Arabic *zafaran*. This crocus flowers in autumn and is a native of Asia and parts of Europe. In Cornwall it grows wild; it was cultivated once at Saffron Walden, from whence the town gains its name, but the plants were destroyed by a fungoid disease and the practice was given up. The Romans used saffron, preferring that of Cilicia (in what is now southeast Turkey) to their own. For a long time it was cultivated in Persia and Kashmir; Mongol invaders introduced it to China. But we use a name of Arab derivation either because it was introduced to medieval Europe from Spain, where the Arabs were growing it in the tenth century, or because it was brought back by returning crusaders from its ancient seat of cultivation in Cilicia (then ruled by the Armenians). Saffron was once used in medicine, but it probably had little effect. It is, however, a poison which the body tends to store up, so it is well to use it only sparingly.

The flower is the birthday plant for 3 February and symbolises marriage. For Christians it is the emblem of charity, and in the language of flowers it warns one to beware of success. When it is said of anyone 'he has slept in a bag of saffron' it means that he is a merry person, since the spice is supposed to be exhilarating. Here it seems to overlap with other species of crocus which are supposed to be the emblem of cheerfulness; also of the courtesan and of illicit love. And yet it was on a couch of croci that Zeus and Hera, the Greek king and queen of heaven, were supposed to have reposed.

There was once a death penalty for those who adulterated saffron; but it still happens since, because of its scarcity, it commands such high prices. The floral parts of the *Scotch cotton thistle* (Onopordon acanthium, *see under* Thistles), and the pollen of the marigold (q.v.) have been used for this purpose in Europe. Another substitute, or adulterant, can be obtained from the flowers of the *golden copper-tip* (Crocosmia aurea), a relative of the crocus found in tropical Africa. A similar colouring matter is obtained from the roots of the Brazilian *wood saffron* (Escobedia curialis).

The *safflower* (Carthamus tinctorius), also known as *bastard* or *false saffron* and *saffron thistle*, provides the most usual alternative. The plant is probably of Eastern origin but has since spread over the Mediterranean region, largely through Arab propagation. It reached China in the second century B.C. Its red florets are used for making a dye, known as *carthamine*, in India, Egypt and the Middle East. For this the plant was cultivated in Egypt from very ancient times; the grave-cloths used for wrapping mummies were very often dyed in it. The flower-seeds, which are usually crushed for the oil they contain, are often roasted as a food, especially in India, where the tender shoots are also used as a pot-herb. The dye is much used in cooking both there and in Egypt, where it acts as a cheese colourant, among other things. One of the Arabic names it bears there, *kurtham*, has given the plant its Latin generic name. The Spaniards, who took the plant to Mexico at an early date, use the flowers in soups and many other dishes, while the Polish Jews were once especially fond of them, adding them to their bread and most other foods.

2. Sword lily (*genus Gladiolus*)

These plants, also known as *corn flags*, are mostly of African origin. Of them Pliny says in his *Natural History* (Book XXI) that they provide 'a pleasant food, one which, when boiled, makes bread more palatable'. However, their use is largely confined to their continent of origin now. The *edible sword lily* (G. edulis) of South Africa has roots which taste like chestnuts when roasted. Elsewhere in the tropics the *spiky sword lily* (G. spicatus) is used by the Lokoja tribe. The Igaras cultivate the *four-day sword lily* (G. quartianus) for its corms, which are pounded in water and mixed with guinea-corn flour. In East Africa the Njelekwa use the corms of the *Zambezi sword lily* (G. zambesiacus). The tuberous rhizomes of a relative, Babiana plicata, known slightingly as *baboon root*, were once eaten by early South African settlers at the Cape.

3. Iris (*genus Iris*)

The word 'iris', deriving from the Greek, signifies that goddess of the rainbow who was Hera's (Juno's) messenger, a part of the eye, and the flower. The sweet-scented roots of at least one species (I. florentina) were once used in perfumery, for medicines and breath-sweeteners, under the name *orrice-root*. Others smell foul, like the *stinking iris* (I. foetidissima), fancifully known as *roast-beef plant*. In South Africa the Hottentots gather the roots of the *edible iris* (I. edulis) and cook them together with the shoots; they are also much sought after by monkeys. In the Mediterranean area the floury root of the *rush iris* (I. juncea) is eaten by the Arabs. It is principally to be found growing wild in forests and damp places, as in Algeria and Portugal, for instance. The *tiger iris* (Tigridia pavonia), also known as *tiger* or *Tigris flower*, has edible chestnut-tasting roots which were eaten by Mexican Indians in pre-hispanic times. The flowers are variously shaded gold-orange, yellow or white, and scarlet-spotted; they are to be found in Central America, Peru and Chile.

The yellow iris is the emblem of May, symbolising flame and passion and the birthday flower for 5 May. In Japan it is the flower of the boy's doll-day festival. The purple iris, however, is the birthday flower for 6 September and symbolises eloquence, hope, light and primeval fire, power and royalty. It was used in charms against evil spirits, dedicated to Hera and considered as being governed by the zodiacal house of Gemini. In the language of flowers it signifies a message. For the Flemish painters it was the royal lily of the Virgin Mary; Memling also gives it to Jesus, and to the royally born St Barbara. In Spanish art it indicates the immaculate conception, The iris is also considered to be the French '*fleur de lys*'.

The Greeks knew the iris as a 'sword lily', like the gladiolus, whose sword-shaped leaves are similar. It symbolises a sword in Japan and is also used to convey congratulations. On the other hand a purple iris is prohibited at weddings because that is the colour of mourning. However, in Japan the *calamus* or *sweet flag* (Acorus calamus), also known as *cinnamon sedge* and *myrtle flag*, *sedge* or *grass* because of its sweet-smelling roots, is often confused with the iris, although it is really a member of the arum family. The leaves and roots are very similar, although the small yellow green flower-masses have nothing in common. The old Japanese name for it (*ayame*) has now been transferred to various irises. The plant is of Indian origin but has become naturalised in marshy places in Asia, Europe and North America. The very young leaves have been used as a salad in the latter continent, while the roots, which are extremely bitter, have been chewed by singers to clear their voice, and pickled in vinegar. They are often used medicinally and in perfumery. They have also been used as a beer-flavouring in Germany.

The Japanese held a very important festival, of Chinese origin, on the fifth day of the fifth month; this has been misnamed the Iris Festival (for reasons given above) but it was in reality the Sweet-flag Festival. At this time rice dumplings were prepared out of a glutinous rice moulded into a cylinder, wrapped in the leaves of reeds and wild rice, and decorated with coloured streamers and seasonal flowers. The festival was intended to ward off disease and the roots and leaves of the sweet-flag (to which were ascribed medicinal properties) were placed on the roofs, worn as a hair ornament, put in bath water, added to medicinal sachets (which also contained mugwort, *see under* Tarragon), added to rice wine as a health drink, and formally presented to the emperor by the court physicians. A Korean poem refers to it:

> *On the sweet-flag festival*
> *I brew healing herbs.*
> *I offer you this drink,*
> *May you live a thousand years.*

Another reference to the celebrations is to be found in the Japanese *Kagero Nikki* diary, in an entry for the year 974.

My brother appeared before we had raised the shutters on the morning of the Sweet-flag festival. 'What's this? You haven't put out the flags yet. You really should have done it last night'.

We got up in great excitement and prepared to open the house. 'Leave the shutters for a while', he said. 'The scent will drift in, and then you can have a good long look'. But we were all up and helping him. It was a fine morning. The wind had changed and the sky had cleared, and the scent of the roots spread quickly through the house. Out on the veranda the boy and my brother had gathered all sorts of plants and were busy with medicine packets. We were rather alarmed at a flock of cuckoos[1]—there really were too many of them this year—near the toilet; it was a bad omen surely. But when two and three flew off singing into the sky the effect was pure delight. 'The mountain cuckoo on the day of the festival' someone murmured (in reference to an imperial love-song), and all of us took it up one after another.[2]

Mallows *(genus Malva)*

Many European mallows were once eaten, especially in Roman cookery, where the leaves were used as a spinach substitute. Apparently they are still a favourite vegetable in Cyprus. Pliny considered them a very powerful aphrodisiac, and they are used medicinally as a remedy for colds and sore throats still, since the leaves abound in mucilage. The flowers may also be cooked as a flavouring – with green peas, for example. Even among country folk, however, the plant gradually came to be considered no more than famine fare. The name derives from the Old English *mealuwe*, an equivalent of the Latin *malva*. Country corruptions of this, such as *malue*, *malus*, *mole* and *maws*, may be compared with the French *mauve*. Further variations on the name include *malice*, *mallard* and *mullers*. *Round dock*, another general name for the plants, would seem to indicate that they were grouped with the docks and sorrels in the countryman's mind.

The leaves of the *running* or *dwarf mallow* (M. rotundifolia) were once esteemed in salads. The species most commonly used, however, was the *common, wild, field* or *high mallow* (M. sylvestris), whose round flat fruits are still eaten by children in some areas. It is the appearance of these which have given the plant many of its popular names: *Bread and cheesecakes*; *bread and cheese and cider*; *butter and cheese*; *cuckoo's* or *old man's bread and cheese*; *cheese flower* or *log*; *chucky* or *Jacky cheeses*; *custard* or *lady's cheeses*; *fairy cheeses*; *frog* or *pick cheese*; *truckles of cheese*; *loaves of bread*; *pancake plant*; *pans and cakes*. The French, too, have similar names such as *frommages* and *fromageon*. Other names celebrate the short opening time of its flower: *fliberty-gibbet*; *good-night-at-noon*; *flower of an hour*. The name *mallow-hock* preserves the old name of *hock* or *hockherb*.

The roots of the *white marsh mallow* (Althaea officinalis), also known as *wymot*, and by the French *guimauve*, are still used occasionally, principally

---

[1] Generally considered unlucky in Japan.
[2] *The Gossamer Years*, trans. Edward Seidensticker. Tuttle, 1964.

in medicine and the sweetmeat that bears its name. They may, however, be used in salads and, according to some, given to babies to chew while teething. The plant is related to the hollyhock and grows in wet places, especially salt marshes. Albertus Magnus explains how one may use it 'if thou wilt seeme all inflamed from thy head to thy feet, and yet not hurt':

> Take white great mallows, or Hollyhocke, mixe them with the whites of egges; and anoint thy body with it, and let it be until it be dried up. And after anoint thee with alom, and afterwards cast on it small brimstone beaten into powder, for the fire is enflamed on it and hurteth not. And if thou make upon the palme of thy hand thou shalt be able to hold the fire without hurt.

In America the Indians eat the pleasant tasting roots of the very nearly allied *poppy mallows* (genus Callirhoe), especially those of the *pimple mallow* (C. pedata) in the west, and the *finger poppy* (C. digitata) in the south. The flowers of the *Brazilian mallow* (Abutilon megapotamicum) are eaten as a vegetable in Brazil, the seeds of its relative A. muticum by nomads in the Sudan. The Queensland aborigines eat the seeds of the *Australian baobab* (Adansonia gregorii) either raw or roasted.

The commonest baobab, a tree member of the mallow family, is known as *monkey bread*, *Indian cork tree* and *Ethiopian sour gourd* (A. digitata), and is to be found all over tropical Africa, especially in the west. It has a bottle-shaped soft-wooded trunk, very broad in relation to the tree's height, and is extremely long-lived; some specimens are thought to be as much as 5000 years old. Its young leaves are used as a sour vegetable, and the seed kernels are prepared in various ways, So too is the acid fruit-pulp, which tastes rather like gingerbread, often used for seasoning corn-gruel and other dishes, and in various drinks. The tree itself is venerated in Africa and the natives hang their good-luck amulets on it. The wood is so soft that Ethiopian bees bore into it and lodge their honey in the hollow. Elsewhere it is artificially hollowed out and dead bodies are lodged in its interior and thus preserved. Often the leaves are pulverised into a product known as *lalo* which is added to all foods in order to inhibit the partaker's sweating.

The *kapok-tree* (Ceiba pentandra) is probably venerated also, in view of one of its alternative names, *god tree*. It is also called *cabbage wood* and, in the West Indies, *silk-cotton tree*. It is probably of American origin and its seeds must have been carried by sea-currents to Africa in remote times. By the tenth century it had been taken to Indonesia, from whence it spread out over South-East Asia and the Pacific islands. The trunk of the kapok reaches a height of 100 feet, and its roots run over the surface of the ground for about the same distance before burrowing into the earth to anchor it. In tropical Asia it is cultivated for the fine downy substance (kapok) which grows out of the ripe fruit. In Java the very young immature pods are eaten, while in some Pacific islands and in West Africa the seeds are crushed, roasted and used in soups.

The widest used of all in the mallow family are the hibisci, of which perhaps the best known member is *okra* (Hibiscus or Abelmoschus esculentus), of African origin and said to have been cultivated in ancient Egypt. It is now grown widely in Africa, the Mediterranean region and the Americas. Its mucilaginous green pod is considered especially good for curries and soups, and may also be eaten as a vegetable. The young shoots and leaves of the plant may also be eaten. The name *okra* is derived from *nkruman*, its name in the Tshi language of Ghana. In Angola the name is *ki-ngomgo*, which entered Portuguese as *quin-gombo* and ended up in the West Indies and southern U.S. as *gumbo*. Arabic names such as *bamyah* (there are slight variations) becomes *bamie* in French. Indians call it *bhindi* and, in their restaurants, *ladies' fingers*, although this is a name applied to many other plants, including grapes, bananas and a kidney vetch. An Australian relative known as *inland roselle* (H. ficulneus) is used in much the same way by the aborigines, although the plant is reputed to cause loss of hair and dermatitis in the sheep that browse upon it.

The better known *roselle* (H. sabdariffa) is also called *thorny mallow* and *Jamaican*, *Indian* or *red sorrel*. The name refers to the sour-tasting edible leaves; there is also a white-leaved less acid variety. The calyces of this plant are boiled with sugar to make a pleasantly sour sorrel drink. They are also used for jellies, chutneys and sauces, while the tender stalks and leaves are used in salads, and as a pot-herb and curry seasoning. Those of the *rose mallow* (H. furcatus) will also do for this. From its place of origin in West Africa it spread into all tropical countries in the Old World; in the seventeenth century it was taken to Brazil and reached Jamaica in 1707.

A less popular member of the same family is the *kenaf hibiscus* (H. cannabinus) which, since it is grown principally for the making of fibres, goes under such names as *hemp mallow*, *brown* or *Deccon hemp*, and *Bimli jute*. Its oily seeds go into the making of soaps, paints, linoleum and lighting oil, but its young leaves serve as a pot-herb. This too, has spread from Africa to Asia. Another plant whose principal use is in the making of fibres is the *tossa* or *long-fruited jute* (Corchorus olitorius), whose relationship to the mallow and hibiscus families is demonstrated by such names as *bush okra*, *Jew's mallow* and *West African sorrel*; one of the African names by which it is known is *krin-krin*. Those plants grown as a vegetable are slightly smaller than the commercial fibre crop. The mucilaginous leaves and young shoots are eaten like spinach. The plant is probably of South Chinese or Indian origin, from which area it has spread via the Middle East to Africa. It is also to be found in Australia, which it reached via Indonesia. In Syria and Egypt, and on some of the Mediterranean islands, it is cultivated and very popular as a vegetable. Elsewhere in Africa other species are gathered, usually wild, and eaten.

Several Australian species of the hibiscus family are eaten. All of these, like several other members of the mallow family, have developed into rather squashy-stemmed trees (cf. the *tree-nettle under* Nettle). The leaves and roots of the *Queensland sorrel tree* (H. heterophyllus) are eaten without preparation by the aborigines. Other names for it are *native rosella* and *green kurrajong* (cf. the *true* or *brown kurrajong* under Potato-beans). The Queensland natives also prize the roots of the *cork-wood* or *cotton tree* (H. tiliaceus), which is to be found growing on the Pacific islands as well, and in times of scarcity will eat the sorrel-tasting leaf-tips.

Milkweeds

Milkweeds are generally named for their milky juice which, in the case of the *silkweeds* (genus Asclepias), may be used as an inferior rubber latex and for chewing-gum. Some of these, which are American plants also going under the general names *swallow-wort* and *silken cissy*, are used for food by the Indians. The buds of the *swamp milkweed* (A. incarnata) are used in soups, those of the *showy milkweed* (A. speciosa), whose young shoots and leaves are also eaten by the Hopi Indians, are boiled as a vegetable. *Common milkweed* (A. syriaca) was brought from America and has become naturalised in Europe; its buds and young green fruits are eaten. Lastly the shoots and pods of the *butterfly weed* or *orange milkweed* (A. tuberosa), also known as *pleurisy, tuber* or *wind root*; the roots have their use in country medicines, as the names indicate, and also serve as a vegetable for the Indians.

The milky juice of several relatives are more or less palatable. The *Ceylon cow-plant* (Gymenema lactiferum) is one, although the juice of its relative, the *Australian cow-plant* (G. sylvestre), has the curious faculty of temporarily destroying the sense of taste, so that sugar feels like sand when chewed. The *South African cow-plant* (Oxystelma esculentum) is similar except for the effect on the taste-buds. The shoots of all these may be used in salads and as a pot-herb. Other African relatives include the *fikongo* (Brachystelma bingere), whose tubers are very popular in the Niger region, and the related B. lineara eaten in Abyssinia. In Western Africa the turnip-like tubers of Raphionacme brownii are eaten raw or roasted by the Fulani and Mandingo peoples, and the roots of the *yakhop* (Xysmalobium cordata) in Senegambia and elsewhere.

The roots of the oriental Metaplexis stauntoni are eaten by the Ainu, those of the *West coast creeper* (Telosma cordata) used for sweetmeats by the Javan Chinese. The flowers and leaves are eaten by the Thais, the young fruits of T. procumbens as a vegetable in the Philippines. Also called *milkwort* or *weed* in Britain is Polygala vulgaris, the seeds of whose relative (P. butyracea) are parched and ground for soups in tropical Africa. The leaves and flowers of the British species (which grows also in Africa and Asia) are eaten as a vegetable or in soup, although bitter-tasting. It is popularly supposed to engender milk in nursing mothers and cows, and the Latin generic name is derived from the Greek for 'much milk'. In Ireland it is known as *four sisters* after the different colours of its flowers. An old set of names cluster round *procession flower*, since according to Gerarde, 'it doth specially flourish in the Crosse or Gang week, or Rogation week (five weeks after Easter); of which flowers the maidens which use in the countries to walk the procession do make themselves garlands or nosegaies'. For this reason it is also called after the various names of that week.

An unrelated milkwort (Glaux maritima) belongs to the primrose family and is also called *black salt-wort* and *sea trifoly*. It grows in Europe, temperate Asia and North America, and is credited with the same milk-producing powers as the above. The young shoots and leaves are eaten in salads, the whole plant in soups and as a vegetable. It has been recommended in times of emergency. Its relative, the *upland shooting star* or *American cowslip* (Dodecatheon hendersonii) grows in northwestern America, where the roots and leaves are eaten by the Indians.

Mustards (*genus Brassica*)

The condiment known as mustard is obtained from the ground seeds of two of the mustard plants. The first of these, *black*, *brown* or *grocer's mustard* (B. nigra), is a Eurasian plant long cultivated in Europe, and which grows as a weed over wide areas of Britain and North America. At first the seeds of this alone were used as a condiment. Nowadays those of *white mustard* (B. or Sinapis alba), also known as *charlock* or *salad mustard*, are added to them. The seedlings of this native of the Mediterranean region are those eaten as a salad together with garden cress (genus Lepidium, *see under* Cress).

Various wild mustard plants also have their uses. *Field mustard* (B. campestris) is grown as a seed and forage crop in India, the seeds being pressed for a cooking oil. There are two varieties of this used, *sarsan* or *Indian colza* (var. sarsan) and *toria* or *Indian rape* (var. toria). By the Thames in England the plant is referred to as *bargeman's cabbage*, which might indicate that it was once used as a vegetable. Certainly the cabbage's remote ancestor, *charlock* (B. arvensis), also known as *corn* or *wild mustard*, and more fancifully as *bread and marmalade*, was so used until recently. In a book of 1727 it is recorded that 'It is called about the streets of Dublin, before the flowers blow, by the name *cornicail*, and used for boiled sallet'. Thus it gained such additional names as *bastard rocket* and *wild* or *field kale*.

Charlock is accounted as a weed with a high nuisance value by farmers. The English Dialect Society glossary records that 'In old Latin leases in the East Riding [of Yorkshire], and doubtless elsewhere, the plant in question is termed *brassica*. Conditions were customarily introduced into such documents in medieval times that the *brassicae* should be duly kept down in the land let'. This led to the Yorkshire dialect word for the plant: *brassics* or *brassock*; the weeding of them was called 'brassocking'. The Irish names are vaguely similar: *prassia, prushus, presha bhwee* (from the Irish *praisseagh buigh* or *phuidhe*, yellow cabbage).

Lastly there is *Indian* or *Chinese mustard* (B. juncea) or *sarepta*, possibly of African origin but taken early to Asia, where it is cultivated extensively, both for its seeds (*rai* in India) and for the pungent leaves. These may be cooked as a pot-herb, provided the water is changed twice; but they are also used for salads in the U.S. The root is also eaten, especially in northern China; it is prepared like celeriac. *Greek mustard* (Sinapis incana or Brassica adpressa) grows in Greece and Turkey. The young plants are eaten in the spring with oil and lemon-juice.

The *tansy mustårds* (genus Sophia) are North American plants largely used by the Indian tribes. This family and the *winter* or *marsh water rockets* (genus Sisymbrium) are sometimes classed together by the botanists into the composite genus Descurainia. The seeds of the *western tansy mustard* (S. incisa) are parched and ground in the west, as are those of the *flixweed* (S. parviflora), once known in England as *surgeon's wisdom* (cf. the French *science des chirugiens*). An old author comments that 'the seeds of flixeweeds dronken with wine or water of the smithes forge stoppeth the bloudy flixe'. Both the seeds and the leaves of the *slim-stemmed tansy mustard* (S. pinnata) are used, the latter either boiled or roasted between hot stones. The tender S. halictorum is cooked whole by the Pueblo Indians of New Mexico.

The *hedge mustard* (Sisymbrium officinale), also called *bank cress*, has dingy yellow seeds which are sharp and strong tasting and a great favourite with birds. It was once cultivated as a pot-herb but is rather tough in its wild state. Other names for the plant include *white charlite* in the west of England, *lucifer matches, singer's plant* (because it was supposed to be good for loss of voice), and *crambling rocket*, a name deriving from *crambe* (kale, cabbage). *Garlic mustard* (S. alliara) was also once used as a pot-herb and in sauces. It comes by the name because it smells like garlic and looks like mustard. Other references to this include *garlic treacle* or *hedgeweed*, *garlicwort* and *poor man's mustard*. Its culinary use brings it the name *sauce-alone* and the old German *sauczkraut*. The frequency with which it is found in hedgerows results in such names as *Jack-by-the-hedgeside*, *Jack-in-the-bush*, *Jack-of* or *run-along-the-hedge*, *Penny-in-the-hedge*. Its curative properties are referred to in several names: *English treacle*, *treacle mustard*, *treacle wormseed* and *wound herb*, and also include some applied to garlic. Other names, most of them old, include *beggarman's oatmeal*, *cardiac*, *caspere*, *eileber* (coming from the Old English), *leek cress* (from another Anglo-Saxon name), *lamb's pummy*, *pick-pocket*, and *swarms*.

A mustard-like condiment is made from the seeds of the unrelated *mustard tree* (Salvadora persica); these are bitterly pungent and also used medicinally. The shoots and leaves are occasionally used as a salad. The tree is to be found in tropical Africa and Asia and goes by such other names as *arak* and *mesuak*. It is called *salt-bush*, since its ashes give a kind of vegetable salt. In Mozambique and elsewhere its twigs are used for cleaning the teeth and it is therefore known as *toothbrush tree*. Although black mustard is common in Palestine and sometimes grows higher than a man, it is considered that this tree is the subject of Jesus' parable of the mustard-seed:

> Whereunto shall we liken the kingdom of God? Or with what comparison shall we compare it? It is like a grain of mustard seed which, when it is sown in the earth, is less than all the seeds that be in the earth; but when it is grown it groweth up and becometh greater than all the herbs, and shooteth out great branches so that the fowls of the air may lodge under the shadow of it.
>
> (Mark 4:30–2)

Nettles *(genus Urtica)*

The tops and very young shoots of the nettle may be cooked; this takes away the plant's stinging powers. The sting is caused by the minute little hairs along the stem which penetrate the skin, releasing acid and causing irritation. The old countryman's saying about grasping the nettle firmly and then it will not hurt you is true up to a point. The trick is to grasp it so that the hairs are held downwards; it is brushing them in an upwards direction which causes the harm. As one proverb puts it, 'He that handles a nettle tenderly is soonest stung'. This, of course, has wider more human implications, as does the saying 'It is better to be stung by a nettle than pricked by a rose'. Worse still, however, is when 'the nettle grows where the rose was expected'. The remark that 'he has pissed on a nettle' is as much as to say that the fellow has 'got out of bed on the wrong side'.

In the language of flowers a nettle indicates spite. It is the birthday flower for 31 October and symbolises cruelty, envy, but also courage. The burning nettle (see below) belongs to 27 January and indicates slander. It's unfair since the leaves are extremely nutritious and much sought after by poultry, like so many other spinach substitutes (*see under* Spinach). From the leaves and stems are also made a deliciously refreshing soup, a light beer, and a supposedly health-giving decoction. An old proverb runs thus in its Scottish version:

> *If they wad drink nettles in March*
> *And eat muggins* [*mugwort*] *in May*
> *Sae mony braw maidens*
> *Wad not go to the clay.*

The plant is supposed to be an aphrodisiac, and love filtres mixed with hellebore and cyclamen were made from it. The Egyptians are credited with the discovery that it can be made into a kind of fibre.

In Anglo-Saxon times nettles (Old English *netel*) were considered powerful medicines. The Nine Herbs Charm lauds it thus:

> *This is the nerb which is called Wergulu;*
> *The seal sent this over the back of the ocean*
> *To heal the hurt of other poison.*

It was used for wounds, sores and swellings, and in a remedy for shingles; the lowest part figures in a potion for the 'dry disease'. One remedy runs: 'Boil the lower part of a nettle in fat of an ox and in butter. Then against pain in the neck you must smear the thigh; if one has pain in the thigh, smear the neck with the salve'. Later on it is used for purging the stomach, towards which 'some boil nettle in water, wine and oil'.

From over thirty species of nettle, Britain contains only three. The commonest and least painful is the *great nettle* (U. dioica), the *stinging* or *tenging nettle* proper. Sinisterly it is known as *Devil's apron* or *leaf*, and *naughty man's plaything*. Other names show a degree of familiarity and even affection: *Jenny* or *female nettle*; *scaddie* (Scotland); *heg-beg, hoky-poky, hiddy-piddgy*. The similarly assonantal *hop-tops* may refer to the fact that the hop can be used and eaten likewise. *Ettle* is an example of the contrariness of the English who add or reassign the letter 'n' at will. (An 'eft' is now a newt, but the 'napron' has become an apron.) The sting of the *small* or *string nettle* (U. urens) is more severe; it is also known as the *burning nettle* (cf. the Dutch *branden-netel*, German *brennessel* and French *ortie brûlante*).

The last, the *Greek* or *Roman nettle* (U. pilulifera), has the severest sting of all, but it is mostly confined to the east of England. It is supposed to be a Roman importation. A legend current in Romney Marsh is that the Romans brought it because they could not get used to the rigours of the climate. When they wanted to get warm they used to strip and lash themselves with bundles of the nettle. In fact there is an old recipe which runs; 'In order that thou may not suffer by cold, take a nettle sodden in oil, smear and rub the body with it, and thou shalt not perceive the cold then on all thy body'. An account from 1713 tells of sadistic gardeners who 'call this plant Spanish marjoram, inducing unwary people to sting their noses smelling it'.

Other nettles are to be found in the U.S.A. To the north there is the *Hudson's Bay nettle* (U. gracilis); to the southeast the *American germander-leaved* or *weak nettle* (U. chamaeroides); and in the centre the *hoary nettle* (U. holosericea). In Japan the young shoots of both the *Japanese nettle* (U. thunbergiana) and the related *Japanese wood nettle* (Laportea bulbifera) are eaten in the spring. In the East Indies region some of the nastiest species appear – horrible stingers such as U. crenulata and U. urentissima, the *devil nettle*. For the first hour there is only the slightest irritation, but it rapidly worsens into a pain like a red-hot poker shooting along the limbs affected; application of cold water only makes it worse, and the pain remains as strong and as irritated by water for up to eight days. A similar sting can be obtained from the huge thick-stemmed *Australian tree-nettle* (U. gigas) which grows up to 120 feet high.

In India they eat the very nutritious tubers of U. tuberosa, either raw, boiled or roasted. There are also edible tubers to be obtained from members of the *flame-nettle* family (genus Coleus) which is really a mint that grows in the East India area. The most important is the *fabirama* (C. rotundifolius), which is also extensively cultivated in the West African region, and is similar to the Kaffir potato (*see under* Potato-beans). Other names for it include such contractions as *fura fura* and *fra-fra potato*, and *daso*. The *patchouli plant* (C. aromaticus), of Indonesian origin and now spread over tropical Asia and taken to the West Indies, serves as a substitute for such herbs as borage or sage. In Indian native medicines a decoction for chest complaints is made from its leaves. Other names for it include *Spanish thyme* and *Indian borage*. Several others growing in Africa and Asia bear either edible leaves or tubers.

The European *dead nettles* (genus Lamium) are also relatives of the mint. In Anglo-Saxon times very little distinction was made between these and the ordinary nettle, and in magico-medical texts one finds both put to much the same uses. Their stalks were used in an emetic and they were ingredients in a salve for various aches and pains. 'The red nettle that grows through the wall of the house' appears in a spell against rheumatism. Both in northern Europe and in North America the tops have been eaten as a vegetable and used as a pot-herb. Collectively the plants are known as *archangels* and *deaf, dumb, blind, dummy, dunce* and *day nettles*. The most popular is the *purple dead nettle* (L. purpureum) also known as *red* or *sweet archangel*. Other names for it include *bad* or *black-man's* (i.e. the Devil's) *posies, lamb's cress, rabbit meat* and *dog* or *French nettle*. The *white dead nettle* or *archangel* (L. album) is next in popularity, and also known as *bee nettle*. Children pick off the white flowers and suck the honey out of the hollowed end, after which practice it has been called *suck bottle, sucky Sue* (in Scotland), and *honey bee, flower* or *suck*. Other names include *lamb's ears, rat's mouths, shoes and stockings*, and the curious *black beetle poison*. There is also the rather rarer *yellow dead nettle* (L. galeobdelon), alternatively known as *weasel's snout*. Various other far less common species would also probably do and are certainly fed to poultry.

Olives (*Olea europea*)

The tree, of which olives are the fruit, is originally from South-West Asia and related to such trees as the ash, lilac, jasmine (the leaves of which the Chinese use as a tea) and forsythia. Its green (unripe) and black (ripe) fruits are used for the extraction of oil and are pickled for eating, soaked first in hot weak alkali to take away their bitterness and then prepared in salt water. The green are the less oily of the two. They are extensively cultivated and to be found as far afield as South Africa, China, Australia and New Zealand. The tree is very hardy and is productive even in extreme age. One tree has been recorded as being at least 700 years old.

Olives have been grown since remote antiquity. They were being used in Crete about 3500 B.C., are mentioned in the Bible, and have been found in Egyptian tombs, although there the first record of them is in the seventeenth century B.C. The olive branch has been considered a symbol of peace ever since the returning dove brought back a leaf from one to Noah's ark, thus signifying that the Flood was over. And because of its productivity it is also a Biblical symbol of prosperity; in the prophecy of Jeremiah, Judah is informed that 'the Lord called thy name a green olive tree' (11:16), while the self-righteous David considered himself like one of these growing in the house of the Lord (Psalms 52:8). When the Romans extended their power into Tunis they taxed that area 300,000 gallons of oil yearly and special conduits for it were built down to the sea.

We derive the word from the Ancient Greek ἐλειϝα which becomes *oliva* in Latin. The oil from its fruits was known to the Greeks as ἐλαιον; *oleum* in Latin. This becomes *oila* in Old French, and thus enters English, to be extended as a name for all oils.

The poisonous evergreen shrub *oleander* (Nerium oleander) is no relation to the olive, but the *oleaster* (Elaeagnus angustifolia) is. This small tree, also known as *wild olive* and *Zakkoum oil-plant*, is of Mediterranean origin and was introduced into the Canary Islands by the Phoenicians. It bears a bitter fruit also used for oil-making. The Romans used to graft branches of this into the true olive tree to improve the stock, and it is to this practice that Paul refers in a passage of his Epistle to the Romans (11:17–24).

Onions and family (*genus Allium*)

1. Onion (*A. cepa*)

The onion and all its relations belong to the lily family. Its use goes back many millenia, for we find it used by the Chaldean, Egyptian and other early civilisations. No other plant was so often portrayed in Egyptian tomb paintings. One has even been found in the hand of a mummy. It was in Egypt, too, that the Jews first seem to have appreciated its virtues; when they sighed for the 'flesh-pots' of Egypt it was for 'the cucumbers and the melons, and the leeks and the onions and the garlick' (Numbers 11:5). By some the onion was regarded as divine, together with garlic; sometimes as too divine to eat even. In India, however, it is considered not fit to be eaten by Brahmins, a tradition which goes back at least as far as the Ordinances of Manu (about the sixth century B.C.) where we read 'garlic, onions also, leeks and mushrooms, are not to be eaten by the twice-born, as well as things arising from impurity'. Similar prohibitions are later laid on those who practise yoga.

For Peer Gynt the onion appears as a tragi-comic symbol of his life's emptiness as he strips off skin after skin in search of a heart (*Peer Gynt*, v). In fact the uncovering of veil upon veil has suggested the onion as a symbol of immortality, eternity and the universe, to others. To dream about it is good luck, and it was used for divining purposes in early England. In the southern U.S. it is burnt in the fire for good luck and carried on the left side to ward off disease. High regard for it is manifested in the approving phrases 'spruce as an onion' and 'to know one's onions'. It is further connected with dreams in the superstitution that one will have a vision of one's future wife if it is placed under the pillow on St Thomas' Eve (21 Dec.). On a more practical level countrymen say that a thick-skinned crop indicates that a severe winter will follow. Everyone knows that cutting up an onion makes you cry; 'to weep with the onion' however, is an alternative way of expressing 'crocodile tears', or the fact that a person weeps too much.

Onions and family

The onion is considered to be of Central Asian origin, and to have been propagated by the Indo-European tribes in their separate migrations. It was known to the Greeks and Romans. A very popular peasant dish consisted of onions, garlic, wild celery and cheese, pounded together with rue in a mortar. It is perhaps for this reason that the Brahmin aristocrats, not to mention the Roman poet Horace, objected to members of the family. It reminded them too much of the early pioneering days before their stomachs grew delicate! One also thinks of its effect upon the breath. In Shakespeare's *Midsummer Night's Dream* Bottom the weaver has to warn his small company of actors to 'eat no onions nor garlic, for we are to utter sweet breath'. In England we find very early mention of 'bread, cheese and onions', still popular in some parts of the country where, also, to eat a raw onion a day is considered a mighty defence against catching a cold. An old proverb has it that 'Onions [or garlic] make a man wink, drink and stink'.

Among the Anglo-Saxons the names for various members of the family are so confusing that it is difficult to tell which they mean by any name. All, however, would seem to enjoy great popularity and to be the subjects of various riddles. Is it, for instance, an onion or a leek to which the following refers?

I was alive and said nothing; even so I die.
Back I came before I was. Everyone plunders me,
Keeps me confined and shears my head,
Bites my bare body, breaks my sprouts.
No man I bite unless he bites me;
Many there are who do bite me.

Part of the fun in Anglo-Saxon riddles is to describe one thing in terms of another, so that there is a level of double meaning running throughout. The above may not have particularly salacious overtunes, but another most definitely has:

I am a wonderful thing, a joy to women,
To neighbours useful. I injure no-one,
No village-dweller, save only my slayer.
I stand up high and steep over the bed;
Beneath I'm shaggy. Sometimes comes nigh
A young and handsome peasant's daughter,
A maiden proud, to lay hold on me,
She raises my redness, plunders my head,
Fixes on me fast, feels straightway
What meeting me means when she thus approaches,
A curly-haired women. Wet is that eye.

Both the above may represent either an onion or a leek, possibly the latter, since it is often connected with love and lechery. One proverb avers that 'Lovers live by love as larks live by leeks', and of a 'dirty old man' it is said that 'like a leek he has a white head and a green tail'. Another example of the riddle concerns a one-eyed seller of onions or garlic:

> *A thing came walking where many sat,*
> *Men wise in mind at the meeting place.*
> *One eye and two ears it had,*
> *Two feet and twelve-hundred heads*
> *Back and belly, and two hands,*
> *Arms and shoulders; one neck*
> *And two sides. Guess what it's called.*

There are many varieties of onion, including a Japanese one which is a foot long. The plant is sensitive to the length of night and day to such a degree that both summer and winter varieties stop growing and will not ripen if planted in the wrong season. The name is derived from the Latin *unio*, which some take to mean oneness or singularity. Another theory is that it is a corruption of *usnio* and linked with the Sanskrit *ushna*, burning or stinging. There are English dialect variants of the name, such as *ingum* and *inning*.

The name *jibbles* refers to many species, but may be connected with the Devon name of *chipple* for a small onion, which will ultimately derive from the French *ciboule* (cf. the German *zwiebel*), used of the *Spring onion* (A. fistulosum), similarly known as *chibol, sybie* and *sybow*. Also called *Welsh* or *green onion*, and *stone leek*, this species comes originally from Siberia and has no bulb but many small scallions. It has been known in Europe since the time of Charlemagne and is very popular in the East. In Japan it is known as *Japanese leek*. A variety of this, known as *catawissa* or *Japanese bunching onion*, is of Chinese origin and was first taken to America, by which route it came to Europe.

The somewhat similar *shallot* (A. ascalonicum), to which the name *scallion* is also applied in common with various others of the family, is also known as *cibbols* (see the related names above). It is very popular in India, but apart from that it is confined to Europe, where it was known in the time of Charlemagne, and even earlier. Pliny mentions it as coming from Askalon in Palestine, which is not true, but the corruption of the name has stuck. It became *eschalot* in French and thus entered English. The difference between this and the Spring onion is that the shallot only has one small undeveloped bulb at the end of its stalk.

Varieties of the onion proper include the *Egyptian tree-onion* (A. cepa bulbiferum) and the *potato onion* (A. cepa aggregatum), also known as *burn* or *underground onion*, and grown more in Eire than in Britain. The *Canada onion* (A. cepa canadense) was introduced into England in 1820;

its very small bulbs are cultivated for pickling (like the tree-onion's) and for cocktail onions.

Many other species are eaten in Asia, especially the Far East. These include A. splendens, whose small bulbs are eaten boiled or pickled in *saké* and soy sauce in Japan, and the *Japanese onion* (A. nipponicum), whose bulbs are eaten in salads by the Ainu. The bulbs of the *old man's onion* (A. senescens) of Europe and temperate Asia are also eaten in Japan, and those of the *Siberian onion* (A. angulare) serve as a winter vegetable in Siberia. Of the Central Asian onions the *akaka* (A. akaka) is sold in the bazaars of Iran, and the *royal salep* (A. macleanii) is eaten from Iran to Afghanistan. The *ruddy onion* (A. rubellum) is used by the hill people of eastern India; there is also the *Himalayan onion* (A. leptophyllum), whose bulbs are exceptionally biting.

2. Leek (*A. porrum*)

Nobody seems very sure where the leek came from originally, though it may have been from the East. Italian leeks were held in high esteem in Nero's time; that emperor was in the habit of eating them several days running every month to clear his voice for singing. For this he was derisively nick-named Porrophagus (leek-eater). The Romans are thought to have introduced leeks into England. They became the Welsh national emblem, worn in the hat on St David's Day, because that muscular Christian instructed his men to wear them in order to distinguish the British fighting against the waves of Saxon invaders. Possibly the misconception that the emblem is a daffodil arises originally from the fact that the flower was once counted as a leek.

The leek is the birthday plant for 9 February. Various popular sayings, usually of a defamatory nature, cluster around it. One is said to be 'green as a leek', and a slight matter 'is not worth a leek' (or a couple of onions). To eat the leek means a humiliation (like eating humble-pie), and possibly arose out of the famous scene in Shakespeare's *Henry V* where Fluellen forces Pistol to eat one after his rude comments on the plant. On the other hand there is an old country saying:

> *Eate leekes in Lide [March] and ramsins [wild garlic] in May*
> *And all the years after physitians may play.*

Leeks, because they are mucilaginous, figure in many soup recipes, especially in the Scottish national broth, cock-a-leekie soup. In fact the word porridge used to denote a thick vegetable soup with a leek base, and was then extended to any thick concoction such as pease or oatmeal porridge. The word probably represents a running together of 'pottage' and the French *porée*. An old Parisian street-cry runs:

Ah, mes beaux poirreau
Qui cuysent en eaue!
C'est un bon pottage
Avec du laictage.

In Cornwall the leek is called *ollick*; another old name is *purret* (cf. the Italian *poretta*).

Among other species of leek, some of which are easily confused with onions or garlic, are to be found, A. lebedourianum a Eurasian plant whose leaves and bulbs are eaten as a salad or cooked as a vegetable in Japan, and the *wild leek* (A. tricoccum) of North America, also known as *ramps*, a name more properly belonging to the British wild garlic (q.v.). Country people used the bulbs for flavouring various soups and other dishes, and the stems served as asparagus, as do those of other wild members of the onion family.

In the Germanic tongues most members of the onion family were called by the name of leek; it is somewhat difficult to sort them out. I list the Old English names below and attempt an identification.

1. *Bradeleac* (broadleek): leek (A. porrum) or broad-leaved garlic (A. ursinum), known as wild leek in Scotland.
2. *Cropleac* (headleek): garlic (A. sativum); cf. *knoblauch* (German), *krunslauk* (Estonian).
3. *Crowleac* (crowleek): Crow garlic (A. vineale) or the bluebell (Scilla non-scripta or nutans) known as crow leek.
4. *Garleac* (spearleek): garlic (A. sativum).
5. *Holleac* (hollow leek): chives (A. schoenoprasum)?
6. *Hwitleac* (whiteleek): onion (A. cepa or var.).
7. *Porleac* (porrum+leac = leek−leek): leek (A. porrum), but used mostly of a young leek or onion; it could therefore apply to the shallot (A. ascalonicum) or Spring onion (A. fistulosum).
8. *Secgleac* (sedgeleek): chives (A. schoenoprasum), also called reed or rush garlic; cf. *bieslook* (Dutch = bulrush leek), *brislauch* (German).
9. *Soteleac* (sweetleek): leek (A. porrum).

3. Chives   (*A. schoenoprasum*)

The Latin name is derived from the Greek for 'rush-leek', the Greek name for the leek, which thus links it with the various Germanic names mentioned above. Our present name comes from the French *cive*, which is a version of the Latin *cepa* (an onion). There are various country versions of the name, such as *civet*, *sives* or *sithes*. The cultivated variety is very nearly related to a wild Alpine variety; others grow widely over the Northern hemisphere. It was introduced into China about 2000 years ago and there regarded as an antidote to poison and a remedy for bleeding. It is, indeed, an antiseptic, since it is rich in an oil containing sulphur, in common with the rest of the onion family. Perhaps the German name *schnittlauch* (woundleek) arose because it was used similarly by that nation.

In England chives have been cultivated from very early times, the tops being plucked and used as garnishings, in salads and as a pot-herb. It has even been added to onions and shallots while cooking in order that they preserve their flavour better. If the tops, which are the only usable part of the plant, are cut regularly more will grow and the stalks will remain tender; chives left too long have a tendency to become tough. The *Chinese chive* (A. tuberosum), widely grown in the East, has longer leaves, flattened towards the tip, and white flowers. It adapts well in Europe.

4. Garlic (*A. sativum*)

In China garlic is of such ancient cultivation that it has an individual ideogram to itself. Elsewhere in the East, though it is very popular with the peasants, the priestly classes tend to frown upon it. Egyptian priests considered it unclean. And those people who had eaten it were not allowed to enter the temple of Cybele, the mother goddess of the earth, and witch goddess of untellable name, worshipped throughout the Near East. The Greeks, however, made much of it, while Hippocrates declared he preferred it to the onion. Criminals were given it to eat in order to purify them of their crimes. Nevertheless, even among them it was connected with Hecate, triple-goddess of the witches, and was buried at crossroads for her. Among the Romans it was fed to labourers to make them strong, and to the army to give it courage, but it was not considered with much favour by the richer classes. Horace waxes furious about it in his third Epode, having been given some by Maecenas at a feast:

> *If ever any man with impious hand*
> *Strangled an aged parent,*
> *May he eat garlic, deadlier than hemlock!*
> *(Ah, what hardy stomachs reapers have!)*
> *What is this venom savaging my frame?*
> *Has viper's blood, unknown to me,*
> *Been brewed into these herbs, or has*
> *Canidia tampered with the poisoned dish?...*
> *Ne'er did such heat of dog-star brood*
> *O'er parched Apulia,*
> *No fiercer did the gift of Nessus burn*
> *Into the Herculean shoulders.*
> *But should you ever wish this jest on me again,*
> *My merry Maecenas, I pray*
> *Your sweetheart stops your kisses with her hand*
> *And lies apart upon the couch's furthest edge.*

Garlic consists of a bulb divided into small segments, generally referred to as cloves. It is regarded as cooling, whether eaten or as an external application. Its antiseptic qualities were early recognised, and it was used in times of plague, and during the 1914–18 war. It is also a digestive stimulant, which explains its popularity in the Southern European regions where appetites grow jaded in the heat. In earlier times it would have been invaluable in helping down the rural repasts such as are described in this book, but there is some truth in Horace's remarks on the toughness of the rustic stomach.

Another poem of the same time as Horace, said to have been translated by Virgil from the Greek of Parthenius, and translated by Cowper during a lucid period in his final madness of 1799, describes another peasant concoction with a cheese base such as is mentioned in the section dealing with onions. A rustic miser enters his garden:

> There delving with his hands, he first displaced
> Four plants of garlic, large and rooted fast;
> The tender tops of parsley next he culls,
> Then the old rue bush shudders as he pulls;
> And coriander last to these succeeds,
> That hangs on slightest threads her trembling seeds.
> Placed near his sprightly fire he next demands
> The mortar at his sable servant's hands;
> When stripping all his garlic first, he tore
> The exterior coats, and cast them on the floor,
> Then cast away with like contempt the skin,
> Flimsier concealment of the cloves within.
> These searched, and perfect found, he one by one
> Rinsed, and disposed within the hollow stone.
> Salt added, and a lump of salted cheese,
> With his injected herbs he covered these,
> And tucking with his left his tunic tight,
> And seizing fast the pestle with his right,
> The garlic bruising first he soon expressed,
> And mixed with various juices of the rest.
> He grinds, and by degrees his herbs below,
> Lost in each other, their own powers forego,
> And with the cheese in compounds, to the sight
> Nor wholly green appear, nor wholly white.
> His nostrils oft the forceful fume resent,
> He cursed full oft his dinner for its scent,
> Or with wry faces, wiping as he spoke
> The trickling tears, cried Vengenace on the smoke!
> The work proceeds: not roughly turns he now
> The pestle, but in circles smooth and slow;
> With cautious hand, that grudges what it spills,
> Some drops of olive oil he next instills.
> Then vinegar with caution scarcely less,
> And gathering to a ball that medley mess . . .
> Obtains at length the salad he designed.
>   (The Salad)

In the thirteenth century there was a very popular sauce of mustard consistency made from garlic, almonds and breadcrumbs, crushed together and bound with a little broth. But by the sixteenth century it was considered as only fit for peasants.

In Anglo-Saxon medicine garlic and other members of the onion family figure widely. They are found eaten with bread in a ceremony against the elf-disease, in a drink against demonic temptations, in an eye salve for a stye, and a poultice for swellings. In the Middle Ages it was considered as a good defence against such manifestations of evil as werewolves, vampires and the witch's evil eye. It even became a good luck charm. The Aymara Indians of Bolivia still carry a piece of garlic about their person when they go bull-fighting. It was also thought to be efficacious against the bites of venomous beasts, as such names as *churl's*, *clown's*, *countryman's* and *poor man's treacle* testify. A 'treacle' was originally a remedy for these, and the names are echoed in the Latin *theriaca rusticorum* and the Greek Θηριακον άγροτων.

Allium, the general name for the family, is really the Latin for garlic, taken up in the French *ail*, the Spanish *ajo* and the Italian *aglin*. It may be connected with the Sanskrit *alu*, which covers any nutritious root. Of the many different species going under the name, some are more like a cross with an onion. They include *oriental garlic* (A. ampeloprasum), which is bigger but less strong than ordinary garlic, and probably that used by the Romans. Peasants still eat it raw. Originally from the East, it is now to be found growing wild everywhere in Mediterranean countries, and is a possible ancestor of the leek. Another large species of Eurasian origin is *giant* or *Spanish garlic* (A. scorodroprasum), a sort of onion-like leek, known as *sandleek* in northern Britain, where it is quite popular. To the Germans it is *rockenbolle* (rock-onion, cf. the Swedish *rackenboll*), a name which has entered French as *rocambole*, and occasionally used in Britain also.

A similar confusion of nomenclature exists with the North American *wild* or *Canada garlic* (A. canadense), also known as *meadow* or *rose leek*, whose bulbs are eaten boiled or pickled by the Indians; and the *drooping flowered garlic* (A. cernuum), alternatively known as *nodding onion* and *lady's leek*, another American species whose strongly flavoured bulbs are pickled or used in soups. The *ballhead garlic* or *onion* (A. sphaerocephalum) is a Eurasian species eaten in the Lake Baikal area of Siberia. Also occasionally cultivated in that area is the *twisted-leaved garlic* (A. obliqum), used as a flavouring in much the same way as garlic. *Daffodil garlic* (A. neapolitanum) and *rosy flowered garlic* (A. roseum) were once widely cultivated in Europe but are now less in favour. However, the *fragrant-flowered garlic* (A. odorum), a species extremely popular in the Far East, is now coming to the notice of European gardeners. This grows wild in

China and the bulb may be eaten raw; the flowers are also edible and used as a pot-herb.

There is a wild English species, the *broad-leaved garlic* (A. ursinum) which was once used in the Spring in sauces and as a pot-herb. It was also known as *bear's, hog's, snake's, common, wild* or *wood garlic; gipsy chipples, gibbles* or *onions; onion flowers; wild leek* or *onion*. A very common name for it was *ransons*, variously transmogrified into *ramsey, ramsden, romings, rosamund, ram's horns* and *buckrams*. It applies to the plant's smell, being derived from Old English *hramsa* (Old Norse *ramse*), a ram, a very smelly beast. Other names also allude to the same fact: *onion stinkers, stink plant, stinking Jenny* or *lilies*. Others include *yamps, badger's* or *iron flower, devil's posy, snake flower, water leek*.

Palms

A palm branch, signifying spirituality, belongs to those whose birthday falls on 17 June. In ancient times it was sacred to Mercury, Venus and orgiastic goddesses. Since it was used as a royal fan in hot Eastern countries it signified justice and royal honour in heraldry. It also stands for triumph; phrases such as 'to bear the palm' and 'to yield the palm' look back to when the palm branch was awarded to winners in the public games during classical times. A fusion of both triumph and royal honour seemed meant by the people who went out to meet Jesus with palms in their hands when he entered Jerusalem riding on an ass one week before his crucifixion. It is this event which has given Palm Sunday its name.

In the old days there was a good deal of ceremonial connected with this particular day; churches were decked out with greenery and processions held. Many parts of Europe, however, were hampered by having no palms of their own and had to substitute catkinned branches of other trees. Thus it has come about that several trees in the British Isles are known by the name of 'palms'. Chief among these are various species of willow, wagon-loads of which were collected from the Thames-side every year and taken up to Covent Garden to sell. On the day before Palm Sunday it was a popular custom to 'go a-palming'; this also took place on the day itself. In the North of England the branches were formed into small crosses and hung about the house.[1]

Because of its close connection with the crucifixion the palm branch has

---

[1] Willow branches were once used as a palm substitute in Russia also. Other popular substitutes in the British Isles include yew and box (used since medieval times), hazel catkins in some parts of Lincolnshire and branches of the silver fir in Ireland.

come to stand for martydom in Christian iconography. It was also the emblem of anyone who had pilgrimaged to the Holy Land; he was thus known as a 'palmer'. Willow and palm branches featured together in the Jewish Feast of Tabernacles, when it was required that the people live for eight days in booths made from them in memory of the forty years wandering in the wilderness. A palm tree was the badge of the Jewish tribe of Manasseh, as it is of the Christian St Ambrose, patron of Milan and one of the 'four fathers of the Roman church'. For the Egyptians the tree was the symbol of the year, each branch standing for a month since it casts off and grows anew a branch each month. And to all the ancient civilisations it was the tree of life and fertility, represented as a phallic pillar issuing flames. It is said that this is what the English maypole is meant to represent.

Also, of course, many people rely on palms for food. The date (genus Phoenix) immediately springs to mind; its fruits form a staple food for many Arabians. The two principal edible items from palms as a whole are sago (which is treated separately since it is also obtained from other trees), and palm cabbage. The latter is generally the terminal bud, which is cooked and eaten like a vegetable. In certain cases, so some say, it is rather like an artichoke. The only trouble is, as happens when obtaining sago as well, the palm is destroyed in the process. In tropical parts where trees are quickly replaced, however, this does not matter very much.

The following is a list of palms used for food, exclusive of those used for sago (q.v.).

1. *Piassava* (genus Attalea). This group of very similar palms from tropical America has now been fragmented, according to the latest experts. The almond-like seeds of the *native almond* (A. or Bornoa amygdalina) are eaten in Colombia; so are those of the Brazilian *urucuri* (A. excelsa or Scheelea martiana), which are also burnt to help coagulate rubber latex. The edible seeds of the *pindova* or *indaja* (A. compta), also from Brazil, are the size of a goose-egg. The young buds of the *cohune* (A. or Orbignya cohune) are eaten in Central America.

2. *Palmyra* (Borassus flabellifer). This tree, also known as *deleb* or *tala palm, brab tree, great fan palm* and *East Indian wine palm*, one of the most used and useful of all, grows in West Africa, tropical Asia and the Pacific. In some places, such as Ceylon, the trees form enormous forests. In many areas the population depends upon them for the necessities of life, food, shelter, fuel and clothing. The leaves have been used since time immemorial for writing instruments, fans, mats, buckets, flutes, and so on. The tree is estimated to have more than 800 uses. The young shoots serve as a potherb, and the fruits are eaten after cooking; the young seeds, of which there are three the size of goose-eggs in every fruit, are also used.

3. *Cane palm* (genus Calamus). These palms are exceptionally slender-stemmed, and seem to have more in common with bamboo. The very

9. PALMS

sugar palm (♂ and ♀ spadix)   sago   palmyra (♀ and ♂ spadix)
coconut

Copyright © 1972 by George Allen & Unwin, Ltd.
All rights reserved. For information, write:
St. Martin's Press, Inc. 175 Fifth Ave., New York, N.Y. 10010
Printed in Great Britain
Library of Congress Catalog Card Number: 77-188546
First published in the United States of America in 1973

AFFILIATED PUBLISHERS: Macmillan & Company, Limited, London – also at Bombay, Calcutta, Madras and Melbourne – The Macmillan Company of Canada, Limited, Toronto

THE VEGETABLE BOOK

An Unnatural History

YANN LOVELOCK

Illustrated by Meg Rutherford

ST. MARTIN'S PRESS   NEW YORK

The majority of ferns belong to the polypody family (Polypodiaceae) which comes by its name in reference to the many branches of its rootstock (the name means 'many feet'). An English herbalist of the sixteenth century, referring to the medical uses to which polypody is put, comments that it 'drieth and lesseth the body'. The budding fronds of the *oak-leaf polypody* (Polypodium or Drynaria quercifolia) are eaten by the Asian poor, and the young fronds of Drynaria rigidula on some Indonesian islands. So also are those of the *golden marsh fern* (Acrostichum aureum), which is also to be found in the Philippines and elsewhere in the Pacific, and in Central America; and those of the *edible spleenwort* (Asplenium esculentum), the British species of which is known as *miltwaste* and *scale* or *stone fern*. In the Philippines these are eaten as a salad or cooked, and are sold in the markets. The name spleenwort was originally derived from the Latin generic name, and later it was thought to be good for diseases of the spleen in sheep.

Other relatives are also used by the poor, mostly in the South-East Asia. The very young fronds of Diplazium esculentum are eaten with rice in Indonesia, and those of the Diplazium asperum either raw or boiled elsewhere. The young shoots and fronds of Pleopeltis longissima are eaten raw or boiled in the Malaysian islands, and both there and in Indonesia the young fronds of Stenochlaena palustris are very much esteemed with rice.

Several ferns not belonging to the polypody group are also used. In some Malaysian islands the sweet young fronds of an *adder's tongue fern* (Ophioglossum reticulatum) are used as a pot-herb, and the succulent fronds of the *oriental water* or *pod fern* (Ceratopteris thalictroides) serve as a spring vegetable in Japan. In some parts of Asia the young shoots and fronds of the *four-leaved pepperwort* (Marsilea quadrifolia) are eaten, but members of this family have their greatest importance in Australia, where the aborigines make a flour by grinding the spore-cases of the *nardoo plant* (M. salvatrix or macrocarpus). Taught by them, the occasional settler and explorer has been saved from starvation by this expedient, but there seems to be some disagreement over how good the flour tastes. According to some accounts it is said to be eminently passable, while others condemn it as not very palatable. Two other pepperworts (or clover-ferns) may also be used for this: *Drummond's pepperwort* (M. drummondii) and *hairy pepperwort* (M. hirsuta). All grow in swampy areas after rain; as the swamps dry the ferns wither, leaving their spore-cases on the ground.

The roots of some ferns are also used in Australasia, such as those of the *net fern* (Gleichenia dichotoma), from which the aboriginals extract an edible starch. The root of the *tara fern* (Pteridium esculentum), another polypody, is an important source of food to the New Zealand Maoris, although the plant is regarded as a troublesome weed by Australians. The tender stalks and fronds of the related *bracken fern* (P. aquilinum) are boiled as a vegetable or used as a pot-herb by the Japanese, and the roots were once eaten by North American Indians. In Britain it is known as *lady bracken* or *female fern* and *adder-spit*. The name *eagle fern* is echoed by the Scottish *ern* (i.e. eagle) *fern*.

Actually the very young shoots of almost any variety of fern may be eaten, either raw or cooked (preferably the latter). The taste is not unlike that of asparagus, although I found the shoots rather bitter and tough; it much depends, I suppose, on the soil and the variety picked. In nineteenth-century England (and doubtless earlier), it is recorded that the young shoots were 'greedily devoured as a substitute for green vegetables'. They are still eaten in rural North America, where they are known as fiddleheads, and in the Far East. The Korean Song Sam-mun (1418–56) is doubtless referring to this practice in a poem written shortly before his execution for plotting the restoration of the deposed king to whom he was minister. He remembers two Chinese brothers who starved to death as a protest against a usurpation:

> *I scan befogged Mt Shou-yang,*
> *Lament the sages, Po I and Shu Ch'i.*
> *They would rather have starved to death*
> *Than pluck the wild bracken here:*
> *Even tho it is an innocent weed,*
> *Does it not grow in the usurper's soil?*

For the Japanese the fern is the emblem of the samurai and signifies honesty. Used in New Year's Day decorations it symbolises hope and prosperity. In the West it also typifies victory over death. As the birthday plant for 24 March it stands for fascination, confidence, sincerity and solitary humility. Because of its rapid growth and spread it is an emblem of colonisers, like the plantain (q.v.). But bracken also represents a serious menace to otherwise good arable land and a great deal of effort is put into ways of exterminating it in Britain. There was once a belief that if gathered at the proper moment it would render a person invisible. Shakespeare, in 1 *King Henry IV*, II, considered that the seed conferred invisibility. It was inadvisable to pick it on St John's eve however, since those who did so had their hats struck off and were kicked about severely.

Another curious plant put to various uses is a cone-bearing fern-ally known as the *horse-tail* (genus Equisetum), a sort of cross between a fern and a conifer which can't quite make up its mind which it would like to be. The tuberous rhizomes of the *meadow horse-tail* (E. pratense) are eaten by North American Indians, and the above-ground portions of the *smooth horse-tail* (E. laevigatum) are dried and made into a mushy porridge by the Indians of New Mexico. The latter plant also goes by the name of *scouring* young unfolded leaves of the *rattan* (C. zeylanicum) are eaten raw or cooked in Ceylon. In Borneo the buds of the *Malacca cane* (C. scipionum) are used.

4. *Pacaya* (genus Chamaedorea). This is another dispersed family of Central American palms. The young flower clusters of Eleutheropetalum sartoni are eaten by the Indians of southern Mexico, and the very young leaf shoots of the *tepejilote* (Edanthe tepejilote). These may be eaten like asparagus, as are those of Neanthe elegans elsewhere in Central America.

5. *European palm* (Chamaerops humilis). This is the only palm native to Europe, and is found growing on both sides of the Mediterranean. It goes also under the names *Mediterranean hair palm* and *dwarf fan palm*. The very young buds may be eaten.

6. *Coconut palm* (genus Cocos). Generally understood by the name is the commercially grown C. nucifera whose fruits are so well known to us, and very important in the diet of some South-Sea islanders. Originally from Malaya, it has been spread throughout the world tropics. Its peculiarity is that it only flourishes well by the sea. A meal is obtainable from the centre of the trunk and the young buds are eaten as a salad or a vegetable. Its edible erstwhile relatives are to be found in South America. The very young buds of the *queen palm* (Arecastrum romanzoffianum) are best liked preserved in oil or vinegar. The *yatay* (Butia yatay) grows chiefly in the Argentine and the abutting districts of Uruguay and Paraguay. Its fruits are used for making a brandy and its young buds eaten.

7. *Guadeloupe cabbage palm* (Erythea edulis). This is used in the West Indies and the Baja California district of Mexico.

8. *Assai* (Euterpe edulis). This is the principal cabbage palm of Brazil. The buds are available in the markets and locally canned. It has been considered as having some export portential.

9. *Shadow palm* (Geonoma binervia). The palm grows in Central America; its flower-clusters are cooked by Mexican Indians.

10. *Sagisi* (Heterospathe elata). A palm from South-East Asia whose buds serve as a vegetable; the nuts are used as a masticatory with betel.

11. *Coquito* (Jubaea chilensis). The tree is Chilean and also known as *Chile syrup* or *wine palm, coco de Chile* and *little coker-nut palm*. These names refer to the fact that palm wine and honey are its products; its coconut-like fruits are edible and crushed to express a cooking oil.

12. *Australian cabbage palm* (Livistonia australis). The buds are eaten raw, cooked or baked by aborigines.

13. *Nibong* (Oncosperma filamentosa). A Malayan palm whose very young leaf-shoots are eaten.

14. *Solitaire palm* (Ptychosperma elegans). This is from New South Wales (Australia), where it is often referred to as *cabbage palm*.

15. *Nikau* (Rhopalostylis sapida). This is also known as *cabbage palm* in

168 / *Parsnips and related roots*

New Zealand, where the buds are eaten by the Maoris. It used to be classed in genus Areca, whose principal member, the *catechu* (Areca catechu) from tropical Asia and the Pacific, provides the betel-nuts used as a masticatory so widely in its area, and whose very young unfolded leaves are used as a vegetable. The leaves of the nikau itself were once much used for thatching Maori huts, but are now giving way to corrugated iron, which they resemble. The vivid red fruits of this palm are the size of large peas and so hard that settlers have been known to use them as shot when ammunition was scarce. But they are much relished by the wild parrots which, unable to find a footing on the exceptionally smooth green trunk, feed upside down while hanging onto a leaf by one claw.

16 *Royal palm* (genus Roystonea). Some of these palms produce sago, and have an edible bud. The *Caribbean royal palm* (R. oreodoxa) is one such. The *Cuban royal palm* (R. regia) of Central America and southern Florida (U.S.) is used mainly for its very young leaf-buds.

17. *Cabbage palmetto* (Sabal palmetto). The leaves of this palm, native to the southeastern U.S. and Caribbean, are used in Easter celebrations, and the bud is eaten. South Carolina is known as 'the palmetto state' since the famous incident during the War of Independence when a stockade made of its trunks, banked up with earth on an island in Charleston harbour, defied the British Fleet in 1776. The naval shot could make no impression on the spongy wood and the attack was called off.

18. *Saw palmetto* (Serenoa semilata). This palm, the seeds of which are important as a food to the local Indians, also comes from the southeastern U.S.

19. *Socratea* or *Iriartrea durissima*. The young buds of this palm, native to Nicaragua and Panama, are boiled by the natives.

20. *Californian Washington palm* (Washingtonia filifera). The palm grows in California and the adjoining area of Mexico. Its buds are roasted by the Indians and its fruits are eaten fresh, dried or ground into meal.

Parsnip and related roots (*Pastinaca sativa*)

The parsnip is a member of the carrot family, and therefore related to other umbelliferous plants, the roots of several of which may also be eaten. The parsnip's starchy root, however, is one of the most nourishing in the whole family, although there are some who complain that it is too sweet. In fact the starch is converted to sugar whenever the root is exposed to the frost. It has a strongish flavour and one is advised not to put it in a stew (especially a vegetable stew) since it dominates all the other constituents. The plant is said to have entered Europe from western Asia, and to have been cultivated from Roman times. But, as in several other cases, there is such a confusion of names in the Latin that it is difficult to say much about its early history. The parsnip is first distinguished from other roots in the time of Charlemagne and, before the introduction of the potato, was very popular as Lenten fare.

The Latin name *pastinaca* is related to the verb *pascare* (to feed). In Old French it is *pastenaie* or *pasnaie*, the latter developing into the modern French *pasnais*. The word enters Middle English via the French and is compounded with *nepe* (a turnip) into *pasnepe*. The Welsh *maip* (a turnip, derived from Old English) has gone towards the formation of *mypes*, another old name for the plant in the west of England. The Latin name (unchanged in Italian) has also contributed to other north European names: *pastinake* (Germany), *pastniak* (Holland, Denmark), *palstemacka* (Sweden), *pasternak* (Russia).

RELATED ROOTS

One of the parsnip's main relatives is the *hogweed* (genus Heracleum). An English species grows as high as 5 feet and is known as *cow's parsnip*, *pig's parsnip* or *bubbles* (H. sphondylium), from the practice of feeding the roots to these animals. *Madnep*, another name for it deriving from the Old English *mede-næp* (meadow turnip), was eventually transferred to the parsnip itself. It may well have been used once as a parsnip substitute, although the plant's close likeness to the hemlock in the same family has given it such sinister names as *Devil's meal*, and *bad, black*, or *dead man's oatmeal*; *ha-ho*, a name it shares with hemlock, supposedly represents the agonised cry of one who has eaten the plant by accident.

A more jolly set of names describes it as *humpy-scrumples*, *limperscrimp* and *rumpet-scrumps*. These may well refer to the stalks, for which several uses have been found, including that of a straw; a Somerset man once spoke of these when he reported 'they drinked up their cider with *wippul-squips*'. The stalks have also been used as candles, pipe-lighters and, among children, served in much the same way as angelica stalks (q.v.), as such names as *spouts* and *squirt-guns*, and the Scottish *bear-skeiters* (barley-shooters), testify. In America the stem of the *American giant* or *cow parsnip* (H. lanatum), also known as *masterwort*, are edible; in Alaska they go by the name of *wild celery*. The Indian tribes in the north also eat the young flowers and the roots. Lastly, the seeds of the *Persian cow parsnip* (H. persicum) are used in pickles in Iran.

The roots of some of the *hog-fennel* family (genus Peucedanum) have played an important part in the diet of Indian tribes in North America.

They include the *chucklusa* (P. canbyi), *tuhuha* (P. farinosum) and *cous* (P. cous). These have sometimes become confused with members of the related *biscuit-roots* (genus Eulophus, Cogeswellia or Lomatium) whose thick bulb-like edible corms are used by several Indian tribes of the northwest. They include the *fennel bread-root* (L. foeniculaceum), *Wyeth bread-root* (L. ambiguum) and *wallowa* (L. circumdatum). The family name comes about because many of the Indians reduce the root to a flour and use it in their soups and for baking.

The *wild parsnip* (Phellopterus montanus) is known by the Spanish name *gamote* in the southwestern states of the U.S. In New Mexico the Indians peel the roots and then bake and grind them for food. The *Japanese corkwing* (P. littoralis) is cultivated in the Far East for its leaves which are used as a condiment and are said to taste like a cross between angelica and tarragon. *Gamote*, however, seems to have been given a new family name, genus Cymopterus, by the botanists, and is thus linked with the aromatic *chimaya* (C. fendleri), whose roots and leaves are eaten in the same area as the *gamote*. Some other members of the family are similarly used.

Root or *parsnip chervil* (Chaerophyllum bulbosum) is quite distinct in use from the salad chervils already treated earlier (q.v.). Several other members of the family share the name *water-parsnip* in common with poisonous relatives of skirret (q.v.). This chervil's root is like a small yellowish-white carrot; it is sweet and floury and considered as one of the most nutritious among vegetables. The leaves, however, are slightly poisonous and are best left alone. The plant was first introduced into England in 1726 but the roots were not eaten until some time later, although in Germany and Holland they made a well-known peasant dish. A larger rooted species, the *Prescott chervil* (C. prescottii), was first noted by Europeans as being eaten in the Ural and Altai regions of Siberia in 1850.

Aracacha (Arracacia esculenta) is also known as *Peruvian carrot* or *parsnip* and grows in the American tropics from Peru to the West Indies. The roots are prepared like potatoes and taste like parsnips, although they are considered lighter, more palatable and nutritious, and easier to digest. The plant bears a number of roots bunched together and looking like carrots. The several attempts to introduce them into Europe have proved abortive; the plants always die, probably because they cannot resist the frost, and our damp climate retards their growth. In South America a preparation for the treatment of stomach diseases is also made from them.

Peas and vetches

1. Common pea (*genus Pisum*)

Peas, vetches and beans were very early standard foods among primitive peoples and ancient civilisations. In the Middle Ages they were regarded as no more than Lenten fare and, but for a rather remarkable change in taste, might have disappeared into modern-day oblivion, as have so many other vegetables. The origin of our pea (Pisum sativum) is somewhat questionable; at one time it was thought to be cultivated out of the *field* or *grey pea* (P. arvense), which grows in a semi-wild state in some countries but is now considered no more than a variety of P. sativum. It would appear to have been this species that was eaten in the Bronze Age (*c.* 3000 B.C.). This plant with reddish flowers and an angular fruit is of Italian origin and is now grown as animal forage. The *Mediterranean pea* (P. elatium), which now grows in the Nile delta and is cultivated in Algeria, is possibly that which was cultivated by the Greeks. Charred remains of it were found at Troy, and it has also been found in Egyptian tombs of the twelfth dynasty. Peas were of long-standing use in India but only appeared in China at a comparatively late date judging from their name, which means 'Mohammedan pea'. The Romans, however, preferred the less bland taste of such substitutes as the chick pea, vetches and lupin-seeds. In the Middle Ages peas were dried and kept against times of famine and shortage; they were preferred to beans and lentils.

In England, even before the Norman Conquest, peas were the principal crop, and their popularity has never waned since. Even the dried pea is still used in such ancient dishes as pease pudding, and until recently the field pea was cultivated in the north under the name of *carlins* and cooked as a special dish either on Mid-Lent or Palm Sunday, also known as Carlin Sunday because of this custom. The dried peas were steeped all night, then fried in butter and sprinkled with rum and sugar.

The pea is the birthday plant for 17 February and is symbolic of respect. To dream of a dry pea is said to foretell a coming marriage, and a green pea signifies perfect happiness. The popular sayings that gather round the pea, however, are not nearly so complementary.

> *Love and pease pottage are two dangerous things,*
> *One breaks the heart and other the belly.*

Is this, perhaps, the reason why 'half a pea a day will serve a lady'? Or is she considered so delicate that she can subsist on this meagre fare, one like the true princess of Anderson's fairy tale who was bruised all over by a dried pea lying at the bottom of a mountainous pile of mattresses?

If a man is said 'to be going into the pease field' it means he is going to sleep, and 'to give a pea for a bean' means a kind action done with an eye to future gain, much as the Biblical proverb is misquoted: 'Cast thy bread upon the waters and it shall come back to thee buttered toast'. Various country rhymes centre upon the planting of the crop:

> *If Candlemas Day [Feb 2] be fine and clear*
> *Corn and fruits will then be dear.*
> *On Candlemas Day if the thorns be adrop*
> *Then you are sure of a good pea crop.*
>
> *David and Chad [March 1–2]*
> *Sow peas good or bad.*

Beware, however, of the terminal date: 'St Benedict [March 21] sow thy pease or keep them in the rick'.

It was not until comparatively late that the peas were actually eaten fresh, but when they were they commanded fantastic prices at first. The variety first perfected in Holland and known to the French as *petit pois*, and to Americans as *turkey pea* and *French canner*, was used fresh and became all the rage in the seventeenth century. Madame de Maintenon comments on their popularity in the court of Louis XIV in a letter dated 16 May, 1696:

> *Le chapitre des pois dure toujours; l'impatience d'en manger, le plaisir d'en voir mangé, et la joie d'en manger encore sont les trois points que nos princes traitent depuis quatre jours. Il y a des dames qui, après avoir soupé, et bien soupé, trouvent des pois chez elles avant de coucher, au risque d'une indigestion. C'est une mode, un fureur, et l'une soit l'autre.*

The British were among the first to develop and catalogue the separate varieties of the pea. Many of the nineteenth-century names display typical Victorian boastfulness and chauvinism: Prince Albert, Victoria, Champion of England, Conqueror, Ne Plus Ultra, William the First, Fillbasket; other names are decidedly trendy: Telephone, Telegraph, Strategem. A popular name for a large variety of pea, possibly the *marrow-fat* used in canned processed peas, was *runcivals*. An explanation of this, probably apocryphal, is that it is a corruption of Roncesvalles, where Roland and Oliver of Charlemagne's army were ambushed, and where the gigantic bones of these dead heroes and their companions were wont to be displayed to the credulous. An early variety of pea used to be cried through the streets in the nineteenth century under the name of *green hastings*. Perhaps this is where the idea of calling one variety William the First came from!

One of the first English varieties, the *sugar pea* (P. sat. macrocarpum), is little heard of now. Both the pod and its fruit may be eaten while still young. If you are prepared to go to the trouble you can make a very passable soup out of the young pods of any variety. In China they use pea-sprouts in the same manner as bean-sprouts, germinating the peas at home and cooking the shoot after about 5–7 days, when it is an inch long. In Burma and some parts of Africa the leaves serve as a pot-herb.

Our word entered Old English as *pise* from the Latin. The plural *pisan* became the Middle English *peason*; but the singular *pease*, like the French *pois*, could also serve as a collective plural, as in 'pease pudding'. As our language underwent further changes the 's' sound was mistaken for a plural ending, and thus the totally incorrect singular form of *pea* was evolved. It is related to the Irish *piosa*, and to our word 'piece' (which is the meaning of the Irish). The root is also to be found in the Sanskrit *pis* (to divide), from which was derived their *pêci*, the fruit of the pea apart from the pod (i.e. a piece of the pea). The same root turns up in the Hindi *pai* and *paisa* which became the Indian monetary units. An *anna* is divided into 4 *pice*, and each pice into 3 *pies*.

2. Vetches (*genus Vicia*)

These are members of the pea family, some of which were used in classical times for the sake of their pulse-like seeds. As late as the sixteenth century a chronicle records that people were using them during times of famine. There are about 150 different species growing in the Americas, Europe and Asia, now mostly used as animal fodder. Both wild and domestic animals find them extremely palatable and they are, besides, valuable as a manure which adds nitrogen to the soil. The name is derived via the Old French *veche*, *vece*, from the Latin *vicia*. It is related to an old European root, *vik*, meaning to bind, wind or coil; one may compare the German *wickeln* with *wicke*, the word for a vetch.

Ervillia or *bitter vetch* (Vicia ervilia) is of South European origin and of ancient culture. Remains of the seeds have been found at Troy, and it is known that the Greeks ate them also. The name comes from the Latin *ervum* (now the generic name of the lentil family), and is *ervo* in Italian. The *common* or *pebble vetch* (V. sativa) grows wild over Europe, North Africa, western Asia, and has become naturalised in North America. The Romans seem to be the first to have used it, both as cattle-fodder and for food. Various names for it are variations on 'vetch' and its alternative, *tare*: *fitch*, *thetch* and *datch*. The seeds are known as *twadgers* and as *birds*, *cats*, *crow*, *mice* or *gipsy peas*. *Podder* is a corruption of the older *podware*; *urles* and *chickling* are two other names. A variety of this, known as *bigpod vetch*, comes originally from Algeria and has fruits as large as peas which are eaten by Arabs. It is also grown for forage and is similar to the *French vetch* (V. narbonensis).

The *chickling vetch* (Lathyrus sativus) belongs to the pea-vine family and is probably of West Asian origin. It grows as far as northern India and was cultivated very early in Southern Europe both as a fodder and for food. Its Latin name *cicercula* (little chick pea) is the equivalent of the present name of *chickling*. It is also known as *grass-pea*. In India it is the cheapest of the pulses, and is eaten by the poor both there and in the Middle East, especially in times of famine. The large white seeds are best parched and boiled, and are made into chapaties, paste-balls and curries. Sometimes the seeds become mixed with those of the common vetch which, if eaten over too long a period, are liable to cause paralysis of the lower limbs. The leaves also are used as a pot-herb.

According to Dioscorides the seedling of the *yellow-flowered pea* or *vetchling* (L. aphaca) was once eaten in Europe, either fried, boiled, or cooked like the lentil. Indians of the American north-west make use of other species. The Chiptewa and Ojibway tribes eat the seeds of the *cream pea-vine* (L. ochroleucus), those in Nebraska the pods of the *showy pea-vine* (L. ornatus), and those in New Mexico the pods of L. polymorphus. The Indians also eat the pods of some of the *milk vetches* (genus Astragalus). Those in Montana eat the pods of the *American ground plum* (A. caryocarpus) raw or boiled; the Crees and others eat those of the *Indian milk vetch* (A. aboriginum). Others in this family grow in the Old World, like the *edible milk vetch* (A. edulis), found from North Africa to western Asia, and whose seeds are eaten in Iran. In that land of sultry romance, also, the ladies of the harem used the sweet yellowish-brown gummy exudation of A. fasciculifolius to give a glossy appearance to their skin.

The seeds of the *deer vetch* family (genus Lotus), to which the English birdsfoot-trefoil belongs, can also be eaten. These include such Mediterranean varieties as L. edulis and L. purpureus and the Arabian *koueh* (L. arabicus). But the best known is the *winged* or *asparagus pea* (L. tetragonolobus) which must not be confused, as it often is, with the Goa bean (Psophocarpus tetragonolobus, *see the section* 'others' *under* Beans). The winged pea is of Mediterranean origin and was introduced into England in the sixteenth century for the sake of its edible young pods which are said to taste rather like asparagus.

3. Chick pea (*Cicer arietinum*)
The *chick pea*, possibly of Caucasian origin, was cultivated in ancient times by the Egyptians, Jews and Greeks. It has been introduced into tropical Africa, Central and South America and Australia. In India it is the most important of the pulse crops and made into cakes, puddings and savoury dishes. The short hairy pods contain seeds resembling either the pea or the lentil, and these may be ground into a flour. But the green pods and tender young shoots may also be eaten as a vegetable. There is a certain problem connected with this, however, since the hairs on the pods exude a sticky

irritating acid; this can be dissipated by spreading a cloth over the plant at night, the acid being absorbed with the dew by the next morning.

The chick pea is often known simply as *gram*, a name by which many other pulse crops are known (much as *dhal*, the name for a lentil, also covers many of them). It derives from the Portuguese *graõ*, ultimately the Latin *granum* (a grain). The Latin *cicer* becomes eventually the French *chiche*, from which the English name is derived. The fruit's likeness to the lentil has caused a certain confusion of names. Lentils were formerly known in England as *chick*, and several Slavic names seem to follow suit: *čočka* (Czech); *soczewica* (Polish); *checheviğa* (Russian). The alternative names of *garavance* and *calavanche* derive through the French from the Spanish *garabanzo*, itself ultimately from the Basque *garau* (corn) + *anzau* (dry). It is also known as *Bengal gram* and *Egyptian pea*.

4. Cow pea (*genus Vigna*)

The *cow pea* is of African origin and is now widely cultivated in Asia, which it reached via Egypt and Arabia. It also spread over the Mediterranean countries and was known to the Greeks and Romans. The Spanish took it to the West Indies in the sixteenth century and it was established in the U.S. by the following century. The dried seeds make a very important pulse crop; they may be ground into meal and used in a variety of ways. They are also used as a coffee substitute. The fresh seeds and immature pods may be eaten, as well as the young shoots and leaves, cooked like spinach or dried and stored for later use, as in Africa. If the leaves are cut back regularly new leaves are produced, as is the case with nettle-tops and chives.

Species include the *catjung* (V. unguiculata), also called *Kaffir pea* and *red bean*, and the related *asparagus* or *yard-long bean* (V. sesquipedalis), also known as *Bodi* or *snake-bean*, which is the most widely cultivated in the East and occasionally grown in Africa. The names for the latter, of course, refer to the very long pods. The *China pea* or *southern bean* (V. sinensis) is also closely related and probably best known as *black-eyed pea* or *bean*. This was introduced into the U.S.A. in the eighteenth century and grows mostly in the south, where the beans grow on vines; in the north the plant is more like a bush. To the Asian Indians this is the *chowlee plant*; other names include *Tonkin*, *Jerusalem*, *cornfield* or *marble bean*.

5. Pigeon pea (*Cajanus indicus*)

The *pigeon pea* is originally from Africa but now widely grown for food in Asia, the Pacific and the West Indies, where it was introduced during the course of the slave trade. It was cultivated in Egypt before 2000 B.C. and remains have been found in twelfth dynasty tombs. It was also cultivated very early in Madagascar and taken to India in prehistoric times. The young seeds are eaten as a vegetable or, when ripe and coloured grey or yellow according to variety, dried and eaten as a pulse or split and used like lentils, as in India. In Puerto Rico and Trinidad they are canned and exported. Other names, some of which only apply to varieties, include *red gram* and *no-eye pea*; *dhal* or *doll pea*; *Angola* or *Congo pea*; *hoary* or *toor pea*. Another name is *cajan*, deriving from the Malay *kachang* for any leguminous plant; *catjung* (see previous section) also derives from this word.

6. Lupin (*genus Lupinus*)

The *lupin* is an unlikely member of the pea family to find in the company of vegetables. Nevertheless the Greeks cultivated it for the sake of its bitter seeds, known to the English as *flat, fig*[1] or *penny beans*, and the Romans actually preferred them to the pea. Nowadays they are fed mostly to cattle, although they are still eaten in some parts of southern Europe. If they are first steeped in water and then cooked much of their acridity is lost, although they are still reckoned fairly indigestible. In Europe the most popular species are the *white lupin* (L. albus) and the *yellow lupin* (L. luteus), also called *Spanish violet* and *Virginia rose*. For centuries the seeds of the latter were roasted to make a coffee-like drink. The ancient Egyptians made use of the *Egyptian lupin* (L. termis). Its seeds once served as a Semitic weight equal to one tenth of a gram. There are several other European and Asian lupins whose seeds may be eaten, as in America may those of the *wild, blue* or *sundial lupin* (L. perennis) and the Peruvian *Cruickshank's lupin* (L. cruickshankii). The *shore lupin* (L. littoralis) from the American north-west deserves special mention, as not only its seeds but also the roots were eaten by the Indians.

7. Broom (*Sarothamnus scoparius*)

The *green* or *Scotch broom* (Sarothamnus scoparius) growing from Europe into western Asia, and introduced into North America, is another strange member of the pea family, and one that is generally accounted poisonous. However, sheep and goats are said to thrive on it; its twigs have been used to flavour beer in their time, while the leaves and buds are preserved in salt and vinegar and eaten in South Germany under the name *brahm* and *geiss kappern*. Other domestic uses include the extraction of brown and yellow dyes from the bark and, of course, the making of besoms from its twigs, from which it derives its English name, the French *herbe à balai* and German *besenstrauch*. Other queer old names include *bannadle* in Wales and Cornwall, and *genest*, from the French, ultimately deriving from the

---

[1] Cf. the German *feigbohne* and French *pois lupin*. German names for the white lupin include many referring to its leguminous affinities: *bitterbohne, flachsbohne, wickbohne, wolfswicke, wolfsscholte, wolfserbse, wolfsbohne*.

old Latin name *planta genesta*. It is this name which became that of the Plantagenet family, after their device, the broom flower, the chosen plant of Brittany which was worn by the pretenders to the overlordship of the house of Anjou. A daughter of this family was married into the English royal family, who eventually succeeded to all the Angevin territories and claims. Louis IX of France founded the *Cosse de Genest* order of knighthood in 1234; the name was taken from the collar on which broom flowers alternated with the *fleurs-de-lys*.

8. Others

Other relations of the pea are eaten by Australian aborigines. These include the young pods of the *Port Curtis yellow wood* (Hovea longipes), a tree-like flowering shrub in the north-east, and the seeds of the *pea-tree* (Sesbanea servicea). The latter originated in Australia and has now spread into South-East Asia and Ceylon. In southern Asia the green pods of the commonly cultivated *agati* (S. grandiflora) are eaten, in addition to the tender leaves and large fleshy petals of its flowers. The leaves of the tropical African S. tetraptera serve as a pot-herb. The yellow flowers of the *Chinese pea-tree* (Caragana chamlagu) are eaten in the north of China, while further north still the young pods of the *Siberian pea-tree* (C. arborescens) serve for food.

The flowers of some others in the pea family are also eaten, including those of some species of the *senna* (genus Cassia), whose purgative properties are well known. But the young pods of at least one of these, the *sickle senna* (C. tora), may be eaten and are known as *shim beans*. The buds of the *Judas tree* (Cercis siliquastrum), so named because it was supposed that the false apostle Judas hanged himself upon it, are eaten and have an agreeably acid taste. They may also be frittered or pickled. A North American species, known to the Indians as the *red* or *rose-bud* (C. canadensis), was early used by the French Canadians. The flowers are either eaten raw in salads, cooked or added to bread.

The *rattlebox* or *castanet plant* (genus Crotalaria) of Central America is sometimes classed under the same generic name as the Judas trees. In Guatemala the leaves and young branches of one species (C. guatemalensis) serve as a pot-herb, and the leaves, flowers and pods of C. glauca are eaten elsewhere in tropical Africa. Another close American relative is the *palo verde* (genus Cercidium), whose Spanish name refers to the distinctive green bark. The seeds of the *blue palo verde* (C. torreyanum) are ground for flour by the Indians of the southwestern U.S.; in the same area the fresh seeds of the *little-leaved palo verde* (C. parvifolium) are made into flour or mixed with *mesquite* meal.

Pignuts and carraways

The name covers a southern European species with an edible tuber, Conopodium denudatum, as it does the many-named English *pignut* (Conopodium majus, alternatively Bunium or Carum flexuosum), also with an edible tuber whose nutty taste is attested to by the name *earth-chestnut* (cf. the German *erdkastanie*). The name comes about, I suppose, because pigs love to root it up. *Hog-nut* and *swine-bread* repeat this theme as does the name *St Anthony's nut*, since he is the patron saint of pigs. Further strains of domesticity appear in the name *dog* and *cat-nut*.

Not content merely with digging them up and eating them, some children used to soak the tubers overnight in water and drink off the infusion in the morning. They were prevented from indulging to excess by a belief that these caused lice to grow in their hair, for which reason the roots were christened *lousy*, and so *Lucy, arnits*. Other unpleasant names such as *dead man's bones, scabby hands, Cain and Abel*, refer to the shape of the tuber when it has the next year's tuber growing out of the side; so also does the Scottish *knotty-meal*.

A whole tribe of variants cluster around the names *earth, ground* and *underground-nut*, influenced by Germanic remnants such as are found in the Dutch *aardnoot* and the Danish *jordnod*: *ernut; jurnut*, developing into *Jocky jurnals, Jack jennet* or *durnils, Job jarlins*; *yarnut*, developing into *yennet, yowe, yornut, ewe yorling*; *arnut*, becoming *hare-nut, hawk-nut, hornecks*; *grunnet*. A Cornish group seems to cluster around *fare* or *vare nut*: *fern nut, faverottes*; two other Cornish names are *killas* and *killimore*. Scotland has the related *curluns* and *gourlins*, which northern forms would seem to have influenced some of the 'earth nut' family of names. *Shepherd's nut* or *drop* are two other English names.

Other plants which are perhaps related to the pignut belong to the carraway family, several of which have quite large edible roots having the taste of chestnuts. The *large earth-nut* (Bunium or Carum bulbocastanum), also known as *tuberous-rooted carraway*, goes under several of the names applied to the pignut. The Arab *talghouda* (B. or C. incrassatum), known to the French as *talruda*, grows in southern Spain and North Africa, especially in mountainous districts. In Algeria the roots are eaten raw, boiled or roasted. In America there is the *edible rooted* or *Californian carraway* (C. kelogii).

The root of the true *carraway* (C. carvi), a native of the Mediterranean which has now spread throughout Europe, is also edible, and may be cooked like a carrot. But it is for the seeds, sometimes called *carvies* (from the French *carvi*), that this umbelliferous plant is cultivated chiefly. Their highly aromatic quality is similar to dill and other members of the family, and they are put to a variety of uses as a cheese-flavouring, in or on bread, in pickles and cooking. They have appeared as an ingredient in aphrodisiacs, and Dioscorides considered them good for 'girls of pale face'. Oil from the seeds is used in the very strong German liqueur called *kümmel*.

Poppy (*genus Papaver*)

The most notorious of poppies is the *opium poppy* (P. somniferum), with white or bluey-purple blooms, from which the drug is made. However the pale yellow oil expressed from its seeds is non-narcotic and is used for cooking; the seeds are 40 per cent oleagenous. Oil is also obtained from the seeds of the *field poppy* (P. rhoeas), also known as *corn* or *scarlet poppy* for the colour of the flower and also since it commonly grows in corn fields. Weed-killer sprays, however, are now making it less common than formerly. The plant is a native of Asia which has spread into Europe. There are now cultivated forms of various shades which go under the name *Shirley poppy*.

The young leaves of the field poppy may be gathered before flowering and cooked like nettles; they have a nutty, rather bitter flavour. The seeds are also used in cooking, sprinkled on bread, in mashed potatoes, in eggs, and in white sauces, cooked noodles and macaroni. The plant is not particularly narcotic but poppy-heads were once used in country medicines for toothache and nervous pains. Various country names testify to a certain suspicion of the plant. It has been called *earache* because it is supposed to cause this if held against the ear; and *headache*, supposedly the effect of its smell or its colour. *Blind eyes* or *men* and *blindy-buff* are other names given in the belief that to gaze on it too long will cause blindness. If the flower is picked and its petals fall this is supposed to indicate that the gatherer is likely to be struck by lightning; it is accordingly named *lightnings, thunderbolt*, and *thunder* or *lightning flower*.

A red poppy is the birthday flower for 10 May and stands for consolation, although a scarlet poppy symbolises extravagance. The poppy also stands for enchantment, witchcraft, evanescent pleasure, laziness, night, sleep, solace and oblivion. To dream of it means amusement and pleasure. It is a Christian symbol of fertility, ignorance and indifference, although carved on the end of church benches and pews it is supposed to designate heavenly sleep. However, its appearance there might have come about by mistake. Originally bench-ends were carved with small figurines known by the French word *poupée* (a doll, cf. the witchcraft dolls known as poppets);

a mistaken interpretation of what this word meant might have led to the practice of carving poppies. In Greek mythology the poppy was an attribute of the love-goddess Aphrodite; its leaves, like those of the bay (q.v.), were supposed to reveal the truth to lovers by their crackling when crushed. It was also the attribute of the earth-mother goddess Demeter and of Hera, queen of the gods. A poppy-head was the attribute of the gods of sleep and dreams, Hypnos and Morpheus. This connection is further underlined by the country names *sleepyhead* and *sleepy weed*.

Many other country names centre upon its appearance at one stage or another. The flower's beauty is marred by the staining greeny yellow juice of the stem; it is therefore called *fair without and foul within*, or *Joan silverpin*, in allusion to an East Anglian term for a single piece of finery ostentatiously worn by slatterns. The red flower has, of course, given rise to a host of names. It is known as *pope*, partly as a play on the name, partly in reference to the papal robe; a Midlander going poppy-weeding used to say he was going 'poping'. Other names include *soldiers* (alluding to the time when uniforms were red), *devil's eye* or *tongue, fireflout* (cf. the German *feuerblume*), *old woman's petticoats, Paradise lily* and *corn, canker* or *gipsies' rose*; also *red Dolly, cap, cup, huntsman, mailkes, nap, rags* or *weed*. *Coch*, the Gaelic for scarlet, gives rise to Scottish names like *cockeno* and *cock's comb* or *head. Collinhood* is another Scottish name. The appearance of the seed capsule has suggested the names *pepper boxes, bull's eyes, cheese ball*, and *golliwogs*. The cup-shaped flower is alluded to in the names *cheese-bowl, cusk* (a drinking cup), and *cup rose*; a dialect variant (*cop*) has suggested *copper rose*. The plant shares the name *guy* or *gye* with other weeds generally found in corn fields.

Potato (*Solanum tuberosum*)

An underground grower, blind and a common brown;
Got a misshapen look, it's nudged where it could;
Simple as soil yet crowded as earth with all.

Cut open raw, it looses a cool clean stench,
Mineral acid seeping from pores of prest meal;
It is like breaching a strangely refreshing tomb:

Therein the taste of first stones, the hands of dead slaves,
Waters men drank in the earliest frightful woods,
Flint chips, and peat, and cinders of buried camps.

Scrubbed under faucet water the planet skin
Polishes yellow but tears to the plain insides;
Parching, the white's blue-hearted like hungry hands.

All of the cold dark kitchens, and war-frozen grey
Evening at window; I remember so many
Peeling potatoes quietly into chipt pails.

'It was potatoes saved us, then kept us alive'.
Then they had something to say akin to praise
For the mean earth-apples, too common to cherish or steal.

Times being hard, the Sikh and the Senegalese,
Hobo and Okie, the body of Jesus the Jew,
Vertigial virtues, are eaten; we shall survive.

What has not lost its savour shall hold us up,
And we are praising what saves us, what fills the need.
(Soon there'll be packets again, with Algerian fruits.)

Oh, it will not bear polish, the ancient potato,
Needn't be nourished by Caesars, will blow anywhere,
Hidden by nature, counted-on, stubborn and blind.

You may have noticed the bush it pushes to air,
Comical-delicate, sometimes with second-rate flowers
Awkward and milky and beautiful only to hunger.

Richard Wilbur's poem is an excellent illustration of the ambivalence of feeling that many Americans seem to have about the potato. Many plants have been introduced into their country from Asia, Africa and Europe, and now flourish there. They have repaid the compliment, and many are the influential commodities that have come from their continent: tobacco, maize, the haricot bean, the potato. But to their minds it would seem that the potato has been the most influential, and yet it is at the same time the least assuming. It is mean and unlovely, and it worries Americans when they think that people see it and automatically think of its continent of origin as if it were *the* representative, the ambassador extraordinary. And yet it baffles them more that this plant is one in which they cannot but have pride; it has saved people from famine and, conversely, the failure of its crop has brought about death, immeasurable misery, and in England a major political crisis, a change in government policy and the fall of ministers. Again, it sometimes seems as if the potato has done more towards seeing a country successfully through to the end of a war than any military intervention. During the First World War few tributes to the troops of any nation equalled that which honoured the potato in a wartime *Punch*: a picture of the lumpy root thumbing its nose at a U-boat commander, with the caption *Tuber über alles*.

How the large-rooted tuber of the potato came about is something of a mystery, and it has even been suggested that it is a mutation caused by some micro-organism or parasitic growth. Wild South American varieties have generally very small tubers; but it must be admitted that cultivation does cause that phenomenon known as gigantism whereby the plant grows to several times its former size in the original wild state. This has been proved by experiments on related species. After four years of intensive cultivation and forced growths the bitter and inedible Mexican S. commersoni and Chilean S. maglia were developed into quite acceptable vegetables. The former grows well in high temperatures and is disease-resistant.

Among other species the Paraguayan S. chacoense, as well as S. antipovickii and S. kesselbrenneri, flourish in high temperatures, while Andean species such as S. juzepczukii are highly resistant to frost, as they must be in that part of the world. They tend however, to taste rather bitter, as many roots growing at such high altitudes and used for food seem to do. Other edible species of potato grow in South America, several of which go under the name *papas criolas* (native potatoes). S. curtilobium, originally from the Andean tableland (Peru–Bolivia), is grown in Argentina. Other tableland species include the *ajanhuiri* (S. ajanhuiri), the *phureja* (S. phureja), and S. mamilliferum. S. goniocalyx is from Central Peru, S. rybini and S. boyacense from Colombia, the *Andean potato* (S. andigenum) from the mountainous parts of both countries. *Fendler's potato* (S. fendleri) grows through Central America to the southwestern U.S. and is eaten raw or boiled by the Indians; the *James potato* (S. jamesii) from Mexico and the adjoining U.S. is treated similarly.

It was probably in the Peruvian Andean region that the potato was first cultivated, its breaking up of the soil thus making the growing of maize possible. Although wild species were to be found in many other parts of the continent, some of which were occasionally eaten, there is no proof that any of the others were cultivated so early. It is already represented on pots of about 800 A.D., thus indicating that it was well known, both a vital and a staple crop. The tuber also influences the shape of the pots, so that some are made in the form of a potato. (By an interesting coincidence potato-pots were being produced by the Staffordshire potteries about the beginning of the nineteenth century, long before the Peruvian pots had been turned up by the archaeologists.) Study of these originals indicates that many superstitions and religious ceremonies had grown up about them, which would seem to indicate a considerable period of cultivation before the ninth century.

By the coming of the Spanish, potatoes and maize were basic foods all along the Andean chain. The red, violet, yellow and white varieties went under a vast number of names, another indication of how long they had been in use. In the Aymara language of the Pre-Incan Indians of Peru and Bolivia, now largely superseded by Quecha, there were eleven names, and nine in the now extinct Chibcha language of Colombia. There was a distinctive preparation of these potatoes, still practised on these and other tubers which tend to be bitter and semi-poisonous: the root is soaked and chilled in a mountain stream so that it becomes sweet, then pounded, dried and preserved for use as a meal.

The first Spanish mention of the potato is in the 1530s, long after other commodities had been recognised and taken to Europe. It was not imported into Spain and Italy until the 1550s. Part of the trouble, of course, was that the plant was a relative of the dreaded nightshade, and bore similarly evil-looking berries.[1] One may add to this that the tubers were at first small, watery and acrid, containing indeed some of the poison for which its relatives were notorious. Such poor things were said by many to be fit only for pigs or peasants; the latter, however, had their own ideas on what was fit to eat.

The poison properties are intensified by being exposed to light, and among those early pioneers foolhardy enough to eat it there were several cases of food-poisoning. Matters were not helped by the habit of many in eating it raw. Not only men but even cattle tended to contract eczema from eating them, and this was mistaken for a kind of leprosy. Improvement came only after people stopped growing new crops from last year's tubers and learned to use seed instead. The recurring blights of the nineteenth century (1840–50, 1870) helped in this by killing off many of the old varieties and leaving the way clear for the new varieties reared from seed. Nevertheless the potato came nearer to being superseded in the nineteenth century than it did at any other time. It is only in the present century that it has really come into its own.

There is great debate about how the potato reached England. Gerarde said Raleigh introduced it from Virginia, but this is highly unlikely. Although it was taken to North America and became naturalised there, it had not reached Virginia at the time. Besides which it appears to have been the Andean potato (S. andigenum), now very common in South America, that was the first species to arrive. It may have been pirated by some British privateer from one of the Spanish boats; at any rate there is no mention of it until the end of the sixteenth century. Raleigh may have grown it on his Irish estates in the Cork area, where he lived close to Spenser while the latter was writing his *Faerie Queene*.[2] One story is told that Raleigh tried potatoes and did not like them, but there is some doubt whether his gardener did not try to give him the berries rather than the root, taking them for a kind of tomato, another of the potato's relations.

In the Highlands and Scottish Ulster there was a great objection to adopting the potato on account of its not being mentioned in the Bible; to the Irish themselves its use was taken to be akin to eating the forbidden fruit of Eden, although they eventually got around this by planting it on Good Friday and having it sprinkled with holy water. Its popularity in

[1] The French and Germans still keep this in mind with their names *morelle tubéreuse* and *knollige nachtschatten*.
[2] Alternatively, potatoes are said to have been grown by him first in the gardens of a group of Elizabethan half-timbered cottages on Castle Hill, Kenilworth; the area is consequently known as Little Virginia.

England was no better. In 1765 an election was fought in Lewes under the slogan 'No Potatoes, No Popery'. In the early seventeenth century it was bracketed with the tomato (the *love apple*) as an aphrodisiac, and was called the *apple of youth* because 'it inciteth to Venus'. For this reason Shakespeare mentions it in *The Merry Wives of Windsor*.

The only reason why the potato has lasted, especially under the many initial disadvantages, is that it was able to prove itself useful in times of famine, and has done so time and again. When the wheat crop failed and there was no bread except at impossible prices, then the plight of the poor was desperate indeed, and many died of starvation. The potato, whether used as a vegetable or to make flour, was able to ameliorate conditions, as it did, for example, during the French famine of 1740, when peasants were reduced to eating ferns and grass-roots. And so essential had it become to the subjected and poverty-stricken Irish that the failure of their potato crop in the 1840s amounted to a national disaster. It was then that the British parliament had to drop the tariff which had kept the price of wheat artificially high and allow the import of cheap foreign wheat. America gave us the potato, and that they may thank for their large Irish immigrant population consequent upon its failure at a vital time.

The memory of that famine persists like a scar among the descendants of its survivors and finds poetical expression in Seamus Heaney's 'At a Potato Digging' (of which only two of its four sections are quoted here):

I

A mechanical digger wrecks the drill,
Spins up a dark shower of roots and mould.
Labourers swarm in behind, stoop to fill
Wicker creels. Fingers go dead in the cold.

Like crows attacking crow-black fields, they stretch
A higgledy line from hedge to headland;
Some pairs keep breaking ragged ranks to fetch
A full creel to the pit and straighten, stand

Tall for a moment but soon stumble back
To fish a new load from the crumbled surf.
Heads bow, trunks bend, hands fumble towards the black
Mother. Processional stooping through the turf

Recurs mindlessly as autumn. Centuries
Of fear and homage to the famine god
Toughen the muscles behind their humbled knees,
Make a seasonal altar of the sod.

III

Live skulls, blind-eyed, balanced on
wild higgledy skeletons,
scoured the land in 'forty-five,
wolfed the blighted root and died.

The new potato, sound as stone,
putrefied when it had lain
three days in the long clay pit.
Millions rotted along with it.

Mouths tightened in, eyes died hard,
faces chilled to a plucked bird.
In a million wicker huts
beaks of famine snipped at guts.

A people hungering from birth,
grubbing, like plants, in the bitch earth,
were grafted with a great sorrow.
Hope rotted like a marrow.

Stinking potatoes fouled the land,
pits turned pus into filthy mounds:
and where potato diggers are
you still smell the running sore.

Another legend in which the potato figures is that which concerns the Frenchman Parmentier in the time of Louis XVI, after whom it was once named by them *la parmentière*. It seems that this gentleman decided that the potato had had enough bad propaganda and needed popularising; in this he even went so far as to persuade the king to wear a potato-flower in his button-hole at court. Meanwhile Parmentier grew fifty acres of potatoes on the wide plain near Paris, isolating it from the outside with huge ditches along which armed guards patrolled. Naturally the local French peasants were persuaded that here was something really worth stealing. They were astounded at how ready the guards were to be bribed or otherwise persuaded to turn a blind eye to such nefarious activities. Little did they know that these same fearsome custodians had been ordered to allow this to happen! Thus Parmentier earned for himself the title of philanthropist and benefactor of mankind. The truth of the matter was, however, that in many parts of France, even if not in this area, potato cultivation was proceeding well. With the benefit of hindsight some are so churlish as to describe this action as just one more folly of a dying regime. The cause for which Parmentier was really campaigning was the substitution of potato for wheatflour in making bread for the poor after the terrible famines of that century. The only result of his propaganda came during the Revolution when the Tuileries gardens were converted into a vast potato field in 1793; during the famine of 1802 even the yards were torn up and the alleys planted with them.

A certain amount of popular lore has grown up around the potato. It is the birthday plant for 13 June and symbolises benevolence. A stolen potato is supposed to be a charm against rheumatism. If one is sliced, rubbed over warts and then buried, the warts are supposed to disappear as the potato decays. The usual American ambivalence shows itself in proverbs and popular sayings concerning it. 'Small potatoes' is remarked of an inferior, mean or insignificant person, much as the English comment 'small beer'; 'small potatoes and few in the hills' is a comment on a trifling matter. Where the English say 'cake-hole' for mouth, the Americans say 'potato trap' (or jaw). On the other hand 'quite the potato' means the correct thing, and a 'hot potato' is someone worth watching.

Like most other popular vegetables, potatoes come in many sizes and colours. There are two main divisions: the floury British potato, and the waxy continental variety, which is superior in flavour, texture and firmness. The British is doubtless well-suited to English cookery, being apt to crumble away into a shapeless mass at the slightest provocation. But it makes the better chips.

Among the waxy varieties of potato are the following: Belle de Juillet, a round variety; the sweet Jaune de Holland; the red-skinned Red Star; the large Yellow Eigenheimmer, and the blue-skinned yellow-fleshed Blue Eigenheimer.

The flesh of some species and varieties run through a range of reds; the Congo, for example, has purple flesh and makes a pretty salad decoration. There is an inedible South American species, the *chapina* (S. stenotomum), which is black; this is cultivated and used for dyeing cloth. Others produce the constituents of textile printing inks and are put to a number of different industrial uses. Among their products are starch, gums, dextrose and commercial glucose, and alcohol. Irish potato whiskey, *poteen*, is so powerful that I have heard of a man who went to a wedding party and did not return for a fortnight owing to his over-indulgence in it.

It is this perverse and illegal constituent of war and revolution of which Americans are somewhat ashamed. It is this poisonous and unpromising root, that has held at bay the spectre of famine nevertheless, in which they feel pride. How are they to reconcile these feelings? It is an embodiment of American dream and myth, it is the poor thing making good. And to Peter Viereck this is very sinister.

> *O vast earth-apple waiting to be fried,*
> *Of all life's starers the most many-eyed,*
> *What furtive purpose hatched you long ago*
> *In Indiana or in Idaho?*
>
> *In Indiana or in Idaho*
> *Snug underground, the great potatoes grow,*
> *Puffed up with secret paranoias unguessed*
> *By all the duped and starch-fed Middle West.*
>
> *Like coiled up springs or like a will-to-power,*
> *The fat and earthy lurkers bide their hour,*
> *The silent watchers of our raucous show*
> *In Indiana or in Idaho.*

'They think us dull, a food and not a flower.
Wait! We'll outshine all roses in our hour.
Not wholesomeness but mania swells us so
In Indiana and in Idaho.

'In each Kiwanis club on every plate,
So bland and health-exuding do we wait
That Indiana never, never knows
How much we envy stars and hate the rose!'

Some doom will strike (as all potatoes know)
When – once too often mashed in Idaho –
From its cocoon the drabbest of earth's powers
Rises and is a star.
       And shines,
       And lours.

A NOTE ON THE NAME OF THE POTATO

The potato, in that it is a comparatively recent discovery which has since achieved great popularity, provides a fine example of how names proliferate and under what influences. The original Peruvian term for it was *papas*, accepted into Spanish but since superseded in Spain. The name continues to be used in South America and the Philippines. To the French the potato was once *la papas des Péruviens*.

At the same time that the potato was discovered other roots were being used in North America, and the names given these tended to collide and coalesce in the potato. The name given to the Jerusalem artichoke, *truffle* or *ground truffle*, also spread to the potato. The Spanish took this up in the sixteenth century, calling it *turma de tierra*, and the Italians followed with *tartufo bianco* and *taratuffli*. To the eighteenth-century French they were *truffes, truffes sèches, truffes rouges, truffières*; in the Lyon area the first name persisted right into the nineteenth century. Versions remained also in other dialects: *tufelle* (Savoy, Geneva); *tufère, tufène* (Languedoc); *tartifle* (the Avignon–Orange region once held by the Pope).

In Europe the roots of the cyclamen were called *earth-apples*, while in America French settlers gave the names *pomme de terre* and *pomme blanche* to the *breadroot* or *prairie turnip* (q.v. under Potato-beans). But by the eighteenth century the former name, and that of *pomette*, had been given to the potato, and it was known as *earth-apple* in England at the same time. In Dutch this became *aard-appel*; *artappel* in Malaya and Ceylon. Other translations appear in different languages: *mere de pamint* (Roumanian); *maaomena* (Finnish); *γηομελον* (Greek); *zamnak* (Czech); *zemniak* (Polish); *zemnyak* (Ukrainian); *sib-i-zamini* (Persian).

The German version of this, *kartoffel*, seems to have been confused with some of the 'ground-truffle' names in its spread. Thus in some dialects it is *tartoffel*; one may compare the Polish *taretofl* of the eighteenth century, and the Icelandic *tartuflur*. *Cartoufle*, the German form, turns up in some French dialects, but generally the German influence is felt in Eastern Europe: *kartofel* (Polish, Russian); *kartoska, kartopha, kartoclo, kartovka* (Russian dialects); *karrofla, karckofle* (Polish dialects); *kartupelis* (Lettish); *krtola* (Serbian); *cartofla* (Roumanian); *kartof* (Bulgarian). It has also been accepted into the Scandinavian countries.

The early French settlers also named the artichoke *poire de terre*. This too was applied to the potato, which was called *poirette*, a form similar to *pomette* of the same time, in the eighteenth century. It was taken up by the Flemish as *grund-birn* and *grond peer*, becoming *cronpire* and *crompire* in Wallonia and Alsace; in the latter district there was the German *grunbirne* (also used of the artichoke) and the dialect version *grumbire* to influence it. In Bavaria it became *gruntbeer*, and in Carinthia *grumber* and *gruntpirn*. Another German version of this was *erdbirne*. The Austrian *krumpir* reappears in Bulgarian, Serbian, Czech, and the similar *krompir* is found in Serbian and Slovenian. Other versions include *grumbir* (Moravian); *gombiri* (Bulgarian); *grumciri, crumpira, crumpena* (Roumanian); *krumpla* (Slovak); *krumple* (Czech); *krumpli* (Hungarian); *krompele, kompery kumpery, kraple* (Polish); *klumberis* (Lithuanian).

Our name 'potato' is one of the least admissable, widespread though it be, for it is no more than a confusing of *batatas*, the Spanish for a sweet potato, with the word *papas*. It is quickly taken up, however, as *pytatws* in Welsh, *patate* in nothern French and Breton dialects, *pataque* in the west of France, *patache* in Anjou and *patata* in Italian. *Patatas* is found in the Philippines and in Indonesian dialects, πατατα in Greek dialect, *batata* in Portuguese Timor and Goa, and *batata ingleza* in the Canary Islands. Similar versions appear in *potatis* (Swedish); *potet(es)* (Danish); *patas, patat* (Senegambia); *lwe-batata* (Kaffir); *matata* (Southern Nigeria); *watala* (Indonesia). Such shortened versions as *'taters* and the Scottish *tattie* are to be found in the British Isles.

There are several other aliases under which the potato goes. The Irish slang *Murphy*, and the English *spud*, for instance. Many, however, centre upon the word for a tuber, or a vegetable used similarly to the potato. Thus the Indo-European *alu* appears in *alu-i-Malkam* (=Malcom's tuber, Persian) and *bilati aloo* (=English tuber, Bengali). In Tibet it is known as 'foreign tuber', and in Japan as 'Holland or Batavian tuber'. But the widest spread is the Greek βυλβος (a tuber), applied to potatoes in the eighteenth century and carried by merchants into the Balkans, where the following versions of it appear: *boulbes* (Yiddish); *bulba* (Ukrainian, Russian); *barabolya* (Ukrainian, Russian); *barabol* (Czech); *baraboj* (Serbian, Bulgarian); *bruboj* (Bulgarian); *garabola* (Ukrainian, Serbian); *gulba, bunba*

(Russian). The potato is compared to the yam in the East, where it is known as *foreign yam* (China), *English yam* (Tibet), *Holland yam* (Malaysia, Indonesia), *Bengal* or *European yam* (Malaya). In China it is also known as *foreign (mountain) taro*, and to the Arabs it is *Frank's taro* (that is to say, a taro used by any European). Finally, in the Indonesian Sunda Islands, it is known as *Dutch cassava*.

A last curious family of names clusters round Czech names for the German state of Brandenburg (*Bramburk*) and for Prussian (*Brambor*), since it was the Prussian nation – who ruled Brandenburg – who introduced the potato to some of the Slavic areas and it was thus named after them. *Brambor* is still the most popular Moravian name for the vegetable, and may be as widespread in other areas as well. Other names deriving from these include *bandraburca* (Rumanian); *mandyburka, gardyburka* (South Russian dialects); *panbowka, perka* (Polish); *peruna* (Finnish).

Potato-beans and other substitutes

The name potato-bean may be said to cover the small tuberous roots of various members of the pea family, more especially *tuberous-rooted wistaria* (Apios tuberosa), which grows in America and is also called *wild bean, mic-mac potato, ground nut, pindar* and *pindal*. The last two names were given it by slaves brought from Africa and derives from the Congolese *mpinda*. Some of the vetches have edible roots that may be eaten boiled or baked. One is the *ground-nut pea*, root of the *tuberous-rooted everlasting pea* (Lathyrus tuberosus) or *tine-tare*. The roots were once referred to as *Dutch mice*. Gerarde comments on this in his *Herbal* (1597), giving the Platt-Deutsch (a North-Western German dialect) as *muysenmet steerten*, 'that is to say tailed mise, of the similitude or likenes of the domesticall mise, which the blacke, rounde and long nuts, with a pece of the slender string hanging out behind, do represent'. Similarly one may use the roots of the *tuberous-rooted bitter vetch* or *pea* (Lathyrus macrorrhizus), also called *heath* or *mouse pea*, and *peasling*. In Celtic areas the roots go under the names *carmele* or *carameile*, deriving from the Scottish Gaelic *cair-mele, corra-meile*, and the Irish *cara-meala*. Other names, such as *gnapperts*, are variations on *knap-wort* (knot-root); and names like *liquory knots* and *wild liquorice* refer to both the appearance and sweetish taste.

North-American Indians make use of the roots of two members of the milk-vetch family (genus Astragalus). Those of the *Canada-milk vetch* (A. canadensis) are eaten raw or boiled by the Blackfoot Indians either in spring or autumn. And in the southwest the Hopi eat those of A. pictus filifolius. The pods and seeds of others in this family are also eaten (*see under* Peas). The tubers of the related *rush pea*, Hoffmanseggia densiflora, are eaten by Indians in the same southwestern area. The roots of one of the cow peas widespread in tropical Africa, Vigna vexillata, are cooked in the same manner as sweet potato in Sudan and Ethiopia, while those of an Australian species, V. lanceolata, are used for food in parts of Queensland, where they are referred to as 'yams'.

There are several other Australian roots which have been similarly christened 'yams', although in many cases the resemblance to that vegetable is tenuous enough. The finger-like clusters of tubers belonging to the *yam daisy* (Microseris scapigera) were once a very important food to many aboriginal tribes in Victoria, some of whom refer to them as *murrnong* (fingers). Similarly named, and once equally important to the aborigines, were the roots of the white-flowered *tree-mallow* (Lavatera plebeia) which are rather like parsnips. The *yam tree* (Brachychiton populneum) of New South Wales, also known as the (*brown*) *kurrajong tree*, is another member of the mallow family; the natives eat the young roots, which are said to taste similar to, but are rather sweeter than, turnips. The erroneously named *native potato* (Marsdenia viridiflora) is in reality the milky unripe fruit of a tree native to Australia, which is eaten raw or roasted by the aborigines. That of the related *doubah* (M. australis) is used in the same way.

The tuberous roots of another member of the pea family (genus Pachyrrhizus) go by the name *yam bean*. The Fijian *wayaka* (P. angulatus), or *short-podded yam bean*, has spread all over the Pacific and South-East Asia from its original home in the Philippines. Its edible tubercules are rather similar to the European rape-root (*see under* Turnip). The *West Indies yam bean* (P. tuberosus) of South and Central America is also known as *sincamas* and *potato bean*, and serves as a very acceptable yam substitute. The seeds of this plant are poisonous and in some places serve as an ingredient for a powder to exterminate vermin. Other relatives in the same area include P. erosus, which is of Mexican origin and now cultivated throughout the tropics of the world, whose young roots are eaten raw or cooked, and in soups. The Central American P. palmatilobus is cultivated for its large turnip-like roots. The yam bean of tropical West Africa (Spenostylus schweinfurthii) is cultivated for both its seeds and the tubers; other species in the same area are only grown for their seeds.

The floury tuber of the *scurf pea* (Psoralea esculenta) was introduced to the early French settlers by the American Indians, who extended its cultivation as far north as it could be made to grow, although it is chiefly found from the Missouri region through the Mid-West into Texas. It was known to the French in early days simply as *pomme de terre* or *pomme blanche*, and is still known by the French *pomme de prarie*. Other names include *Missouri bread root, prairie turnip* and *Cree potato*. In 1846 it was introduced into France as one of the very many potato substitutes, none of which met with much success. Other American species are eaten by the Indians, especially in the deserts of the south-west, and in Mexico. The large roots of the *beaver bread* (P. castorea) are ground into a flour for porridge or bread-making. So too are those of the *skunk top* (P. maphitica), especially by the Indians of Utah, who also eat them cooked or raw. The Luiseño Indians of California use them to extract a yellow dye.

A last potato substitute is the wild tuber known as *omime root* (Plectranthus ternatus) which grows in the southern half of Africa and is preferred before all others by the natives. The name comes from the Malagasy, since it originated from that region, although it goes by a host of others. The plant belongs to the cockspur or spur-flower family. While the root is rather watery, it may be used exactly like a potato. It was spread very rapidly by the French over their former tropical possessions, to the great relief of the colonial administrators who, by all accounts, found the lack of potatoes one of the hardest burdens that the white man had to bear. Edible relations include the southern African *Hausa* or *Kaffir potato* (P. esculentus) and the Indian *country potato* (P. tuberosus).

Purslanes

There are several related families of plants, all of which go under the name purslane. The best known is Portulacca oleracea, a native of Persia, where it was used 2000 years ago. It has now become a world-wide common weed, especially in the U.S. where it is also known as *pussly* and considered a menace – the more so because it retains moisture and thus germinates long after it is pulled out of the ground. The leaves and tender tips can be used as a pot-herb, or raw in salads, and are thus used in country districts of the world, and by the Asian poor. The plant was known to the Greeks, cultivated by the Romans, and adored by the Arabs in Medieval times, when it was known to them as 'the blessed vegetable'. In Europe it was cultivated from the fourteenth century. Of the two varieties the green-leaved has been used the longest, but the golden or large-leaved, specially developed in the seventeenth century, is the better for cooking. Its name derives from the Latin *portulacca* through the Old French *pourcelaine*; the softening of the 't' is sometimes explained as an assimilation of Pliny's name for the plant, *porcillaca*.

The leaves of this purslane are said to taste like watercress, and are both nutritious and antiscorbutic. Australian aborigines not only collect the leaves but gather the plants for their seeds, which they wash and grind or eat raw. In northeastern America the seeds are said to be used for bread. The method is to collect the plants into heaps, turning them over at intervals so that they dry and the seeds fall out. Queensland aborigines also gather the seeds and turnip-like roots of P. napiformis, those of South Australia the tap-roots of P. intraterranea. In the Pacific the shoots of P. lutea are eaten; poor Asians eat those of P. quadrifida and the leaves of P. tuberosa. The largely ignored tuberous roots of the Brazilian *sunplant* (P. grandiflora) are edible; the Indians in the southwestern U.S. use P. retusa as a vegetable.

The purslane most favoured by the North American Indians is the *white mountain rose* or *bitter root* (Lewisia rediviva). Its dark-coloured roots are white and floury inside and resemble arrowroot when boiled; they have great vitality and are capable of a large yield when cultivated. From it the Californian Indians make a flour known as *spatlum* (or *spatulum*) which is surprisingly nourishing. It is said that one ounce provides sufficient for a meal. The plant grows along the Pacific coast as far as British Columbia, and inland into the desert regions.

The leaves of the *Indian lettuce* or *water chickweed* (Montia fontana) are eaten in North America, and in some mountainous areas of France and elsewhere in Europe. The plant is usually to be found in sandy soils by the side of ponds in warm areas. Other names for it include *water blinks* and *blinking chickweed*. It is closely related to *winter purslane* (M. or Claytonia perfoliata), alternatively known as *miner's* or *Spanish lettuce*. The plant is originally from Central America and has spread as far north as the Vancouver district in Canada, to which it was introduced in 1790. The leaves have antiscorbutic properties and may be eaten raw or as a spinach substitute. The same would seem to be true of *Cuban spinach* (C. cubensis) and *Chinese chickweed* (C. sibirica), also known as *Siberian purslane*. The tubers of the latter are also eaten in Siberia. The roots of the *Spring beauty* (C. virginica) are highly valued by North American Indians, and those of the *Carolina spring beauty* (C. caroliniana) are recommended in times of scarcity. In Alaska the Eskimos eat the roots of C. acutifolia.

Other purslanes are eaten in Central America. *Samphire* or *seaside purslane* (Sesuvium portulacastrum) is eaten in the West Indies and North America. The *fameflower* (Talinum patens), of Central American origin and known as *puchero* in Mexico, may be cooked like purslane; the leaves serve either as a vegetable or for seasoning. In the southwestern U.S. the Indians cook the roots of the *orange fameflower* (T. aurantiacum), and the *Guyanese purslane* (T. crassifolium) is eaten in the American tropics and East Africa. *Pot-herb fameflower* (T. triangulare) is of central African origin and has now been spread to the American tropics and South-East Asia, flourishing best in warm humid climates. It is also known as *waterleaf, Surinam purslane, Lagos bologi* and *bologi spinach*.

Rock purslane (genus Calandrinia) is closely related to winter purslane. Of its 200 species about three-quarters are found in the western areas of the Americas, but it also grows all over Australia, even in the most arid areas. It is of particular importance there since it stores up moisture and suffices cattle when there is no other water available. The aborigines bake the succulent leaves of the *parakilja* (C. balonensis) with bark as a food; the plant was once eaten by the early settlers. In Western Australia, C. polyandra is eaten. The Californian *desert rock purslane* or *red maids* (C. caulescens var. menziesi) is used as a pot-herb or garnish.

Lastly there is *Cape spinach* (Emex australis), also known as *double-gee*, which once served as a spinach substitute in South Africa. From there it was taken to Australia where it flourishes so well on sandy soils, and especially on the beaches, that it too has been classed there as an exceptionally troublesome weed.

Radish (*genus Raphanus*)

The radish is another member of the mustard family. In its wild form (R. raphanistrum) it is practically indistinguishable from charlock (*see under* Mustard); both are very common and widespread inland plants. Many of its names are shared in common with charlock, but it also known as *wild* or *jointed charlock*, *wild radish*, *red root*, and in the north as *runch*. Another more pungent ancestor is the *sea radish* (R. maritimus) to be found growing widely in warm maritime areas. The name comes from the Latin for a root, *radix*. It is also known as *rabone*, from the Spanish *rabano*, and in Scotland as *reefort*, from the French *raifort* (charlock).

Radishes were eaten in Egypt from the beginnings of civilisation. Herodotus records an inscription he saw there which said that builders of the Great Pyramid ate enormous quantities of a radish called *gurmaia* together with onion and garlic (*History* II. cxxv). Pictures of them have been found in the Egyptian tomb paintings and at Pompeii. In the Middle Ages the radish was known more in the south of Europe than in the north. It is recorded from the thirteenth century onwards but, though it was used in England, it is not mentioned for the purposes of cookery before the sixteenth century. However, it seems to have figured largely in Anglo-Saxon medico-magic, where we find it used for a variety of purposes. It appears in concoctions for shingles, madness, demonic temptations and possession; in a poultice and a drink for pains in the right side; in salves for headache, pain in the joints, eye-ache, wens (warts), weakness in all the limbs, and the 'dry disease'. Eaten beforehand it is accounted a remedy against any poison. In more modern times we are advised what to do in cases of 'heaviness of mind: give to eat of radish with salt and vinegar; soon the mood will be more gay'.

In Europe the species of radish used, R. sativus, may be eaten either raw or cooked. It does not appear to have occurred to many that these roots are very pleasant cooked, and that cooking removes the hotness which some find unpleasant. The several sizes of the radish are usually classified as: round or turnip-shaped; long; large (winter-variety); and olive-shaped. The *Bavarian radish*, which is served in southern Germany to be eaten raw with the beer, is a good cooker. So also are the black *Spanish radish* (about 3 inches in diameter), and the large *China rose*, originally brought to Europe by the Jesuit missionaries. The leaves also may be cooked or used in soups. In India they cultivate a variety known as the *cabbage radish* for the leaves alone; its large spongy root is very seldom eaten.

Radishes are very widely used in the East, and there are varieties up to 3 feet long. There is a mention of them in China in 1100 B.C. The Japanese radish known as *daikon* (R. acanthiformis) accounts for 25 per cent of the Japanese vegetable crop. It is a long white variety most frequently used pungently pickled. Another radish of Japanese origin, frequently pickled, is the Javan radish known as *mougri* (R. caudatus). It is also known as *rat* or *serpent-tailed radish*, and was one of the weird vegetables pioneered in Europe in the nineteenth century by those indefatigable experimenters Pailleux and Bois. Nevertheless it is still confined chiefly to Asia and grown for its seed-pods (8–12 inches long) which may be eaten either fresh or pickled. The root is not very good, but may be used as a turnip substitute.

Kenko, the fourteenth-century Japanese monk and diarist, records a curious story in which one of these Japanese radishes figures:

> Once, in Tsukushi province, there was a certain Prefect of Police whose name escapes me. Believing in the marvellously health-giving properties of the *tsuchione* [a large variety of radish], he had for many years eaten two broiled every morning. One day enemies attacked his residence on every side when there was nobody by to defend it. Two warriors suddenly appeared who fought without regard for their lives and put all to flight. 'Gentlemen', the completely astounded man asked them, 'since I have never had the honour of seeing you before, might I ask you who you are who have fought so remarkably?' They replied, 'We are those in whom you have put such confidence for so many years, the radishes you have eaten every morning, come now to serve you'. Saying this, they disappeared. Such is the virtue of profound faith.
>
> (*Tsurezuregusa*, section 68)

Rampion *(Campanula rapunculus)*

A wild native of Britain and related to the harebell, rampion was once more widely used than it is now. It is to be found in Asia and North Africa as well as in Europe. The white sweetish roots may be eaten raw, or cooked like radish, and the young leaves will do either in salads or cooked like spinach. An Elizabethan recipe suggests the roots 'boyled and stewed with butter and oil and sprinkled with black pepper'. Evelyn describes their flavour as nutty and considers them more nourishing than radishes. Although they were certainly eaten earlier, it is not thought that they were cultivated before the Middle Ages. So highly were they considered at one time that it was said one of the torments of Hell was having to do without them. They fell out of favour, like so many other roots, after the potato became popular. The English name is thought to be a nasalisation of the French *raiponce*; there is nasalisation also in the German *rampunzel*. (Grimm's fairy-tale of this name hinges on the theft of rampion from a magician's garden.)

In the north of England the young shoots of the *great bell-flower* (C. latifolia), known locally as *wild spinach*, were eaten once. Other names for the plant include *white foxglove*, *hask-wort* (because, like several others in the family, it was used for sore throats by country people), and *gowk's hose* in Scotland. The leaves of C. versicolor of southern Europe and Turkey are occasionally used in salads, and as a vegetable. The young leaves of the related Indonesian Lobelia succulenta are eaten with rice, but most other members of this family are poisonous.

In Japan the roots of several in the *gland bell-flower* or *lady-bell* family (genus Adenophora) are eaten. Those of A. communis are boiled or used in soups; those of A. verticillata are eaten principally by the Ainu, fresh or dried. The Ainu also roast or eat raw the roots of another relation, Codonopis ussuriensis. In Siberia the roots of the *broad-leaf ladybell* (A. latifolia) and the vanilla-smelling A. stylosa are eaten by the native peoples. The leaves of all these are sometimes added to soups.

Reeds and rushes

1. Gromwell reed *(Coix lachryma jobi)*
This plant is a relative of maize, of Asiatic origin and used for food in India, South-East Asia and the Philippines. It has been introduced into

Europe and is now naturalised in Spain and Portugal. The hard white tear-shaped seeds are sometimes used as beads (as are those of several of the Asiatic reeds following), but also form an important part of the diet of some peoples although, because of their low gluten content, the flour obtained from them has to be mixed with a portion of wheat flour for baking. The name of the plant is given in reference to the seeds, deriving from the Old French *gromil* or *gremil* (from the late Latin *gruinum milium*, millet-seed). Another popular name given both to the seeds and the plant is *Job's tears*; the French name them *larmes, larmilles, larme de Job*, and the Germans *thränen-gras, Hiobsthränen, Mosesthränen, Marienthränen* and *Christus-thränen*.

2. Flat sedge (*genus Cyperus*)

Among the 600 species of this plant is to be found the *Egyptian papyrus rush* (C. papyrus) from which the famous paper-equivalent was made, and among which the infant Moses was concealed. Several others have almond-shaped tubercules which are said to taste like chestnuts. The Indians of North and South America eat those of the *bearded flat sedge* (C. aristatus), while the *yellow nut-grass* (C. esculentus) has been cultivated there since early times, although its chief use on that continent now is as a pig food in the southern U.S. This plant, also known by its Valencian name *chufa*, and as *tiger nut* or *earth almond*, grows wild in southern Europe, Asia and Africa. It is very popular, and cultivated as a food, along the sea-coast of Ghana.

The mild form of ginger known as *galingale*, much used in Mediterranean cookery, comes from the root of another in the family, sometimes known as *sweet cypress* (C. longus). The name 'galingale' derives ultimately from the Chinese *Ko-liang-kiang* (mild ginger from Ko or Canton); this name passes through Persian to become the Arabic *khalangan*. In Medieval Latin this became *galanga*, in French *galangue* and in Middle English *galangal*. A similar substance is obtained from *greater galangal* (Alpina galangala), *lesser galangal* (A. officinarum) and *East Indian galingale* (Kaempferia galanga). The latter is used by the Sinhalese as a stimulant.

3. Spike sedge (*genus Eleocharis*)

This plant is related to flat-sedge and other bulrushes. The nut-sized tubercules, known as *water-nuts*, formed on the root of a Far-Eastern species, E. tuberosa, may be eaten either raw or boiled. Martinus Martini mentions it in his *Novus Atlas Sinensis* (1665) in connection with a curious Chinese belief that 'if you put a copper coin in your mouth at the same time as the fruit, you will break it with your teeth as easily as you do the fruit, and you will reduce it to an edible pulp by marvellous force of nature such as I have often made proof of myself'. Among other edible species are the

10. REEDS AND RUSHES

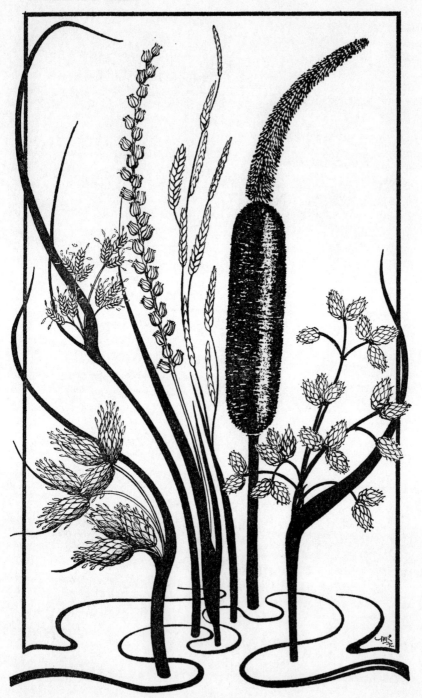

(*From left to right*) seaside clubrush, sweet cypress, sea arrow-grass, water manna grass, reedmace, common bulrush

Pacific E. esculenta and the Australian E. platiginea or sphacelata, whose spherical tubers are eaten raw by the aborigines.

4. Water manna grass (*Glyceria fluitans*)

This plant is commonly found floating on water (from which habit it is known as *float grass*) and in boggy ground. Its sweet grains have given it such names as *sweet* or *sugar grass*, *manna croup* and *Poland manna*, the last because both the Polish and East German poor once used it for food, as did the Indians of North America.

5. Ditch reed (*Phragmites communis*)

This plant is a member of the grass family common to all watery or marshy places in the Northern hemispheres from the temperate to the tropical regions. The tall close-set stems, ranging from 5 to 10 feet high, are of great importance as a soil binder, and in the conversion of marshy to dry ground. The seeds have been used for food by the poor, and the roots have served as such for North American Indians. Among the many names by which it goes the more common include *reed, cane* or *water grass*, and *great, bog, marsh, loch, common* or *water reed*. Others are often found in different dialects; *bennels* is a Scottish name applied to them and the mats woven from them which are used as partitions or laid across rafters to form an inner roof. This will also explain the name *twill* (used of woven fabrics). *Doudle* is the Scottish name used mostly of the partially decayed roots, from which children make a musical instrument similar to the oaten pipe of Greek pastoral poetry. Hence the name *empty-rods* used of the dried grasses, and *windle-straws*, a name which also covers dried grasses in general. *Goss* is a Cornish name generally regarded as a corruption of gorse, a misnomer which probably refers to their spiky nature, elsewhere referred to in such names as *spick, spears* and *spire*. *Lesch* is a name from the fenny districts of England deriving from the French *laiche* (a rush). *Flaggers* (cf. the French *flagière*), *hangel* and *leerspole* are other names applied to it and rushes in general, as are *sheggan* and *seife* (pl. *seaves*), northern forms of *sedge*.

6. Bulrush (*genus Scirpus*)

These plants are also relatives of flat and spike sedge (see above). The *common bulrush* or *lake clubrush* (S. lacustris) grows in the pools and marshy places of the Northern hemisphere. In the Californian area the local Indians eat the thick fleshy root-stock as a vegetable, or alternatively dry it in the sun and powder it into a rather fine-tasting flour. The seeds and young shoots are also eaten, while the rushes are woven to make mats, baskets, and a kind of coracle known as *balsas*. Similar usage in Britain has given the plant such names as *panier-rush* and *chair-platt*; *bolder* is another name referring to both it and chair-bottoming. *Bass* is a name given to a rush-woven horse-collar in Cumberland, alternatively known as *bumble* in Yorkshire, both of which are referred back to the bulrush. Amongst other strange-sounding names are the English *bent*, a name applied to all dried sedges (cf. the German *binse*) and the American *tule*, which derives from the Nahuatl Indian *tollin* (a rush). Others include *spurt-grass, son's brow, chaix* and *frail rush*. The last of these refers to the fact that it is hollow, a fact also alluded to in the Old English *holrisc* and Middle English *holrysche*. The name bulrush itself seems to be a corruption of *pool-rush* (cf. the Old English *ea-risc* and French *jonc d'eau*).

Other edible species include the *seaside clubrush* (S. maritima), which has starchy tubers, and the *tuberous bulrush* (S. tuberosus), whose tubercules are eaten in the Far East. In North America the roots of the *Nevada bulrush* (S. nevadensis) are eaten raw by the Cheyenne Indians, while several other tribes use those of the *alkali bulrush* (S. paludosus) similarly. Alternatively they are ground for flour, and the pollen is also mixed with it for bread-making.

7. Giant bur-reed *(Sparganium eurycarpum)*
This plant is to be found in the Northern hemisphere in those areas with a cool or cold climate, and also in Australia and New Zealand. The roots have been eaten by North American Indians. The name comes from the fuzzy bead-burr at the end of the reed, which makes it easily recognisable. The bur-reed belongs to the same family as the more splendid reedmace (see below).

8. Sea arrow-grass *(Triglochin maritima)*
The plant, also known as *shore pod-grass*, is a native of the north temperate zone; the very young leaves have been recommended as a vegetable in times of emergency. The rhizomes of its Australian relative, T. procerum, are eaten by the aborigines.

9. Reedmace *(genus Typha)*
The reedmace is also often called a bulrush, and fits better one's mental impression of what it looks like: a reed-spike crowned with a dark woolly cylinder. The name reedmace is accounted for by the fact that it was this which was supposed to have been given Jesus as a sceptre during the crowning with thorns (cf. the French *roseau de la Passion*). Many other names of the English species, *broad-leaved reedmace* (T. latifolia), refer to its main characteristic: *baccobolts*; *flaxtail*; *blackamoor*; *black boy, caps, puddings* or *sticks*; *blackie toppers*; *bulrush catstail*; *cat-o-nine-tails*; *cat's spear*; *chimney sweeps* or *flue-brushes*; *devil's* or *holy poker*; *lance-for-a-lad* (obviously referring to a child's game); *candlewick* (cf. the French

chandelle); *torch reed* or *water-torch*; *dunse down*; *sootipillies*; *marsh pestle* (or *beetle*); *fairy woman's* or *frog's distaff*. Other names link it with the common bulrush: *bull segg* or *wand*; *sow's brow* (reminiscent of *son's brow*); *mat reed*, indicating it had similar uses. The name *Mary's tears* probably refers to the legend behind the name reedmace. And *levers* derives from the Old English *loefer*, which covered flags and rushes generally. *Dod* and *goss* are two other curious names.

Broad-leaved reedmace grows in Europe, northern Asia and America. It has edible rhizomes and its stem is sometimes pickled, or else the inner portions are baked and peeled. The root-stocks of two Eastern species are also used: the *narrow-leafed catstail* or *small reedmace* (T. angustifolia), known in India as *reri* and to the Australian aboriginals as *balyan*; T. muelleri, known as *cumbungi* to the latter. They are astringent and starch-filled and their main use is in cases of dysentery, although the aborigines use them for flour. The young flower shoots can be eaten as a substitute for asparagus. *Elephant grass* (T. elephantina), so named because it is a favourite of that animal, bears an abundant mealy pollen which is used for making bread in India and New Zealand. Care should be taken, however, since it is exceedingly inflammable.

Salsify (*Tragopogon porrifolium*)

Salsify is a member of the chicory family and, in the early spring, may be used as a substitute for it. The roots also are eaten, cooked like carrots, and are said to taste rather like asparagus; others suppose them to taste like oysters, from which belief it gains its name *vegetable oyster* and *oyster plant*. In the thirteenth century it was gathered wild and eaten in both Germany and Italy. It is the Italians who seem to have been the first to cultivate the plant early in the sixteenth century. In England it was grown for its ornamental value long before anyone thought of eating it.

The name, also spelt *salsafy*, derives from the Italian *sassefrica*, of doubtful origin; it has been suggested that it develops from the Latin *saxa-frica* (stonebreaker), but why it should be thus called no one has explained. The name came to England via the French *salsifis*, with which one may compare *salsifi* (Spanish) and *sersifim* (Persian).

Other English names are *purple goat's beard*, *star of Jerusalem* (another play on the word *girasole*, as in the case of the Jerusalem artichoke), and *nap-at-noon*. The last refers to the fact that the flower closes early, a characteristic which it shares with its cousin, *meadow salsify*, *yellow goat's* or *buck's beard* (T. pretensis), also known as *Joseph's flower* (cf. the German

Josephsblume), *shepherd's clock, go-to-bed-at-noon, noontide, noon-flower, sleepy head, odd-man-out, paint brushes, maiden's prayer, stoat's meat* and *yellow succory*. The roots of this also were eaten either fresh or preserved in sand for winter use. The young stalks were cut into lengths and boiled like asparagus.

Samphires

The name samphire is a corruption of the French *herbe de St Pierre* and reappears in a number of spellings and pronunciations: *sampire, sampier, samfa*. There is a similar contraction of the Italian *herba di San Petra* into *sampetra*. The plants going under this name are dedicated to the fisherman saint because they are generally to be found growing by the sea or in salt marshes. The best known in England is *sea* or *rock samphire* (Crithmum maritimum), an umbelliferous plant that once enjoyed great popularity, especially in the eastern counties. Its thick fleshy leaves have a high iron content and are thus good for skin complaints such as acne; they may be eaten boiled or as a salad from July to September, and have a taste rather like asparagus with the salt tang of seaweed. They are also eaten as an appetiser or pickled and have a high regional sale. Other names reflect the original, as in *St Peter's wort* and *Peter's cress*. It is also known as *sea fennel, crest marine, pierce-stone* and *paspar* (in Scotland).

Marsh samphire (Salicornia herbacea), a member of the goosefoot family and thus a relation of spinach, is a curious leafless plant whose fleshy jointed stems taste like aparagus when boiled. In Lincolnshire it is the preferred samphire and also pickled. In Cheshire it was once sold under the name *sampion*; it is still occasionally to be found in European markets. It also grows in the treacherous Suffolk mud-flats, one of the few plants that can survive there; and because it is so inaccessible is considered rather a treat. This and several other species, all going under the common names *crab* or *frog grass, sea-grape* and *saltwort*, were once reduced to ashes for making glass; it is therefore known as *glass-wort* (*glaskraut* in German). Other names include *pickle-plant* and *swy*. Similarly used are the Mediterranean *lead-bush* (S. fruticosa), which also serves as camel fodder in North Africa, and *Australian saltwort* (S. australis).

Prickly samphire (Salsola kali) is related to the above and used in very similar ways. The tender young shoots may be boiled and are recommended as an emergency food. Its relationship is underlined by such similar names as *prickly saltwort, bastard sea-grape* and *barilla plant*. *Barilla* is the Spanish name for the soda-ash used in the making of glass and soap. The

Arabic name for it, appearing in the plant's Latin designation, is *kali* (it is the same word as reappears in 'alkali'). Seaweeds were often burnt in a similar way to obtain their chemicals, those used being referred to as *kelp*. This plant has thus been named *kelp-wort*. It is known, in addition, as *eestrige*, *sea-thrift* and *Russian thistle* (because of its spiked branches). The *European barilla plant* (S. soda) grows on the sandy shores of the Mediterranean and in Asia. In Japan it is cultivated and the young leaves and stems are eaten boiled. *Asparagus samphire* (S. asparagoides) is another oriental relative; the young plants are boiled in spring and summer.

Golden samphire (Inula crithmoides) is a member of the daisy family and less aromatic than the others. It was once sold in Covent Garden market. *Jamaican samphire* or *bushy sea oxeye* (Borrichia frutescens) is a West Indian relative. *West-Indian* or *maritime saltwort* (Batis maritima) is also known as *Jamaican samphire*. It is a shrub commonly to be found on the sea-coasts of tropical America, the West Indies and Hawaii, and was once used for soda-ash. Its salt-flavoured leaves are used in salads.

Scorzonera and burdock (*Scorzonera hispanica*)

Scorzonera is also known as *Spanish salsify*, or *black salsify*, because of the colour of its roots, the rind of which must be removed before cooking. It grows throughout Southern Europe (except, apparently, in Sicily and Greece), and as far as Siberia; it was introduced to general European culture from Spain round about the eighteenth century. The English diarist, Evelyn, recommends soaking and then eating it raw, or else stewed with marrow, spice and wine. But it is equally good baked, boiled or fried. The name derives from the Spanish *escorza nera* (black bark), but at one time it was thought that it had something to do with the Spanish for a viper (*scurzo*) and was therefore believed to be a remedy against snake-bite. This explains its other names in English, *serpent-root* and *viper's grass* (*schlangengras* and *vipernwurzel* in German).

The legend has grown up that the African serf of a certain Cerverus Leridanus discovered it growing in Catalonia; seeing some reapers bitten by a viper, and remembering that the juice was an old native remedy in his homeland, he used it to save them and thus spread the knowledge of its use. It was later employed as a remedy against small-pox, and it was only slowly that people began to use scorzonera as a vegetable rather than medicinally. Another native belief is recorded by the Mexican poet Alfonso Reyes who says the Tarahumara Indians consider it good for those who feel cold (see *Yerbas del Tarahumara*).

Gentle scorzonera (S. mollis) is a Mediterranean species widely cultivated in Sicily and used principally in sweet dishes and preparations. Its chocolate-scented flowers are used in salads. S. schweinfurthii grows in the desert areas of North Africa, where various tribes use the root for food. The Kalmuks, a Mongol race from Central Asia, eat those of other species.

In Japan a very acceptable substitute is greatly cultivated; this is an edible variety of our burdock known there as *gobo* (Lappa major var. edulis). But the roots of our own *great burdock* may be eaten, although it is best to cook them in two changes of water. They are still used in that most delicious of fizzy beverages, 'Dandelion and Burdock'. The shoots, also, may be eaten like asparagus. Apart from such northern names as *bourholm*, *eddick* and *flapper-bags*, most of the burdock's other popular names refer to its sticky burrs. These are known as *sticky* or *beggar's buttons*, and *clot*, *clod*, *cockly* or *hurr-burrs*. A children's game, similar to that played with goose-grass, is to throw them onto the backs of people and thus make fools, or 'cuckolds', of them. Thus the burrs are also known as *cuckolds* or *cuckles*, *cuckold's buttons* and *cuckoldy busses*.

Seaweeds

Seaweeds have a multiplicity of uses and are a largely untapped source of food, high in vitamins, protein and iodine. This last is a conditioner of the thyroid glands and can usually be obtained via the water supply, except in some districts, such as Derbyshire in England. Here there is a tendency for a disagreeable condition sometimes known as Derbyshire goitre, unless it is supplied some other way, usually from a special cooking salt containing iodine.

In the British Isles it is mostly the old Celtic and Gaelic inhabitants who still continue to eat seaweed, although it was once eaten over a much wider area, including in England itself. Other island or seafaring peoples in Europe also use it, but it is most widely eaten in the East and in Polynesia. The earliest reference to seaweed's use as a food is in Chinese poetry of about 600 B.C. where those species in use are often referred to as 'sea-vegetables'. The ancient Polynesians used to cultivate 'sea-gardens' of edible algae, and certain species are specially cultivated in Japan, where seaweeds and their products make up 25 per cent of the national diet. Over 100 species are eaten in the Pacific area, either processed into made-up foods, as a base for soups and condiments, as a salad, or mixed with other vegetables. Cooked rice, for instance, is often wrapped in various types of seaweed.

11. SEAWEEDS

blade kelp bladderwrack carragheen
         purple laver

More and more investigation is being made into possible commercial uses for seaweeds, and into their edible properties. (The subject is known as phycology; a seaweed eater is known as a phycophage.) The following are a selection of seaweeds most commonly eaten.

1. Murlins *(genus Alaria)*
One may eat the stalk and fertile fronds of the brown murlins. The most popular seems to be *tangle* (A. esculenta), described as 'the best of all the esculent algae when eaten raw' by the *Oxford English Dictionary*. In Japan it is made into a processed food similar to *kombu* (see below) and known as *sarumen*. The weed is counted a delicacy in Ireland, and in Scotland is eaten by man and cattle, where the edible parts are referred to as *bobbyn*. Other Scottish names are *henware* (explained, perhaps mistakenly, as a corrupted form of *honeyware*, a name more accurately applied to Laminaria saccharina), *dabberlack* and *badderlocks*. The last two names are perhaps corrupted forms of Baldur's locks, a reference to the Norse god; or they may have been coined for the Old English *baldor*, signifying a prince or hero. In the North of Scotland it is called *piersil* or *purcil*, while in the Orkneys the Irish name 'murlin' is corrupted to *mirkles*; it might also represent its being fused with another local name, *keys*. The more prosaic English refer to it as *wing-kelp* and *sea-ware*.

In Japan and elsewhere various relatives of the murlins are eaten, in particular the *trumpet seaweeds* (genus Ecklonia), otherwise known as *horn plants*. Among these are *arame* (E. bicyclis) *kurome* (E. kurome), and *broad-fronded horn plant* (E. latifolia). These are often processed into kombu-like preparations. *Wakame* (Undaria pinnatifida) is another of the family found all along the Asian Pacific coast, eaten with vinegar, boiled or roasted, and specially processed in China and Japan.

2. Kombu *(genus Arthrothamnus)*
These are large brown *oarweeds* related to genus Laminaria (q.v.) and processed into a special food by the Japanese and Koreans. The weeds are dried, then boiled, compressed, dried again, and finally shredded or powdered. Different methods of cutting produce a range of products: white or black pulpy kombu; shredded kombu; filmy kombu; hair kombu; sweet-cake kombu. The complicated and painstaking preparation is undertaken both commercially and privately.

3. Irish moss *(Chondrus crispus)*
This is to be found in great abundance on the Atlantic coasts of Europe and America, and is gathered commercially in the Channel Islands, Connemara in Ireland, and in the Hebrides. It is high in iodine and

potassium. Byproducts are used in ice-cream, orange-juice, plastic bindings, treated cotton wool, beauty preparations and fertilisers. One medical preparation is used in the treatment of stomach ulcers. But Irish moss can also be used as a food, principally in the making of blancmange. A Hebridean crofter told me how this may be prepared. Having collected the purple weed, one must expose it until it is dry and a purplish white. Then steep it in warm milk into which an egg has been whipped and leave it overnight. The weed dissolves and the preparation sets. Some flavour it with orange juice, nutmeg, vanilla, sherry (in Ireland) and rum (in Canada). It is said to be good for tuberculosis and other lung diseases. In Ireland the dried weed can be bought in some grocers shops. The Irish name of *carragheen* is one by which it is also known in some parts of Britain. *Pig's wrack* is another name by which it goes in Ireland, since it is there boiled up with potatoes and meal and fed to pigs. At one time it was sold in Covent Garden under the very apt name of *oak-lungs*, descriptive of its tough, pink and branchy aspect. Less apt is another English name, *pearl moss*.

4. Mirume (*Codium mucronatum*)
This is a small, dark green seaweed that grows prolifically on the rocks and in the shallow coastal waters of Japan, where it is eaten when young. Other members of the family are to be found in the Mediterranean and on the English coast. Mirume is referred to in a particularly beautiful love-poem by the eighth century Kakinomoto Hitomaro, 'the saint of poetry', in the *Manyoshu* anthology (II.135).

> *In the sea of Iwami,*
> *By the cape of Kara,*
> *There amid the stones under the sea*
> *Grows the deep-sea mirume;*
> *There along the rocky strand*
> *Grows the sleek sea-tangle.*
>
> *Like the swaying sea-tangle,*
> *Unresisting would she lie beside me –*
> *My wife whom I love with a love*
> *Deep as the miru-growing ocean.*
> *But few are the nights*
> *We two have lain together.*

There are several other species used in Japan and elsewhere in the East, including C. lindenbergii, eaten fresh or dried and salted in Japan, and C. tomentosum, which is sold in Malayan markets; this the Japanese often eat with soy sauce and vinegar. C. tenuus is eaten in the Philippines, and C. **muelleri in Hawaii, pounded together with chilis.**

5. Enhalus (*Enhalus koedigii*)

Enhalus is a totally submerged marine plant which grows in the tropical water of Asia and the Pacific. Some eastern peoples eat the ripe fruits, which float to the surface and may be roasted.

6. Bladder-wrack and Sargasso-weed (*genus Fucus*)

Some of these brown seaweeds are eaten in Scotland, especially F. esculentus, whose edible ligaments, in common with those of the murlins, are known as *bobbyn*, probably in reference to the little bladders. (Elsewhere the seed-pods of several plants and trees are known as 'bobbins'.) Members of this genus are also known as *tangs*, a name by which they go in Germany also. In Scotland the *knob-tang* (F. nodosus) is known to the children as *sea-whistle*, since the fronds may be cut transversely near the end to make whistles. In the Hebrides they eat a species known as *sea-oak* or *Lady-wrack* (F. versiculosus), boiled in water with 'red fog', a kind of sea-moss. Among other names for the weed, *sea-bottle* refers to its nodules which, when extended, lighten its normally dark colouring; in this state it is known as *strawberry ware* in Scotland. In northern England it goes by the name *black tang*. There are also several Cornish names: *lig* or *liggan*: *oreweed*; and *blade ore* in the Scillies. 'Ore' appears to be the southern form of the northern 'ware' (cf. the French *varèche*). A Pacific species, F. fuscatus, is eaten in Alaska, and members of the related genus Turbinaria are eaten raw or as pickles in Malaysia and Indonesia.

Sargasso-weed (genus Sargassum) is a near relative of the bladderwracks. The similar noduled appearance has given it its name, which derives from the Portugese *sarga* (a grape). It is especially abundant in the waters of southern Australia and, of course, in the West Indian Sargasso Sea. Here the dense carpets of sargasso-weed floating on the surface impeded the progress of the early wooden sailing-ships, and it was regarded as a considerable menace. Various species are used by the Japanese for food, and one in particular, S. enerve, which turns green when dried, is used by them in New Year's Day decorations. The *sea lentil* (S. vulgare), a species common in both the Pacific and Atlantic oceans, is eaten in the Philippines, and *limu kala* (S. echinocarpum) is used in several different ways in Hawaii.

7. Dulse (*Iridaea edulis*)

This edible weed is best known in Ireland, from whence its name stems, as does the variant *dillisk*, deriving from the Irish for water-leaf: *duillisg* (leaf) + *uisge* (water). *Pepper dulse* (Laurencia pinnatifolia) was once eaten in Scotland but never gained great popularity. The name refers to the fact that it has often (though not always) a hot biting taste (cf. the German *pfeffertang*). Other species are found in the Mediterranean, and in the

Pacific and Indian oceans, where they are eaten by various islanders. They are sold in Hawaiian markets under the name *limu lipeepee*.

8. Blade kelp (*genus Laminaria*)

Many species of this brown seaweed, whose wide leaves are named *devil's aprons* in Scotland and the north of England and *fyams* in Ireland, grow prolifically in the colder waters of the temperate zones. Several are made into the standard Japanese food *kombu* (q.v.), while one, L. potatorum, is eaten by some Australian aborigines.

Cuvy or *cairn-tangle* (L. digitata) is occasionally eaten, as by the Russians and Chinese, for example. Its main use, however, used to be for the production of carbonate of soda which was retrieved from its ashes. (Most of the seaweeds once used for this purpose were referred to generally as kelps.) The old kelp-gatherers used its stalks as clubs, from whence it gains the name *sea-clubs*. And because the old fronds are washed ashore after spring storms, it has been called *May* or *drift weed*. Other names include *slack marrows*, *dead man's toes*, *cow's tail*, *sea girdles* or *wand*, and *sole-leather*. There are also a crop of Celtic names. It is the Orcadians, who have a differentiating name for every kind of seaweed they collect, who call it *cuvy*. Other Scottish names are *slatenhara* and *hangers*. In Ireland it is known as *burro*, a name also applied in derision to a tall gangling person.

Sweet tangle, *honeyware* or *sugar wrack* (L. saccharina) is so called because it is covered in a whitish powder which is sweet to the taste. Like other members of the family its frond is long and ribbed, giving it such names as *sea-belt* and *ribbon weed*. It is also referred to as *poor-man's weather glass*, since it is generally this species which is hung up as a kind of barometer: the fronds go dry and brittle in dry weather, turning soft and sticky at the approach of rain. Perhaps it is because of this prophetic quality that the American Kwakiutl Indians of the north Pacific coast use kelp as a charm in summoning the wind.

Several other species, including *Japanese kelp* (L. japonica), are used in Japan to make the processed kombu already mentioned. In powdered form this is added to sauces, soups and rice. The weeds may also be eaten as a vegetable, pickled or candied; in one of its states it is an ingredient added to *saké* (rice-wine), in another it may be used as a tea substitute. In the Pacific area is to be found *bull kelp* (Durvillea antartica), common on the Australasian coasts and in Antarctic waters. This is eaten by the aborigines and New Zealand Maoris, and on the other side of the South Pole is sold in the markets of southern Chile; there it is eaten as a vegetable or used in soup. *Bladder kelp* (Nereocystis leutkeana), or *sea-otter's cabbage*, is found on the Pacific coast of North America, where orientals eat its succulent parts.

9. Nostoc (*genus Nostoc*)

Nostoc (the name was coined by the alchemist Paracelsus) is a blue-green alga which grows in moist places, often clinging to rocks, and is made up of filaments united into a jelly-like spherical or lobed colony. Its simple organisation has changed little over the last 500 million years. Some species are edible, including the Chinese *yuyucho* (N. commune), a land form also eaten in Java and Equador. This is to be found in Britain, where its shining gelatinous appearance once aroused the inhabitants to suspicions that flying saucers had landed, or at least that they were confronted with star-fragments of extraterrestrial origin. It is, comments one old account, the 'stinking tawney jelly of a fallen planet, or the nocturnal solution of some plethorical and wanton star'. This belief has given it many of its names: *Fallen stars*; *spittle of the stars*; *star jelly*, *shoot*, *slime*, *slubber* or *slutch*.[1] It was also known as *scoom* (scum), *Will-of-the-wisp*, and *lock-lubertie* in Scotland.

10. Purple laver (*genus Porphyra*)

This red seaweed was once eaten under the above name in England, and is still used in Korea, China and Japan. When grilled on toast it tastes, so some say, like oysters. All species are high in protein. Among those eaten are the European P. umbilicalis and *marine sauce* (P. vulgaris), which is specially cultivated in China. Best known in Ireland and Scotland, however, is *sloke* (P. lacianata), also known as *slake kale* and *sloukawn*. It is generally eaten stewed very hot with a little water. At the turn of the century, it is said, there were few homes without a sloak-kettle, a container with a lid and a long handle which was used to pass the seaweed round the table.

In South Wales, where the weed is known simply as *laver*, it can be bought from fishmongers and butchers, and was once obtainable even as far inland as Hereford. The weed gathered from the Gower coast is not enough for the local Welsh demands, and the rest is gathered from Cumberland and Stranraer. Nevertheless its popularity is waning. In 1965 126 tons were sold, which may be compared with the 212 tons sold annually in the late 1950s. In Japan, however, it is specially cultured, and about 250,000 tons are sold annually. Some of this is sold fresh, but most of it is dried in the sun and formed into sheets 8 × 6 inches, known as *amanori*. The Japanese method of cultivation is to sink a bundle of bamboos in the sea off-shore until a crop of young weeds have established themselves on them. The bundle is then transferred to specially enclosed nurseries where the weeds flourish in the less salty water. (It is usually the species mentioned above which is also used in the Chinese restaurants of North America for making seaweed soup.)

[1] German names of the same tendency include *schleimling*, *himmelsblatt* and *sternschnupfer*.

Among other species used by the Asian Chinese are P. nerocystis, and the highly prized *tsu choy* (P. suborbiculata), which is used in New Year feasts. P. perforata is found along the Pacific coasts of North America, especially California, where it is used by the local Indians. This is also used by the American Chinese and is even exported to Formosa. In Hawaii *limu luan* (P. leucostica) is considered a great delicacy; this species is to be found not only in the Pacific but in the Atlantic and the Mediterranean.

11. Shell dillisk (*genus Rhodymenia*)
Shell dillisk used to be sold in Scotland and Ireland and is still eaten there and in New England, U.S. The preferred type, from which the general name derives, was one that grew on the rocks near low-water mark and was frequently covered with mussels. The commonest species, which may be eaten either raw or cooked, is R. palmata, also known as *sheep's head* and, in Scotland, as *sou-soell*. This is eaten by a peculiar fast-disappearing species of mountain sheep in northwestern Scotland, and probably in Norway as well, so they name it *sheep-weed*. It is also eaten in Iceland among other places. Fishermen are said to chew dried or partly dried strips as a masticatory, and it is supposed to be good for sea-sickness. On the Siberian Kamchatka peninsula it is used for making an alcoholic drink. *Craw (crow) dulse* (R. ciliata) is another species sometimes eaten in Scotland.

12. Sea cabbage (*genus Ulva*)
The two species usually eaten are U. lactuca and U. latissima, which seem to share their names in common. It is a green laver which was once sold in London, and eaten in South Wales under the name of *laver bread*. (The Welsh *llaven* is reputed to mean black butter.) It is also known as *slake kale, green sloke* and *sea liverwort* (the last being something of a misnomer influenced by the name *laver*). Another name, once popular, is *oyster-green*; the seaweed was reputed to grow near oyster-beds and was much in demand by oyster-women in the market for decking their wares. *Sea lettuce* or *cabbage* is a name given it for its appearance as much as for its use as a vegetable.

In Asia and the Pacific these and several other species are eaten. Chinese fishermen gather and sell them both as a vegetable and as a fever medicine. The Hawaiian *limu lipahapaha* (U. fasciata) is cooked into a gelatinous mass resembling agar-agar (see below). Thinner-fronded relatives (genus Enteromorpha) are also eaten in the same area; for instance the Hawaiian *limu eleele* (E. prolifera), and several others. Although they are also to be found in European waters I have not come across any mention of their being eaten. In Japan *aonori* (genus Monostroma), another relative, is harvested and used either as a salad or relish.

13. Other Asian seaweeds
Of the various coloured seaweeds, by far the most popularly used are the red algae, the least popular being the blue-green species, most of which are poisonous. Besides nostoc, however, the *weedy kuo-yi* (Brachytrichia quoyi) of the Pacific and Indian oceans is eaten in China and Japan. Among the brown species *hijiki* (Hijikia fusiforme) is especially favoured with many references to its being gathered for food in early Japanese poetry. Several members of the Mesogloia family (genera Cladosiphon, Heterochordaria, Mesogloia, Phyllitis) are eaten, either fresh or dried in the sun. Red seaweeds belonging to the genus Grateloupia are eaten as spring and summer vegetables in China and Japan. The related Halymenia formosa is used in the Philippines, and Cryptonemia decumbens in Polynesia. Among other unrelated species, Catanella impudica is sold for food in markets along the Malay peninsula, and Acanthophora specifera is eaten in Indonesia and the Philippines. Digenea simplex is used in China and Japan both as a vegetable and a medicine for worms.

14. Seaweeds of Hawaii and the Pacific
Various seaweeds going under the Hawaiian name of *limu lipoa* (genera Dictyota, Dictyopteris) are eaten in the Pacific. These are branched and generally to be found growing on rocks; they range in colour from yellow-brown to olive-green. Among the edible green seaweeds are the *water-net* (Hydrodictyon reticulatum), which has a world-wide distribution, and various members of the *Cladophora* family (genera Chaetomorpha, Cladophora, Pithophora, and the related Stigeoclonium). *Limu kobu* (Asparagopsis sanfordiana) is a much used red seaweed which must be first pounded and soaked in fresh water to remove its bitterness. It is then salted and eaten as a salad or relish, or else cooked mixed with other seaweeds. It also figures as an ingredient in the condiment known as *inomena*. Alternatively it is powdered, pressed into balls, and packed with salt in barrels to be shipped to other areas.

15. Agar-agar and substitutes
Agar-agar, whose name derives from the Malay, is a gelatinous substance similar in effect to the product of Irish moss. That obtainable from health-food stores in Britain comes in translucent strips or as a white powder, and is mostly imported from Japan. There it is known as *kanten* and largely derived from the *tongusas* (genus Gelidium), also known as *agarweeds* and *Japanese* or *Chinese moss*. In areas where the seaweeds are readily obtainable they are often boiled fresh, or dried in the sun after washing in fresh water and kept until needed, the nutritious jelly obtained from them being used as an accompaniment to the meal. In China, G. divaricatum is often

boiled with vinegar or else sweetened. Acanthopeltis japonica is a red seaweed in the same family used for the same purpose.

During the last war when supplies of agar-agar were cut off, attention was turned to other seaweeds which might serve as substitutes. These included the already-mentioned Irish moss and members of the genus Gigartina in the West. Another Asiatic substitute was to be found in Ceylon or *Jafna moss* (Gracilaria lichenoides), alternatively known as *sea-moss*, from the tropics. A species of this, G. compressa, is used in Japan, and another, G. verrucosa, commonly found along the southern Australian coast, in Australasia. The Hawaiian *limu manauea* (G. corronopifolia) is often used locally in the Pacific, especially in soups. Other Australasian agar substitutes include members of the genus Pterocladia, also found in Japan, species of which are used for food, and those of the genus Eucheuma are very widely used, of which E. speciosa is the *Australian jelly plant*.

Other types used in Japan are the genus Gymnogongus and genus Ceramium. In the Malayan seas Mastocarpus klenzinaus serves both for food and agar. Two families, however, are especially popular as substitutes. Those belonging to the Hypnea family furnish the Hawaiian *limu huna* (H. nidifica), the Indonesian H. cenomyce, and in China H. musciformis, used in spring, and H. cervicornis in the summer. Related to this family is the Malaysian genus Gelidopsis, which is highly prized in some areas, and Sphaerococcus gelatinosus. In the other family is found the Hawaiian *limu akiaki* (Ahnfeltia concinna), *Sakhalin agar* (A. plicata), used in Russia and Japan, and the Russian *Black Sea agar* (Phyllophora rubens). It might be pointed out that all of these are red algae.

Agar is now produced in South Africa, Australia, New Zealand (whose product is finer than that of Japan), U.S.A., Korea, China, the Philippines, Morocco, Spain, Portugal, France, Denmark and Britain. The product of the last two countries is inferior to others. Of these countries Spain is now the second largest producer after Japan and a major exporter, since little is used for home consumption. Manufacturing began in 1940 and production is now in the region of 1000 tons a year, 50 per cent of which is exported to the U.S., Britain and Germany. Other purchasers include Japan itself, Poland, Russia, Finland, Italy, Holland, Czechoslovakia and the Argentine.

Spanish agar is manufactured particularly from Gelidium corneum; this and other seaweeds are gathered on the Atlantic and Cantabrian coasts, the Canary Isles and the Sahara coast. 60–75 per cent of that collected is sorted out on the shore after it has been deposited by autumn and winter storms and sent to drying plants, after which it is processed. Others are collected from rocks at low tide; from May to August frogmen bring the weeds up from the sea bed, working from ships specially equipped for this. The best quality is gathered in this way, but the cost is very high.

16. Future developments
Since the last world war phycologists have been looking into the possibilities of mass-growing protein-rich plankton such as the green freshwater *Scendesmus* and, especially, *Chlorella*, and then processing it into a flour, in order to supply the rapidly growing food needs of a mushrooming and increasingly hungry world population. These microscopic unicellular algae might well be the new foods of the near future.

Skirret *(Sium sisarum)*

Skirret is a sweet-rooted plant of East Asian origin which is still largely used in Japan and China. It was once highly valued in England during the sixteenth and seventeenth centuries but fell into disuse after that. The Roman emperor Tiberius, during whose reign the area across the Rhine was frequently penetrated by Roman armies, is said to have received tributes of this root from the Germans. The plant only grows well in cool climates and was not generally known in Europe until the fifteenth century, when it was much cultivated. Since the plant has now largely returned to its wild state it is rather tough and woody, although some cultivation would soon breed this out. It may be cooked in all ways like scorzonera (q.v.).

The name derives from the Dutch *suikerwortel* (sugar root), a form of which seems to appear in the old name of *skywort*.[1] In Scotland it is known as *crummock*, from the Gaelic *crumag*. The green stalks of another species, S. helenianum, are known as *jeelico*, a corruption of the name angelica, to which it is related, and are eaten raw on the Atlantic island of St Helena.

Wild skirret or *Spring carrot* are names given to the edible roots of the *silverweed* (Potentilla anserina), a member of the rose family. They are recommended as an emergency food in time of want, and the plant is to be found all over the world in the northern and southern temperate zone. Some of the older dialect names obviously refer back to a time when this plant was eaten. They include *moors*, deriving from the Old English *moru* (a root), and associated more particularly with the turnip; *blithran* and *briclane*, deriving from the Gaelic *briosg*, *briog* (brittle), and comparable with similar names for marsh woundwort (*see under* Artichokes); *sweetbread*, in reference to the carroty taste; and *bread-and-butter*, a name given to several plants by children who dig them up, pluck, and nibble. Here one might note that Scottish children, whose liking for pignuts (q.v.) has already been mentioned, used to treat similarly the roots of the related

[1] The German names are *zuckerwurzel* and *zuckerrübchen*; the French is *chirouis*.

marsh cinquefoil (Potentilla palustris or Comarium palustre) which they knew as *bog* or *meadow nuts*.

Other names for the silverweed include *argentine* (fairly obviously) and several perversions of that word; *wild, white, dog's* or *goose tansy*; *fern buttercup*; *fish bones*; *fair days* or *grass* (because it is a summer plant opening to its fullest extent when the sun is shining); and *goose* or *gander grass*, because it was fed to poultry. The leaves of several other species of this plant are used as a kind of tea substitute by such widely separated peoples as various tribes in Russia and Siberia, and Indians and Eskimos in North America.

Sorrels and docks

The true sorrels (genus Rumex) belong to a family which also includes such unlikely members as rhubarb and buckwheat. They are not related, however, either to the wood sorrels (genus Oxalis) or to the mallows (genera Malva and Hibiscus), some of which also go under the name of docks. The name derives from the French *surele* which is based on the word for 'sour'. Most languages incorporate some reference to the acid quality of the leaves into their names: *sauerampfer, sauerklee, sauerserf, säuerling* (German); *zuring* (Dutch); *suran* (Welsh); *acedera* (Spanish); *acetosa* (Italian). [In Sanskrit the adjective *amla* (acid-tasting) was used of the yellow-flowered wood sorrels.] The modern French *oseille* derives from the Latin *oxalis*, itself drawn from the Greek ὀξύς, which means sour. Oxalic acid is a poison found in more or less quantity in the wood sorrels and, among other places, in rhubarb leaves. For this reason it is unwise to eat the latter, although it is said that they were once used, like so many others in the various sorrel families, as a spinach substitute.

1. French sorrel and family (*genus Rumex*)
Nowadays, when one mentions sorrel in connection with cooking, it is generally taken as referring to *French, Roman* or *buckler-leaved sorrel* (R. scutatus), the least bitter of the family, which has superseded all others in popularity. In Europe it has been cultivated since the fourteenth century, mostly in France and Italy, An English traveller named Lister, passing through these regions in the seventeenth century and noting how widely it was grown, suggested it might well replace lemons as an antiscorbutic in the English navy.

The plant had actually been introduced to England the century before.

Until that time the leaves of the *English, garden* or *dock sorrel* (R. acetosa) had been in use, either accompanying spinach or by themselves. The plant is still cultivated in the temperate zones of America, Europe and Asia, and used as a vegetable and salad. Many of the plant's country names testify to its use as a condiment and salad, and to its taste: *green sauce, snob* or *snow,* and *London green sauce*; *sallary, sellery,* or *sollery, salad* and *sallet*; *sharp dock, sarock, sooracks* and *sorrow*; *sour dock* or *docken, grass, grabs* or *sabs*; *sour leek*. The name *arrow-pointed meadow sorrel* is self explanatory; *hunter's meadow sorrel* is thought to contain a corruption of 'hundreds' thus referring to its many seeds, which also give it the names *brown sugar, gipsy baccy, donkey's oats, tea, hundreds and thousands,* and *Tom Thumb's thousand fingers*. Other names include *cock* or *cow sorrel*; *cuckoo* or *gowke meat*; *gobbelty-guts*; *ranty-tanty* (Scotland); and *canker root,* possibly given because the roots were used in a preparation for cankers and sores. Some of these names it shares with the near-related *sheep sorrel* (R. acetosella).

The name 'dock' covers all members of the family in Britain, as do such old names as the Northern *doodykye* and the Irish *phorams*. They appear variously in Anglo-Saxon medicine as emetics, in a bonesalve for headaches and other pains, and a poultice for chicken-pox. The seeds were fed to elf-shot cattle. It is doubtless well-known that the leaves may be used for rubbing on nettle stings. Behind this practice lies the old proverb 'nettle in, dock out' used to indicate erratic and changeable behaviour (cf. Chaucer's *Troilus and Criseyde,* IV. 461). It comes from the charm that used to be repeated while applying the leaf to the affected part:

> Nettle in, dock out,
> Dock in, nettle out;
> Nettle in, dock out,
> Dock rub nettle out.

A last bit of country lore is that the dock may be used to restore dried tobacco, or to keep tobacco from going dry without tainting it. (Cabbage and lettuce leaves are sometimes recommended, but these give it a bitter taste.) One poem about dock leaves appears in an anthology of Yorkshire poetry:

> Docks, the dumb beasts heave up their backs
> To feed with bitter leaves of teeth
> About my yard;
> The rough wind strokes their gentle pulp.
>
> While green-salves burgeoning with fresh
> And eager perfumes blot and block
> Their seaweed sky
> Tart juices sluice and sour bones grate.

> Sorrel-lyes, samphire, dulse, absinthe,
> Filter through seeds of grubby marl
> The innocent,
> The absent madness of their breathings.
>
> The mucous cattle have come home;
> Tenderly drooping, listen, they lip
> Their acrid port
> With berry-ripe murmuring.
>
> It is earth articulating
> Factories of timbered vowels
> Taut as a cord caught;
> Testing the taste of tongue and teeth.

Many of the docks were actually cultivated, or gathered and eaten in earlier times, and the practice still continues occasionally. The *spinach dock* (R. patientia) is used in the early spring and known in Europe as *English spinach*, and goes by such names as *patience* and *patient dock*. It comes originally from Persia and Turkey and was used as a food in classical times. In England it was cultivated in most eighteenth-century gardens. The name patience is derived from the Latin for a dock (*lapathium*) which in Italian was eventually split into *la pazzio*. This, however, was mistaken for *passio* and applied to Christ's Passion, from whence grew up the custom of eating it at Passiontide.

Another dock whose cultivation dates back to classical times is the *blood-veined dock* (R. sanguineius), known to the Anglo-Saxons as *red dock*. Other names for it include *blethart, dragon's blood* (cf. the German *drachenblutkraut*), *bloodwort*, and *bloody dock* or *patience*. The *curled dock* (R. crispus), of European and North Asian origin, and introduced into the Americas and New Zealand, is used as a vegetable and a pot-herb. So also is the *blunt* or *broad-leaved dock* (R. obtusifolius). This has been named *butter* or *batter dock* from its being used to wrap up butter in the old days. Because it was also used medicinally it became known as *doctor's medicine*, while the Scottish name, *smair dock*, refers to the practice of rubbing it on stings, whether of nettles or bees. Its green seeds give it the name of *celery seeds*, and its vigorous growth that of *land-robber*, while children call it *cush-cows* from their game of milking the leaf by drawing the stem through their fingers.

The same function of wrapping butter was performed by several other plants, amongst them the *alpine dock* or *monk's rhubarb* (R. alpinus), most common in the mountainous districts of Central Europe, the Balkans and Caucasia. The object was to preserve unsalted butter during the hot summer months. The young leaves were used both as a salad and a vegetable, and considered good for the ague. There are also edible African species, including the *Abyssinian spinach* or *rhubarb dock* (R. abyssinicus), which is now cultivated in the Congo. The young leaves are used as a pot-herb and the powdered roots impart a brick-red colour to butter. The *bladder dock* (R. vesicarius) of North Africa and the Orient is cooked as a vegetable by the Bedouins. *Rosy dock* (R. roseus), also known as *African spinach*, used once to be considered only a variety of the above.

There are a host of edible American species, from the *Arctic dock* (R. arcticus) of Siberia and Alaska, whose leaves are eaten fresh, soured and in oil by the Eskimos, to the *Brazilian dock* (R. brasiliensis). In between these extremes the other edible species not so far mentioned are to be found in the West, where their use is restricted almost entirely to various Indian tribes. The most popular is the *canaigre* (R. hymenocephalus), occasionally cultivated for its tuberous root, which is used in tanning and as a source of a yellow dye. The Hopi and Papago Indians use it for colds and sore throats. The stalks serve as a rhubarb substitute and the mildly laxative leaves as a vegetable. The leaves of the *Mexican dock* (R. mexicanus) and *western dock* (R. occidentalis) are used by many tribes; those of R. berlanderi are cooked with the fruits of the prickly-pear cactus by the Indians of Arizona. Both the leaves and stems of the *mountain sorrel* (R. paucifolius), a northwestern species, are eaten by the Klamath Indians of Oregon. The name should properly belong to Oxyria reniformis, a very similar relative found in Arctic and alpine habitats, whose very young leaves are eaten raw or in soups. Those of the *alpine mountain sorrel* (O. digyna) are used by the Indian tribes of Alaska.

2. Knotweeds (*genus Polygonum*)

The knotweeds are relations of the dock and grow in much the same areas of Europe, Asia and North America. They are characterised by jointed stems, after which they are named, and minute greenish flowers. Perhaps the best known in Europe, certainly in the way of food, is *bistort* (P. bistorta). Its roots are very astringent, but after repeated washings lose their bitterness; a nutritious flour can then be obtained from them. In Russia this is mixed with wheaten flour to make bread.

The plant also bears the name *poor man's cabbage*, and *Pencuir kale* after the region of Ayrshire where its leaves were used to make broth. At one time it was confused with the *patience dock* (see above) which was eaten at Passiontide. This will explain some of the names it shares with that dock, such as *passions* or *passion dock*, as well as the name *sweet* or *gentle dock*. Its use at Easter will also explain some of the perversions of its Medieval Latin name, *aristolochia*. Beside the obvious *astrologia* it is also known as *Oysterloyte* (with variations) and *Easter* or *waster ledges*. In the eighteenth century it was eaten in Cumberland under the name of *Easter mangiant* (from the French *manger*, to eat), especially in a concoction known as *'erb* or *yearb puddin'*. This consisted of groats (or grits, coarse cracked wheat or the edible part of oat kernels), nettles, chives, bistort leaves, and those of the great bell-flower, cooked together in a linen bag.

The name bistort, taken from the Latin and meaning, as one of the names coined for it puts it, *twice-writhen* (cf. the French *renouée*), refers to the twisted roots, which are used in astringent medicines. By the doctrine of signatures, these roots, reminiscent of snakes, were good for the snake-bite; they give the plant such names as *English serpentary*, *greater snake-weed*, *adderwort* and *dragonswort* (cf. the German *drachenwurz*, *natterwurz*). In America they are added to soups and stews by the Cheyenne Indians. The white magician, Albertus Magnus tells us, finds a more curious use for the plant.

> This herb, put in the ground with the leafe called three-leafe grasse [trefoil], engendereth red and greene serpents, of which, if powder be made and put in a burning lampe, there shal appear abundance of serpents. And if it be put together under the head of any man, from thenceforth he shal not dreame of himselfe.

Several Asian species have their various uses. P. odoratum is cultivated in Indo-China as a seasoning; in Japan and China the *giant knotweed* or *Japanese fleece-flower* (P. cuspidatum) is cultivated for its roots, much used in Chinese pharmacy, from the bark of which a yellow dye is extracted. The slightly acid young shoots may be prepared like asparagus, as are those of the *Sakhalin knotweed* (P. sachalinense) by the Ainu. These people also pound the fruits of P. weyrichii in a mortar and mix them with millet. The sharp and acrid leaves of another Japanese species, P. maximowiczii, are cooked as a vegetable.

The young shoots of the *big-rooted lady's thumb* (P. muelenbergi) of America are eaten by the Sioux, while the Klamath Indians of Oregon and Washington use the seeds of the *Douglas knotweed* (P. douglasii) for flour and as a food. Also in western America the tender stems of the related *desert trumpet* (Eriogonum inflatum) are used as a salad before the plant has flowered. The leaves of E. corymbosum are boiled with cornmeal by the Indians of Arizona.

3. Wood sorrel and family (*genus Oxalis*)

Some members of this very bitter-leaved family have, or have had their uses as sorrel or spinach substitutes. The Asian poor still continue to gather the leaves and stalks of the *creeping yellow-flowered wood sorrel* or *lady's sorrel* (O. corniculata) and the *upright wood sorrel* (O. stricta). In South Africa those of the *Cape wood sorrel* (O. zonata) and the *goat's foot wood sorrel* (O. compressa) are used, and in the U.S. the *nine-leaved wood sorrel* (O. enneaphylla) and O. frutescens.

The leaves of the *common wood sorrel* (O. acetosella) were once cultivated as a salad, and are still used as such by Icelanders in the spring. Some, in fact, believe that it was this plant which served the Irish as famine fare under the name *shamrock* (from the Irish *seamrog*); the name was variously corrupted into *shamrogues, shamroots* and *shame-rags*. Nowadays that name is usually applied to a member of the clover family, but in the last century wood sorrel was called by this name in parts of Britain and Ireland. It is known besides as *cuckoo's, hare's, lady's, sleeping* and *sour clover*. Like the clover it has a three-lobed leaf; and the name of *good-luck* would seem to indicate that the finding of a four-leaved variety, as with the clover, is considered fortunate.

The plant goes under a variety of other names, some of which it shares with other types of sorrel (genus Rumex, q.v.). Among the more singular are those that refer to its food-like qualities; it is called *God's meat* and *God Almighty's bread and cheese*, but it figures more humbly as *bird's food* and *cuckoo's cheese and bread* (cf. the French *pain du coucou* and German *kuckkuksbrot*). It is known in addition as *bread and milk, butter and cheese* or *eggs*, and *bread and cheese and cider*. Why it should be so distinguished by God is perhaps explained by its appearance between Easter and Whitsun 'when Alleluja was woont to be sung in churches'; by this cry of rejoicing it is known both in Britain and elsewhere in Europe.

The wood sorrel's early appearance connects it with that herald of spring, the cuckoo, from which it gains the name of *cuckoo sorrel* and many other variations; the bird was supposed to clear its voice by eating the leaves. The white flowers of this plant seem to shine out in the damp shady woods which are its favourite habitat; from this it gains such names as *candle of the woods, evening twilight* and *fairy bells*. Other names include *cups and saucers, hen and chickens, hearts, laverocks, oxys, stubwort, saltcellar, sleeping beauty, sookie-sourack, sour Sally, Whitsun flower*, and *woman's nightcap*.

The young leaves and flowers of the Spanish-American *oca* (O. tuberosa) are used in soups and as a pot-herb, and the flowers serve as a kind of vinegar substitute in salads. It is cultivated, however, for the sake of its egg-sized tubers, which are an important staple in Mexico and the Andean states. *Oca* is a Spanish adaptation of one of the plant's Indian names; in Columbia it is known as *ibia* and in Mexico as *papa extranjera* (foreign potato), obviously because it has been introduced and is not native to that region. The tuber, especially the red variety, is acid-flavoured when fresh; the less acid yellow variety, however, is the least favoured. After they have been left in the sun to wrinkle and dry they become floury and sweet.

Spanish writers first mention *oca* in the sixteenth century. It was brought to England and Germany in about 1830, and shortly after to France; during the potato blights of the 1840s it was seriously considered as an alternative for cultivation, but attempts had been abandoned before the century was out. Two other species from Mexico, the *rosette oca* (O. deppei) and the *four-leaved oca* (O. tetraphylla), were introduced into England in 1827, and six years later into France. About these there were conflicting reports, some saying their flavour was more delicate than a young carrot's or asparagus, others that they were uneatable. They now seem to have little more use than as ornamental plants. But the bulb of the *buttercup wood sorrel* (O. cernua) would still seem to be under cultivation in southern France and North Africa.

4. Butterbur *(genus Petasites)*
The butterbur, whose main European species also goes by the name *batterdock* (P. vulgaris), is a plant generally to be found growing in swampy ground; its general appearance is rather like rhubarb. This was yet another plant used for wrapping butter, as the names suggest. Whether it is edible I have not been able to find out, although the old Yorkshire name of *poison rhubarb* does not sound encouraging. The leaves of some American species are said to be used as a vegetable, although their main use amongst the Indians seems to have been as salt, after they had been reduced to ash. The leaves of the northwestern P. frigidus, however, are eaten by the Alaskan Eskimos, while the boiled stalks of the *Japanese butterbur* (P. japonicus) serve in Japan as a spring and summer vegetable. The slightly bitter flower-heads are also boiled or preserved in vinegar.

Spinach, orach and blite

1. Spinach *(Spinachia oleracea)*
Spinach, a member of the goosefoot family, comes originally from Persia. The Arabs thought very highly of it, naming it 'the prince of vegetables', and it was they who introduced it into Spain in about the elevenht century. But it may have been the Crusaders who brought it back to Christian Europe in the thirteenth century. In Persia itself the first mention of cultivated spinach is in the fourth century, but its actual use must have been much older than this.

Our name derives ultimately from the Persian *ispanai*, which became *isbanakh* and *isfanakh* in Arabian dialects. It may have been either the former or the latter that produced the Hindu *isfany*. However, the Arabian name eventually became corrupted to *sebanakh*, and it would seem to be this which lies behind the early European names: *spanachia* (Late Latin); *espinaca* (Spanish); *espinache*, *espinage* (Medieval French); *spinach*, *spinage* (English). Some of the sound changes, at least, may be accounted for by the name's corruption with forms of the word for a thorn: *spinus* (Latin); *espino* (Spanish); *espine* (Medieval French).

The original spinach did indeed look like a 'spiny plant'. That the English variety is leafy and succulent is accounted for by the fact that the plant form altered when it reached colder climates. The leafy variety is also noted as being heat-resistant and therefore does not run to seed immediately the temperature rises, as the lettuce does, for example. (In fact, lettuce in this state gives one an idea of what the old spinach looked like.) Spinach is a highly nourishing food, rich in calcium, and vitamins A and C. This fact began to be appreciated in the U.S. during the 1920s, and it was consequently pushed commercially with great enthusiasm, one of the selling points being its truly remarkable effect in the adventures of Popeye the Sailorman!

The history of spinach, then, has been that of an innovation to the West which, like the potato, has triumphantly ousted a whole range of leafy herbs which were once used in its place; and that despite the fact that it does not differ from them in boiling away to nearly nothing however lightly cooked. Amongst those displaced were many of its relatives in the goosefoot family, next to be considered.

2. Goosefoots (*genus Chenopodium*)

Mercury goosefoot (C. bonus henricus), best known by the name *good king Henry*, was once very widely cultivated as a vegetable and pot-herb both in Europe and in North America, and still flourishes as a weed in both continents. Its broad, succulent, bright green leaves are known as *wild spinach* still, but it is advisable to cook wild varieties in a double change of water to avoid their bitterness. The young budding shoots, once known as *Lincolnshire asparagus*, serve as an asparagus substitute. The high regard in which it was once held is emphasised by the old saying.

> *Be thou sick or whole*
> *Put mercury in thy koole* [cabbage pot].

The praise is awarded not only for its edible, but also for its medicinal properties. This is further underlined by such names as *all-good* and *tola-bona* and the European equivalents such as *toute bonne* (French), *allgut* (German), *algoede* (*Dutch*). Possibly it gains the name mercury from the Latin god of medicine, other variations on which are *English, false* or *wild mercury, marguerry* and *marterry*.

The curious name *good king Henry* probably derives from the European equivalents such as *goeden Henrik* (Dutch) and *guter Heinrich* (German). The phrase is used, apparently, of a good fellow. Where the 'king' part of the English name comes from no one seems to know. Among other country names for the plant, *smiddy leaves* is awarded because of the belief that it is usually found growing near a smithy, and *sowbane* reflects the strange belief that pigs will die after eating it (a sixteenth-century writer claims that in 'base Almaigne' it is *seu* or *schweines tod*); but it hardly fits with many other names for plants in the family which indicate an animal preference for the leaves: *Friar's pot-herb, flowery docken* and *mutton dock* all refer to its edible qualities and link it with other plants used for much the same purpose. The name *blite*, which was used to cover both docks (q.v. *under* Sorrel) and goosefoots, derives via the Latin from the Greek for 'insipid'. The name *goosefoot* itself springs from a fancied resemblance of the leaves to that part of the bird's anatomy. Good king Henry is, perhaps, aberrant in shape, being known as *shoemaker's heels*.

Several other names are held in common by members of the goosefoot and related families, such as *orach, meals*, and the Scottish *milds*. A saying from Scotland bids one 'boil myles in water, and chop them in butter, and you will have a good dish'. *Fat hen*, another common name, is usually explained as referring to the thick succulent leaves, found useful in fattening poultry. *Lamb's quarters* also covers several in the family, although more strictly it is reserved for *white goosefoot* (C. album), and it was under that name that the plant was once sold in Ireland in the Spring.

The last mentioned, also known as *white* or *frost blite, white orach* and *grey kale*, has mealy white leaf-stems once much used as a vegetable, and still occasionally cultivated in Europe and America. The leaves and young tops have also been eaten by American Indians and the seeds used for flour and gruel. Seeds have been found in European Iron-Age settlements and there is evidence that the plant was a cultivated crop even then. In Asia it is still used as a vegetable and pot-herb, and there too the seeds are much relished. Other names for the plant such as *mutton chops* or *tops* link up with the name *lamb's quarters*, but also with good king Henry's *mutton dock*. A further connection with the docks exists in the name *more smerewort*. *Pigweed* is a name that might seem to contradict the belief that the plants of this family are inimical to pigs, but it might only refer to the fact that it is frequently to be found on manure and rubbish heaps, a habit testified to and deplored by such alternatives as *dirty Dick* or *John, midden* (which might have developed out of the Scottish *milds*), and *muck, mud* or *dirt weed*. Still other names include *bacon weed, confetti, smooth tongue* and *Johnny O'Neele*.

Other spinach substitutes include *nettle-leaved goosefoot* (C. murale), *amaranth goosefoot* (C. amaranticolor) and *red goosefoot* (C. rubrum). The last is also known as *French spinach* and by the contradictory names *pigweed* and *sow* or *swine's bane*. *Strawberry blite* or *spinach* (C. or Blitum capitatum) is so called after its edible fruit. From ancient times until the Middle Ages the plant was gathered wild, and cultivated, eventually becoming despised as mere famine fare. *Australian spinach* (C. erosum) was once eaten by early explorers. It was probably after this that the Fat Hen Plains were named. The so-called *Mexican goosefoot* (C. ambrosoides) grows from the temperate to tropical zones, and seems to have been introduced into America from the Old World. The Chinatect and Mezatect Indians of Mexico use the leaves as a condiment and in soups. The plant also goes by the names *demigod's food* and *wormseed* (the last because it, in common with one or two other species, is specially cultivated for a medicinal oil which is used to treat worms).

The majority of American species, however, are valued more for their seeds than their vegetable uses. Chief among these is *quinoa* (C. quinoa), whose name derives from the Peruvian Indian (Quechan) *kinua* or *kinoa*. It was cultivated from before Incan times by the Andean peoples, and by the time of the Incas was more widely cultivated than any other vegetable except the potato. The plant was regarded as sacred, and in the planting ceremony at the beginning of the growing season the first furrow had to be opened with a golden implement. The seeds, which are known as *petty rice*, are still a staple among some tribes. They may be boiled, toasted as tortillas, mixed with wheat flour in bread, and used in soups and porridges. They have the same amount of protein as maize but fall short in carbohydrates. Many attempts to grow it elsewhere, however, have failed by and large. The tender stems and leaves are best used as a salad. It was for these that the nineteenth-century attempts to grow it in Europe were made but, for all the experts thundered in its defence, it never attained popularity because the leaves were small and gummy. Under ideal conditions quinoa grows from four to six feet tall, the flat seeds being coloured black, white or red.

Cañihua (C. pallidicaule) is cultivated to a lesser extent in the Andean area and has the advantage of flourishing at even higher altitudes than *quinoa*. The seeds of *Fremont's goosefoot* (C. fremontii) are roasted or ground for flour by the Klamath Indians of Oregon, and elsewhere in northwestern America. Several other desert tribes mix the seeds of the *slim-leaved goosefoot* (C. leptophyllum) with corn-meal and salt. The plants are also eaten either raw or cooked. In Arizona the Indians use the seeds of the related Monolepsis nuttaliana for meal, and the roots are washed and cooked with fat and salt. In Central Australia the aborigines grind the seeds of C. rhadinostachyum. In Mongolia and Siberia the seeds of the related *soulkhir* (Agriophyllum globicum) are very important to some of the peoples.

3. Oraches   (*genus Atriplex*)

Most plants in this family, a relative of the goosefoots, go under the name *orach*, or variations upon that name such as *arache* and *orage*. It derives from the French *arroche*, a corruption of the Latin *aurago* (golden herb, a name by which they are also known). Their taste and choice of alkaline soils has given them the name *salt bush*, and the strong scent that of *musk weed*. Garden orach (A. hortensis), alternatively known as *mountain spinach* and *butter leaves* (they too were used as a wrapping), a native of Europe and Siberia, is reckoned to be among the oldest cultivated plants. It was extremely popular during medieval times, and was added by Italians to pasta, much as they add spinach to their green pastas nowadays. There are varieties with red, green and white leaves.

Another edible species from the same area is the *spreading* or *fat hen orach* (A. patula), whose varieties include the *halberd* or *broad-leaved orach* (var. hastata), known as *fat cabbage*; the *trailing* or *delt orach* (var. prostrata), last named for its triangular shaped leaves; and *upright*, *spear* or *oak-leaved orach* (var. erecta). In Ceylon the poor gather the leaves of A. repens, and the Gosuite Indians of Utah eat the seeds of the *four-winged salt-bush* (A. canescens), a shrubby plant growing from north-western America to Mexico. Other species are regarded as foods to be used only in emergencies. The silver-leaved *sea purslane* or *shrubby orache* (A. portulacoides) is mentioned in the Bible as famine fare: 'For want and famine they were solitary, fleeing into the wilderness in former times desolate and waste; who cut up mallows by the bushes and juniper roots for their meat' (Job 20:4). The Hebrew *malluwakh* is here mistranslated, and in the Revised Version is changed to 'salt-wort'. Its leaves were once pickled and eaten as a relish in England.

The *broad-leaved sea-purslane tree* or *Mediterranean salt-bush* (A. haulimus), growing on alkaline soils in Africa and usually fed to cattle, is consumed in times of want by the Tuareg. The related Haloxylon salicornicum grows in saline places in Iran and Afghanistan, where the local peoples eat the young stems in time of famine. Other obscure relatives include the *desert seepweed* (Dondia suffrutescens), a shrubby North American plant whose leaves are cooked by the Indians of California and Arizona, and Arthrocnemum glaucum, a Mediterranean shrub whose leaves are added to garlic porridge in Greece. The leaves of another species (A. indicum) are eaten as a salad in India.

SPINACH SUBSTITUTES OF UNRELATED FAMILIES

As has already been mentioned, there are many spinach substitutes. Those whose use is incidental to others are mentioned elsewhere (*see under* Colocasia for example), while large classes like amaranths, nettles, purslanes and sorrels, have sections to themselves. Lesser classes appear below. Taken altogether it should be possible to have spinach or a substitute all the year round.

1. Chickweed (*Stellaria media*)

The plant is so called because poultry is very fond of it, a fact born out by other names such as *hen's inheritance, chicken's meat, chicken weed* or *wort*, and *chick-wittles* (i.e. victuals). It is also named *passerina*, after the French for a sparrow, presumably because these birds show a preference for it. The plant is to be found growing in temperate regions all over the world, and is capable of withstanding severe cold. It has even been found blossoming under light snow in mid-winter, which will explain such alternative names as *winter weed* or *winter chickweed*. Its small white flowers grow at the top of a tangle of green leafy stems, thus suggesting the name *starwort*, which is echoed in the Latin of its generic name. Others include *arvie* (from the Danish *arve*), *muruns, gosk, flewort, craches* and *mischievous Jack*.

Chickweed would once seem to have been more widely used than one would have expected. A Latin poem, supposed to have been translated from the Greek by Virgil (*for this see under* Onions), mentions a peasant who grows the plant in his garden and then takes it to Rome to sell, leaving the more rustic herbs for his own repasts! An eighth century poem from the *Manyoshu* anthology also mentions the plant as being eaten, presumably by country people.

> *Over there on the Kasuga Plain*
> *I see wreaths of smoke rising –*
> *Are the young girls*
> *Boiling starworts*
> *Plucked from the spring-time fields?*

It is best to use the plant when young. The raw leaves reminded me of rather mild cabbage to taste. An old recipe advises one to 'take chickweed, clythers [burdock], ale and oatmeal and make pottage therewith'. This is probably one version of the ale pottage to which marigolds (q.v.) were added in the seventeenth century.

Various other relatives may be used in the same way as chickweed; for instance, the *sandworts* (genus Arenaria), so named because they prefer sandy soils, whose young leaves are sometimes used as a salad in North America. These include *thymeweed* (A. serpyllifolia) and *chickweed sandwort* (A. trinervis). *Sea chickweed* or *purslane* (A. peploides) is also used, and the leaves are eaten soured or in oil by the Alaskan Eskimos. The young leaves and shoots of the *bladder campion* (Silene vulgaris or inflata) are added to soups in North America, while in Europe the young shoots are cooked like asparagus and are said to taste rather like fresh young peas. Most of the popular names for this plant centre upon the flower-bladder (which children amuse themselves by popping in much the same way as unopened fuschia flowers are treated): *thunderbolts*; *snappers*; *clapweed*;

rattle-bags; *bull rattle*; *cow cracker* or *paps*; *blether* (i.e. bladder) *weed*; *white bottle*; *birds eggs*. Other names include *adder and snake plant*, *sprattling* or *frothy poppy*, *behen* (the German name) and *catch-fly*.

Corn-cockle (Agrostemma githago), also known as *bastard nigella* and *wild savager*, is a European relative introduced into North America. Its seeds are poisonous, which is more than annoying, owing to its likeness to nigella (*see under* Pepper) whose seeds are used as a condiment. The young leaves have been eaten as a vegetable and with vinegar in times of want. *Corn spurrey* (Spergula arvensis) is generally used for fattening cattle, and has been named *franke* after the stalls in which cattle used to be shut up for this purpose. Its seeds, however, may be used for making bread. It is also known as *yarr*, *yawr*, *yur* and *yarrel*, all shortened forms of yarrow; both plants look rather like milfoil. The Latin name is a shortened form of *asparagula*, a diminutive form of asparagus, which plant it is also thought to resemble. Other names include the Irish *granyagh*, *lousy grass* (cf. the German *läusegras*), *farmer's ruin*, *make-beggar*, *cow's quakes* and *sandweed*.

2. Pokeweed (*genus Phytolacca*)

The most commonly used member of this family is the *American pokeweed* or *nightshade* (P. decandra), once very popular but now regarded as little more than a troublesome weed. The roots are poisonous but have medicinal properties. The young shoots and leaves may be cooked like spinach or asparagus. The young berries, dark purple and juicy, were once used for colouring purposes and, in Southern Europe and especially Portugal, are still used in wines. Various names still record such usages: *dyer's grapes* *red-ink plant* or *inkweed*, and *crimson berry plant*. They also had a certain medicinal value, especially in veterinary medicine, if we are to believe the seventeenth-century name of *pork physik*. *Garget*, another old name meaning the throat or a throat infection, might also indicate a primitive use but, on the other hand, may merely be a reference to its colouring. Its common name derives from *pocan*, by which it was known to the Indians; variations upon it include *scoke* and *coakum*.

The plant, and related species, are to be found in warm areas of North America, the Caribbean, Europe, Asia and Africa. *Spanish calalu* (P. octandra), otherwise known as *West Indian foxglove*, grows in the Central American region, and is eaten like spinach. The leaves and shoots of the *Venezuelan pokeweed* (P. rivinoides) are also eaten in South and Central America. *Indian poke* (P. acinosa) and the Chinese *food poke* (P. esculenta) are occasionally cultivated in Asia, although they are less tasty than American species. In East Africa the young stems and leaves of the *Ethiopian poke* (P. abyssinica) are eaten.

3. Spanish needle (*Bidens pilosa*)

This plant is a member of the daisy family found growing in all warm parts of the globe. In South Africa, at least, and probably elsewhere, its leaves are eaten prepared like spinach or cooking lettuce. In the Philippines it is used as an ingredient in a local rice wine (*tafei*). The American name of *railway beggarticks* is graphic as a description of its burred appearance and one of its favourite habitats. An alternative name, *New Zealand cowage*, is a misnomer since the name 'cowage' strictly belongs to the very different Stizolobium pruritum (*see the section* 'Others' *under* Beans).

4. Basella (*genus Basella*)

Red basella (B. rubra) first became known to Europeans in 1688, when the Dutch governor of Malabar noticed the natives eating it. It can be found growing in most tropical Asian countries and was introduced into England early in the eighteenth century, where it is alternatively known as *red Malabar nightshade* and *East Indian spinach*. The latter name also covers *white Malabar nightshade* or *country spinach* (B. alba), known to Indians by the Hindi name *poi*. The name *basella* derives from the Singhalese. *Large-leaved Chinese basella* or *vine spinach* (B. condifolia) was brought to Europe in the nineteenth century and is preferable to both of the foregoing because of its large leaves and abundant growth. It has, however, a laxative effect and is not very nourishing. Two other edible species are known as *Japanese spinach* (B. japonica) and *Chinese spinach* (B. lucida).

5. New Zealand spinach and iceplants (*Tetragonia expansa*)

This heat-resistant summer vegetable was first discovered by Sir Joseph Banks, sailing with Cook in Queen Charlotte Sound, to the north of South Island (N.Z.). It was an extremely gratifying find at a time when antiscorbutic plants were at a premium on long sea voyages. It was first brought to England in 1770, and by the nineteenth century was being used widely both in Europe and the U.S.A. It is now to be found growing all over the Pacific, and in Chile, China and Japan. To the Australians it is sometimes known as *warrigal cabbage* (*warrigal* being an aboriginal word for the dingo, later extended to mean anything wild by the whites). As the leaves tend to taste rather bitter it is advisable to cook them in a change of water. *Australian spinach* (T. implexicoma) is also known as *Victoria bower spinach* and *Tasmanian iceplant*.

The true *iceplants* (genus Mesembryanthemum) are relatives from the southern hemisphere, now introduced to warm areas in the north. They are named after the shimmering silvery dots that cover their leaves. Alternatively they are known as *fig marigolds* after their edible fruits, and in Australia as *mid-day flowers*. The most commonly eaten as a vegetable, especially in the islands of the Indian Ocean opposite Africa (Reunion, Mauritius, etc.), is the *frost* or *diamond plant* (M. crystallinum). The *heart-leaved fig marigold* (M. cordifolium) is also edible though not so good. In Australia the aborigines bake the leaves of the *karkalla* (M. aequilaterale), also known as *pig's faces*.

Sweet potato and other creepers (*genus Ipomea*)

Sweet potato is the tuberous root of a tropical vine (I. batatas, formerly Batatas edulis), related to convolvulus and the morning glory, and bearing purple flowers. The root, which now grows in all warm areas of the world, is large, thick, sweet and mealy, entirely replacing the use of potato in some regions. In Europe it is too much trouble to grow it compared with the potato, although there has been large-scale cultivation in the U.S.S.R. since 1930, and people in the U.S. are particularly fond of it. There are several hundred different kinds of this vegetable, but they divide broadly into two main groups: those with dry, hard, yellow-fleshed tubers, and those whose tubers, yellow or white in colour, have a sweet, soft watery flesh. There are others whose tubers are pink, purple and brown. In the West Indies the most popular variety has a round red-skinned tuber. The leaves, also, are used in soups and stews as a pot-herb in Africa, Indonesia and the Philippines.

The plant is of doubtful origin, but it would seem to be another of those which have travelled from the East to America via Polynesia, and taken from there to Europe. Many names are similar, such as the Peruvian Quechan *cumar* compared with Polynesian and Maori names: *kumala* (or *gumala*), *kumara*, *umara*. On the other hand it must be of very ancient culture in America since there are different names for it in the separate language groups: *apichu* (Peru); *camote* (Yucatan); *maby* (Caribbean islands); *batata* (San Domingo). The last name was originally applied to our potato in England, and from it was coined the word 'potato'.

Columbus brought the sweet potato back from his first voyage, and it was cultivated in Spain, Portugal and Italy during the sixteenth century. From Mexico the Spaniards took it to the Philippines, and from there it reached China in 1594 and Japan in 1698. Meanwhile the Portuguese took it to their settlements in Africa and Asia. John Hawkins was the first to take back examples to England from Venezuela in 1565; less than a century later it was being grown by the colonists in Virginia. Louis XV was very fond of it, and had it cultivated in France for himself. Later on it was

12. ROOTS AND TUBERS

Sweet potato
yam bean   arrowroot   cassava
heath spotted orchid
Jerusalem artichoke   ulloco   ysaño

prepared for Josephine, Napoleon's wife, who was a creole and had acquired the taste for it during her childhood. With the Bourbon restoration, however, it fell out of favour again. Sweet potato is now grown throughout the tropics, and especially in Africa, in parts of which, and in the East, it is ousting yam-cultivation.

Parts of other creepers in the same family are also eaten. The roots of I. tiliacea or fastigata, known locally in the West Indies as *wild potato*, are sometimes tuberous and occasionally eaten in South America. It is this plant which has been suggested as a possible parent of the sweet potato. The large succulent roots of the *jicama* or *bejuco blanco* (I. or Exogonium bracteata) are eaten by the Indians living on the Pacific coast of Mexico. The roots of another species, I. mammosa, are eaten in Indonesia and the Philippines. The young shoots and leaves of the *water glorybind* (I. aquatica or reptans), also referred to as *water spinach*, are fried in oil or cooked like spinach in Asia. The plant grows everywhere in the waters of southern China and elsewhere, and is cultivated in some countries, such as Ceylon. In Ceylon also the young flowers of the *moon creeper* (I. bona-nox or Caloncytion aculeatum) are eaten as a vegetable. This plant is cultivated in tropical America and elsewhere as a decorative plant for the sake of its large, fragrant night-blooming flowers.

The *Madeira vine* (Boussingaltia basselloides) is an unrelated creeper belonging to the goosefoot family, of tropical American origin and found growing from Mexico to Chile. Its edible tuber was once suggested as a possible substitute for the potato during the nineteenth-century crop-failure, but it was found to be too mucilaginous. In the same area the *mountain rose* or *coral vine* (Antigonum leptopus) is also to be found growing. This has edible tubers of a nut-like flavour which are eaten by the natives. And in Columbia and Venezuela they eat those of the *ulloco* (Ullocus tuberosus), which is also known by such local names as *melloco*, *uyusa*, and *oca-quira*.

In Australia the *Burdekin sorrel-vine* or *treebine* (Cissus opaca), a relation of the grape-vine, has tuberous roots which weigh up to 10 pounds. Although these are rather pungent the aborigine prepare them in the same way as water melon, after first immersing them for a long time in water. Those of another relation, Cayratia clematides, are treated by them similarly. Besides the actual grapes, the leaves of the grape-vine (genus Vitis) are edible as a vegetable; their delicate flavour is considered set off at its best when they are stuffed with unseasoned rice. These are well liked in the vine-growing areas of Europe, Asia and America.

Thistles

1. Tassel burs *(genus Carduus)*

However prickly a plant may be, if it has sufficiently tasty parts it has been eaten by man, especially in earlier times when stomachs seemed to have been stronger and one did not know where the next meal was coming from. The only thistle that still retains great popularity is the globe artichoke (q.v.), but there are several more which taste as good, if not better; and some which are still very popular in their native areas. The archetypal thistle, so much so that it is known simply as *thistle, bristle thistle* or *tassel bur*, is of the genus Carduus whose stems are still eaten (probably peeled first) in the manner of asparagus by country folk. In America those of the *Indian thistle* (C. or Cirsium edulis) are eaten by the Cheyenne and other Indians, while several other tribes eat the roots of the *wavy-leaf thistle* (C., Cirsium or Cnicus undulatus).

The roots of several other thistles related to the above may also be eaten, as for instance the European *tuberous-rooted plume thistle* (Cirsium or Cnicus tuberosus), which may be stored through the winter and is recommended as a food in times of emergency. American Indians use the roots of many others belonging to the genus Cirsium: the *Drummond thistle* (C. drumondii), *western thistle* (C. occidentale), *Virginian thistle* (C. virginianum), and C. scopulorum. In the East the Japanese eat the young leaves of the *Japanese blessed thistle* (Cnicus japonicus) as a vegetable. Those of a Eurasian relative, the *meadow distaff* (Cnicus or Cirsium oleraceus), are non-prickly and eaten like cabbage in some areas, such as Russia and Siberia. The plant is commonly to be found on wet peat-moors and in boggy woods. An alternative name is *pot-herb thistle* (cf. the French *chardon des potagers*). The Germans name it *cabbage thistle* (*kohldistel*, etc.).

2. Carline thistles *(genus Carlina)*

Carline thistles are so called because of a tradition that Charlemagne had a vision in which an angel gave him the roots of these to use medicinally in time of plague. They are to be found in Europe, North America and western Asia, and are especially common on chalky ground. The flower receptacles of the *wolves'* or *smooth carline thistle* (C. acaulis) are eaten in some European mountainous regions in much the same manner as artichokes. In the Alps they are known as *chardousse*, in the Cévennes as *cardavelle*, both names deriving from *carduus*, from which the French *chardon* (a thistle) also comes. To the Germans they are 'wild artichokes'. In the Mediterranean region the roots and flower-head of a close relative, the *pine thistle*

(Atractylis or Circelium cancellatum), may also be eaten boiled and seasoned, and are accounted very good.

3. Centaury *(genus Centaurea)*
Plants belonging to this family generally go under the names *knapweed* or *centaury*. Several resemble thistles, and all belong to the very large family (Compositae) which includes thistles, but most have more in common with the aster family. Some grow in desert regions and are a valuable food for camels (who can eat all sorts of thorny plants). The *star thistle* (C. calcitrapa), also known in Britain, where it is rare, as *caltrop*, is generally to be found in the Mediterranean region, and is distinguished by its brilliant rose-purple flower-head. The young leafy stems are eaten in Egypt. Another species, C. eryngoides, also grows in Egypt and Arabia, where its leaves are used as a salad by the Bedouin.

4. Sea holly *(genus Eryngium)*
Members of this family, although they have a thistle-like appearance, and in many cases are called thistle, are in fact umbelliferous plants. The best known is probably the *sea holly* (E. maritimum), common to the sea-coasts of Europe, whose very young roots (referred to as *ringo-roots*) may be eaten like asparagus. Those of the inland *snake root* (E. campestre), known in France as *chardon Roland* or *roulant*, are occasionally eaten candied; but they serve as a vegetable among the Kalmuks. In the Midlands of England this plant also goes by the names *Watling Street* or *hundred-headed thistle*, and as *Dane weed*. According to Daniel Defoe it gained this last name because of a local belief that it sprang from the blood of Danes killed in battle, or during the great massacre. In tropical America the offensively smelling roots of the *stinking thistle* (E. foetidum) are used as a condiment.

5. Hacub *(Gundelia tournefortii)*
Originally from Asia Minor, where it grows wild, this belongs to the same family as *scolymus* (see below), and its roots are considered the best among thistles. The buttons, leaves and stems may also be eaten and are said to taste like artichokes or asparagus. When cooking them it is advisable to add lemon-juice to the water, otherwise the green parts are liable to turn black.

6. Cotton thistle *(genus Onopordon)*
Among these is one of several thistles which claims to be the original *Scotch thistle* (O. acanthium). It grows in Europe and North America and is often to be found 'grown to the height of a man among the corn'. Its very white leaves have given it such names as *argentine* or *silver thistle*; it is

also known as *down* or *oat thistle*, *Queen Mary's thistle*, and *thistle-upon-thistle*. The roots, flower-head and (peeled) stem are edible. Both the French and the Germans refer to it as wild artichoke. The flower-heads of the Central European *Illyrian cotton thistle* (O. illyricum) are also eaten.

7. Scolymus (*Scolymus hispanicus*)
This golden thistle, known also as *Spanish salsify* and *Spanish oyster plant*, is cultivated in Spain for the sake of its roots, which some say are as good as its namesake (q.v.), others say are better. It grows wild all over southern Europe and had a certain amount of cultivation in the south of France during the last century, where it is also eaten wild. The fibrous centre, which is uneatable, must be removed either before or after cooking. One difficulty which mars this plant's usefulness is that it is extremely prickly; nevertheless the young leaf stalk and the mid-part of the leaf is occasionally used, cut up as a pot-herb, added to omelettes, etc.

8. Milk thistle (*Silybum marianum*)
This plant, now naturalised in some parts of the U.S., was once cultivated in Europe for its edible roots, leaves and large purple head. The leaves at its base are white-spotted, traditionally explained as being splashes of the Virgin Mary's milk. It is therefore known also as *blessed* or *Lady thistle* and *Lady's milk*. There are French and German variations on these names; the latter also call it the wild artichoke.

9. Sow thistle (*genus Sonchus*)
There are some very rough and prickly species of this thistle, also known as *swine thistle*, or simply as *swinies* in Scotland. The first century Greek Dioscorides says the leaves of the *milk thistle* (S. oleraceus) are edible; these are not prickly, as the alternative names *smooth* or *soft thistle* testify, and in northeastern America and Europe they are eaten as a salad. The name milk thistle comes about either because of the white-spotted leaf markings or because of its milky juice. Several names insist on one or another of these: *sow bread* or *milk*, *Virgin's milk*, *milky dashel* or *tassel*, *dindle*, *dickles* (all corruptions of 'thistle'). Names such as *hare's thistle*, *colewort* or *lettuce*, are explained by the old belief that the hare eats the leaves to recruit its strength, or else recover from summer madness; the thistle is also the *hare's palace* 'for yf the hare comes under it he is sure no best can touche hym'. Other names include *hog's mushroom*, *scent bottle* and *St Mary's seed*. The Australian *dune sow thistle* (S. megalocarpus), which is juicy and lactiferous, has been eaten by explorers of the continent when they have been short of food, and in Indonesia the leaves of

the tropical *sow-thistle tassel-flower* (Emilia sonchifolia) are eaten with rice and in soup.

Tomatoes   (*Solanum lycopersicum or Lycopersicum esculentum*)

The word tomato is a Spanish rendering of the Nahuatl (Mexican Indian) *tomatl*. In the New World, where it was discovered, the Indians ate it long before any European dared to try the experiment; indeed, it was not accepted in some parts of Europe and the U.S. until as late as the nineteenth century because of its connection with the nightshade family. Its bright shiny colours – red, orange, yellow and white – were most suspicious. Nevertheless, its importation to Europe antedates the more profitable products of the New World: the potato, maize and tobacco. In about 1550 the foolhardy Italians were actually growing it to eat, and the rest of Europe slowly followed suit, observing that the nation was not noticeably diminished from the effects of this experiment. It was being grown in England by 1580, largely as a curiosity at first.

Tomatoes, like most other fruits, come in assorted shapes and sizes. There is the *cherry tomato* which is to be found growing wild in Ecuador and Peru and is possibly the ultimate ancestor of all cultivated tomatoes; it is small and sweet, and grows in clusters, and is now grown all over the world. The *fruit tomato* is a very juicy, almost seedless variety; the *Italian plum*, sweet and coarse-centred; the *oxheart* is large, and coloured pink or yellow; the *pear tomato*, so named for its shape, is red or yellow. The smallest, a separate species S. pimpinellifolium, is the *current tomato*, or *German raisin*.

The tomato is supposed to have originated in Peru, and now grows wild over most of South America; it will propagate happily in any hot country. The fruit was cultivated before the arrival of the *conquistadores*; indeed it was one of the principal crops of Mexico. Upon its first arrival in Europe it went under the name *wolf's peach* or *Peruvian apple*. Soon, no doubt with the old myth of the golden fruit in the garden of the Hesperides in mind, tomatoes were named *golden apples* (cf. *goldäpfel*, German; *pommes d'or*, French; *pomi d'oro*, Italian). But with the growth of a belief in their aphrodisiac qualities they were re-christened *love apples* (cf. *pommes d'amour*, French; *liebes apfel*, German). Similarly, they were known as *mad* or *rage apples* at a time when these words had overtones suggesting wanton or amorous disorder.[1]

---

[1] Another German name, *Paradiesapfel*, might have been influenced by any of the alternatives above and coalesced with the myth of Paradise.

RELATED SPECIES

Cannibal's tomato (S. anthropophagum) is to be found in Fiji; its berries look like tomatoes and are cooked into a kind of sauce. But the fruits have little to do with the true and original use of this plant, from which it gains its exciting name. The Fijians believed that the prowess of a strong and respected enemy would be transferred, absorbed into oneself, if he was killed and eaten. Unfortunately he was found very hard to digest and the reveller was apt to suffer for two or three days afterwards. Experiment showed that the addition of vegetables aided the process. Thus it came about that the meat was wrapped up in the leaves of the cannibal's tomato together with those of two local trees, the *ramoon* (Trophis anthropophagum) and the *tudana* (Omolanthus pedicellatus). The whole was sprinkled with salt and roasted over heated stones. This was eaten with a fork and not allowed to touch the hands since it was believed that contact with the meat engendered diseases of the skin that could be passed on by contact with the flesh of children, and the Fijians were very fond of their children.

Ethiopian nightshade (S. aethiopicum) is a native of Africa whose leaves are used as a pot-herb; the young immature fruits are cooked as a vegetable and used for seasoning.

S. agrarium is a native of Brazil whose fruits are eaten; a decoction of the leaves is used in cases of gonorrhea.

Children's tomato (S. anomalum) is occasionally cultivated in tropical Africa. The fruits are used in sauces and as a condiment.

The *Kangaroo apple* (S. aviculare or vescum), also known as *Australian nightshade*, comes originally from the Australasian region, and is now occasionally to be found growing in European gardens as a decorative plant like many other solanaceous plants and as was the tomato once. The Maoris and Australian aborigines cultivate it for the big mealy berries – yellow, orange or violet, according to the variety – which may be eaten either raw or cooked. They must, however, be gathered when fully ripe since immature fruits taste acrid and burn the throat. Maori names include *poro-poro* and *kohoho*; aborigines refer to them as *gunyang*. In Australia the name *kangaroo apple* covers yet another edible species, S. laciniatum.

S. diversifolium is a prickly shrub of Central American origin, whose young fruits are important to West Indian negroes, eaten with salt as a relish.

Randa or *ranto* (S. ellipticum) are names given to the fruit of this plant by the natives of Central Australia, who especially esteem it. Elsewhere on the continent the fruits of the *quena* (S. esuriale) are eaten.

Gilo (S. gilo) is of African origin and has been introduced into South

America. The immature fruits may be cooked as a vegetable or for seasoning, as may those of the South-East Asian S. ferox.

Cocona (S. hyperhordium) is a native of the upper Amazon whose red and yellow edible fruits are at present on commercial trial in Puerto Rico, and elsewhere.

Afghan thistle (S. hysterix) is a prickly plant which can be found growing in South and Western Australia. In food-scarce areas the aborigines remove the prickles and seeds from the berries, pound up the remaining pulp with the root-bark of eucalyptus shrubs, and make cakes from the mixture.

Indian morel (S. indicum and S. xanthocarpum), also known as *yellow-berried nightshade*, bears edible fruits and leaves which are gathered by poor Asians. The plants also grow in Africa, where the immature fruits are cooked as a vegetable and for seasoning, as are those of S. incanum. In Indian native medicine the roots of S. indicum are used for chest complaints. Both the leaves and fruits of S. macrocarpum may be eaten; the plant is of African origin and has since been introduced into Malaysia.

Melon shrub (S. muricatum) is a native of South and Central America, introduced into Europe from Ecuador in the nineteenth century. In that country its fruit (the *melon pear*) goes by the Spanish names *pepino* and *guayavo*, and is very popular eaten either raw or cooked. It tastes rather like an acid-flavoured egg-fruit.

Black-berried nightshade (S. nigrum) grows throughout the Pacific and is an extremely popular ornamental flower, rejoicing in such names as *petty morel, solanberry, sunberry, quonderberry, houndsberry* and *garden huckleberry* or *nightshade*. The leaftops may be used as *calalu* (see below), but there is some doubt among the botanists about the berries. The general concensus of opinion is that they tend to be toxic when young and green, but may be innocuous when fully ripe and blackish coloured. Perhaps they are best left alone, or at least used very cautiously.

Branched calalu (S. nodiflorum) is another native of the Pacific now growing widely in the West Indies. The tops are edible and, cooked like spinach or in soups, have a light bitter flavour. They are very highly thought of in hot countries. In the West Indies the name is applied to other plants which are eaten when the above is rare; they include *Spanish calalu* (Phytolacca octandra *see under* Spinach) and *prickly calalu* (Amarantus spinosus, *see under* Amaranth). Tinned *calalu* is obtainable.

Olive tomato (S. olivare) is a native of tropical West Africa, growing as far south as the Congo. The fruits are eaten by the natives.

Olombeh (S. pierreanum) also grows in the West African region; its fruit is eaten by the tribes of Gabon and elsewhere.

Cut-leaf nightshade (S. triflorum) is a native of western America, growing from Canada to Mexico, and used by various tribes of Indians. The ripe fruits are eaten raw, boiled or ground and mixed with chili and salt; they are also added to bread.

Turkey berry (S. torvum) is the West Indian name for a plant widely spread through the tropics whose young fruits may be cooked as a vegetable or used as a seasoning. The young shoots are eaten raw or cooked in Java.

The above, of course, are only a very brief selection from over 1200 species, several others of which are eaten as a dessert, still others being poisonous. The leaves of at least one, S. inaequilaterale, are smoked by the natives in the Philippines, thus reminding us that the tobacco plant is another useful relative of the tomato.

Other families have various uses. The *ground cherry* (genus Physalis) bears sweetish fruits generally used as desserts, as do some plants belonging to the genus Solanum. *Vegetable tomato* or *mercury* (genus Cyphomandra) is a sub-tropical fruit-bearing shrub of American origin used for centuries among the Peruvian Indians. That mostly used is the *La Paz tomato* (C. betacea), also known as *New Zealand tree tomato*, whose light-brown egg-shaped fruits have a musky acid taste. The tough skin and seeds are generally removed first, the pulp being eaten raw or cooked. C. hartwegi bears reddish berries, similarly used, and sometimes to be found sold in the markets of Colombia, Argentina and Chile.

Another relative is the *prickly box* (genus Lycium), whose leaves are eaten as a vegetable and the young shoots like asparagus. In America the Indians of Arizona and California use the raw or dried berries of the *desert thorn* (L. europaeum) in soups and porridges. The plant also goes by such names as *box-thorn, asses' box tree, squaw bush* and *wolf berry*. In Chile the berries of L. humile are used, and in Asia the *Chinese box* or *tea plant* (L. chinense) serves as a vegetable.

Turnip and other worts (*genus Brassica*)

The turnip is a cultivated root that has been developed from the *field cabbage* (B. campestris or asperifolia), which grows wild in sandy soils, especially near the seaside in northern Europe. On the banks of the upper Thames it is called *bargeman's cabbage*; elsewhere it is, or was, known as *wild* or *annual turnip, keele* and *kalewort, suumer rape*, and simply as *yellow*. It was also called *wort*, in common with many other plants in the cabbage family. 'Wort' derives from the Old English *wyrt*, meaning a root, but having the idea about it of growth, since it is formed from the verb *wyrþian* (to grow, cf. the Sanskrit *vridh*).

The noun 'wort' was gradually narrowed down to any cultivated plant or pot-herb, and then applied especially to the *brassicae*. Under cultivation several forms developed, one of which was *rape* or *colerape* (B. rapa), a slender rooted plant cultivated for its seeds, which go into the making of rape-seed oil, and are also used as bird-seed. The greens are used as a forage crop for sheep and pigs. The word 'rape' has an interesting ancestry going back into the prehistory of the Indo-European languages. The Sanskrit root-word *rap* has about it the sense of swelling or bigness, and thus it came to be connected with many developed or cultivated roots. The Greek ῥαπυς and ῥαφος give us the Latin *rapa* (a root) and *raphanus*, the generic name for a radish (q.v.). The French may get their word *rave* from the Latin. The Irish is *raib*; Old High German *raba* becomes German *rübe*, and *rofa* in the Scandinavian languages; Old Slavic *repa* is *rjepa* in Russian.

The turnip is used as a forage crop as well as a vegetable. Usually one thinks only of eating the large root, but the leaves may be eaten too, and are found on sale in shops under the name *turnip-greens*. The turnip is usually designated as B. napus, and seems to share many of its popular names with the rape. *Colza* is usually understood to mean rape; it is derived from the Dutch *koolzaad*, meaning cabbage-seed. *Dowball* is another name for it, and *keblock*, which is obviously connected with *smooth cadlock*, a name for the turnip. The name *pottage herb* is the turnip's, as are *nape, navet,* and *gentle* or *garden navew*.

With reference to the last names we must look at the ancestry of the name 'turnip' which, by the way, used often to be mispronounced as *turmit*. The Old English name *næp* comes from the Latin *napus*, as does the Gaelic *neip*. The prefixed *turn* seems to mean 'ambiguous'; nobody seems to know why it was used, and it may turn out to have some other meaning, or even be some other word. With the development of the language we get the form *turnep*, and then *turnip*. Something similar has happened in the case of parsnip (q.v.). The Old English word obviously gives us *nape*, and the forms *navat* and *navew* come from the French, where *naveau* is applied to the field-cabbage, *navet* to the turnip, and *navette* to rape. Yet another name for turnips is *knolles*, which is allied to the Dutch *knol* and Danish *knold*, meaning a tuber.

Turnips need a cold damp climate to come to perfection, and grow well in sandy or gravelly soil where nothing else prospers. This is summed up in the proverb, 'turnips like a dry bed with a wet head'. For this reason they were a very important food crop for the early European races. Turnips were the principal source of food for the ancient Gauls and Germans, being filling, although, like the carrot, of no great nutritional value. Even in comparatively modern times they continued a staple crop for what remained of the old Celtic races, as in the Limousin region of France. There is an epigram by Rabelais written in the local dialect which runs

> *Se la rabiola et la castagna*
> *Venount a manqua*
> *Lou pais es rouina.*[1]

They were also cultivated by the Greeks and the Romans, being especially important to the latter in their early rustic days while they were struggling for supremacy with the surrounding Latin races. A favourite story concerns Curius Dentatus, consul in 290 and 275 B.C., a Roman hero and victor in their wars for survival, who had retired to his farm after the performance of his civic offices. Here ambassadors from the Samnites, whom he had previously defeated, attempted unsuccessfully to bribe him to their side while he was humbly roasting turnips over a fire. These he preferred to golden treachery.

At a later time, when they had fallen out of favour, one threw turnips at people as an insult, in the manner of bad eggs and tomatoes later. A country tradition was for a girl to give a young man a turnip when she tired of his favours. Thus the saying 'she has given him (cold) turnips' means a man has been cold-shouldered and jilted. Another proverb which looks with no happy eye upon the root is an equivalent of the adage about getting blood out of a stone: 'There is no blood to be got from a turnip'. But in the disruption of the Middle Ages they enjoyed a high reputation and only came to be displaced in popularity by the potato. An old Parisian street-cry runs:

> *Quand je fus mariée rien n'avais;*
> *Mais, Dieu mercy, j'en ai pour l'heure*
> *Que j'ai gagné à mes navetz.*
> *Qui veut vivre, il faut qu'il labeure.*

There are many varieties of turnip of all shapes and sizes; their colours range from red through white, yellow and grey, to almost black. A red variety, eaten either raw or in pilao in Central Asia, is called *gongoulou* in Kashmiri and *chalgam* in Turki. In England they have become increasingly popular as a forage crop since the eighteenth century. The four principal varieties are long (three times longer than broad); tankard or spindle shaped (twice as long as broad); round or globe-shaped; flat (broader than total length).

The *swede* or *Swedish turnip* (B. napobrassica or rutabaga) is so nearly allied to the turnip that many do not distinguish between them. In some areas both names are applied with equal indifference to the orange swede, although its flesh is of quite different and milder taste. In *Sons and Lovers*,

[1] 'Should turnips and chestnuts fail, the land's ruined.'

set in Nottinghamshire, we learn from D. H. Lawrence that 'swede-turnip' was one of Mr Morel's favourite vegetables. *Swaddie* is another name for the root. In the U.S. they are generally known as *rutabagas* after their name in Swedish dialect, *rotabagge* (red bags). This is preserved in the English *rootybaker* and *root rams*.

Swedes have found their poet in Edward Thomas, whose sympathetic and mysterious vision of them follows:

> *They have taken the gable from the roof of clay*
> *On the long swede pile. They have let in the sun*
> *To the white and gold and purple of curled fronds*
> *Unsunnned. It is a sight more tender-gorgeous*
> *At the wood-corner where Winter moans and drips*
> *Than when, in the Valley of the Tombs of Kings,*
> *A boy crawls down into a Pharaoh's tomb*
> *And, first of Christian men, beholds the mummy,*
> *God and monkey, chariot and throne and vase,*
> *Blue pottery, alabaster, and gold.*
>
> *But dreamless long-dead Amen-hotep lies.*
> *This is a dream of Winter, sweet as Spring.*

Cabbage turnip (B. ol. caulo rapa or gonglyodes) is an edible-rooted variety of the cabbage proper. It is also known as *knol-kohl* (cf. *knolles*), *kohlrabi* and *Hungarian turnip*. It is possibly the vegetable referred to by Pliny as 'Corinthian turnip', but there is no real evidence of its being known in classical times. There are further ambiguous references in Charlemagne's time, but it was certainly known in fourteenth-century France. In 1558 it is mentioned as having recently been introduced into Italy and Germany. In the latter country and in India it is very popular, but in England it is used little except as animal-fodder. People only began to grow it in their gardens in about 1850. But it is a particularly useful vegetable in that it can withstand both frost and drought. Furthermore one can also eat the enlarged stem like a cauliflower, and the leaves like cabbage.

Lunary or *moonwort* (Lunaria annua) has an edible rampion-like root which must be used before the development of the flower. The flat seed-cases are round and penny-sized, and have given the plant many of its popular names. Since the seeds can be seen through the pods it is called *honesty*; it was popularly supposed that where the plant abounded the local husbandmen must be possessed of singular virtue. The likeness of the cases to coins gives rise to *money-flower* (cf. the French *monayère*); *money-in-both-pockets* (because the cases are on both sides of the stem); *penny flower* (cf, the Dutch *penninck bloemen*); *dollar plant*; *shillings* and *silver plate*. The French continue the analogies with their *medaille de Judas* (cf. the German *Judas silberlinge*).

Gerarde comments on its appearance that 'the innermost skin where on the seed doth hang or cleave is thinne and cleere shining like a peece of white Satten newly cut off the peece' in explanation of such names as *white satin, silks and satin,* and *satin leaves.* Older more mysterious names were *bolbonac, shawbubbe* (apparently from the German *schabab,* of doubtful application) and *pricksong-wort.*

One of the moonworts was of great importance to the ancient alchemists; unfortunately plants going by that name abounded at that time, as a seventeenth-century writer points out in bemusement: 'There are so many herbes called by the name of *Lunaria* that it would make any man wonder how so many should be called'. It is, however, the birthday plant for 28 July, symbolising bad fortune and forgetfulness. But in the language of flowers it signifies that one is dreaming of love. And since there is an old belief that it will open the lock on any door if it is put in the keyhole, so it is believed that it will open a heart that is closed.

Violets *(genus Viola)*

Violets were once cooked along with other herbs in medieval times; whether it was only the flowers or the leaves as well is not made very clear. Certainly the flowers were used in a sweet pottage for which the recipe runs: 'Take violet flowers and boil them, then press and grind small and mix with either almond or cow's milk; add rice flower or *amidon* [wheat flour that had previously been steeped and then strained and dried in the sun], and add sugar or honey'. The same basic method is used to make similar dishes in which figure either roses, primroses or hawthorn flowers.

Most violet roots are poisonous, although the roots of some are said to be added to soups in North America. The *early blue* or *palmate violet* (V. palmata) may be used thus. The root is known as *wild okra* to the negroes; it is mucilaginous like its namesake, and seems to have been used in much the same way. Two related Brazilian plants, the *lobolobo* (Alsodeia physiphora) and the *rinorea* (Rinorea castanaefolia), are eaten like spinach, especially by the black population. In England at one time the seed vessels of the *wood violet* (V. sylvatica), alternatively known as *blue, dog, pig* or *hedge violet* and *gowk shoe* in Scotland, were eaten by children under the name *pig-nuts.*

The violet symbolises constancy, modesty, humility, secrecy and mourning. It is the emblem of Rhode Island, Illinois and New Jersey, and was chosen by the exiled Napoleon as his. For the Christians it was the emblem of Christ, the Virgin Mary, St Fina and confessors. In the East, however, it plays a less passive part. The Koreans name it 'savage flower' because it blooms early in spring on any kind of soil. It is symbolic of intelligence and sprightliness in Japan and represents the male sexual organ in India.

In classical myth the violet appears as the attribute of the love-goddesses. The flower is said to have sprung from the blood of Attys, a handsome Phrygian shepherd beloved of Cybele, the primeval Asian earth-goddess. When he proved unfaithful to her she threw him into a state of madness in which he castrated himself and then changed into a fir tree. Alternatively, it was produced by the sweat of Io, a girl beloved of Zeus who was changed into a cow by the jealous Hera and tormented by a gadfly which drove her from land to land.

Water lilies

1. Water shield (*Brasenia peltata*)
The *water shield*, known in North America as *deer food*, is a large-leaved water lily found on all continents except Europe. The fresh young leaves are eaten seasoned with vinegar in the spring by the Japanese.

2. Prickly water lily (*Euryale ferox*)
The *prickly water lily*, also known as *gorgon plant*, is occasionally cultivated in the Far East for its edible seeds, which may also be ground into flour. In addition the young root and shoots are also eaten, although they are not very highly thought of and are generally classed as fit only for peasants.

3. Lotus (*genus Nelumbo*)
Members of this family make up the sacred lotuses of the East, but are not to be confused with the mythical lotos of Tennysonian fame, the eating of whose fruit brought forgetfulness of one's former life and a surrender to ever-present languor. The plant is probably of Indian origin but its yellow, white, red or blue flowers have been held equally sacred in China, Japan and Egypt. The plant is emblematic of the sun, and hence of light, life, fecundity, creation, resurrection and immortality. It also stands for androgyneity, the self-created and self-existent, prosperity, concord, peace, silence, purity, steadfastness and exaltation. In the East it is worn as a good luck amulet. In the West it stands for estranged love in the language of flowers, and it is the birthday plant for 9 November.

The five petals of the flower symbolise the five stages in the wheel of life: birth, initiation, marriage, rest from labour, and death; and also the perpetual cycles of birth and death in reincarnation. Since the seed-pods, buds and flowers are found simultaneously on the same plant it stands for the three stages of existence; past, present and future. It springs from the mud, and yet leaf and flower are undefiled, even by the water beneath them, thus suggesting man's spirit rising above worldliness and adverse surroundings, and therefore boundless possibilities. The flower opens on the water facing the sun, emblematic of meditation and entry into Nirvana (for the Buddhist). The flower follows the sun, seeking, as it were, the highest and purest of human possibilities.

There is a further connection with the sun in that the sun-god is represented as being born from the bloom in several religions; it is also represented as the ark in which the sun crosses the underworld on its nightly voyage from West to East. In Chinese art the plant is often shown with ribbons emanating from it as symbols of the sun's sacred rays. In Egyptian myth Horus, god of the new-born sun, rose from the lotus as it expanded its leaves on the face of the primeval deep. It is also the throne of Isis, goddess of the moon, her sister Nepthys, goddess of night, and their brother Osiris, god of the setting sun and husband to both; Horus is his son by Isis. The flower, emblem of Upper Egypt, also stands for royalty. A red lotus is the emblem of India, and a blue lotus stands for celestial and earthly love, that which produces and that which is produced.

There are also several sexual connotations, the bud standing for the *lingam* (male organ), the blossom for the *yoni* (female organ), and a lotus opening and rising for the sexual act, in India. In Zoroastrianism a lotus and flame is emblematic of the union of fire and water, and thus of the sexual act and of creation. It is also the symbol of the tree of life. For the Greeks, however, the flower is obnoxious to Aphrodite (goddess of love), but serves as a couch for Hera (Juno, queen of heaven) and as a boat for Heracles.

In China the lotus is the symbol of Ho-hsien-ku, the only woman among the eight immortals. (She corresponds to the Japanese Kasenko, and is called upon to assist with household management.) The flower is also symbolic of summer, female beauty, harmony in marriage and progeny. For the Japanese, however, although it is the flower of truth, it also stands for death and therefore must never be used in festive decorations. But the lotus does have its place in flower-arrangements in representing the three stages of existence. The past is designated by a partly decayed or worm-eaten leaf, or by an open leaf with seeds; the present by a shining open leaf (known as a mirror leaf) or by a half-opened leaf and flower; the future by a curled leaf or closed leaf and bud.

In Korea the lotus is known as 'the flower that speaks' or 'daughter of the sun'. According to Hindu myth Brahma sprang from the lotus in Vishnu's navel and the world is pictured as a lotus floating in a shallow

13. WATER PLANT

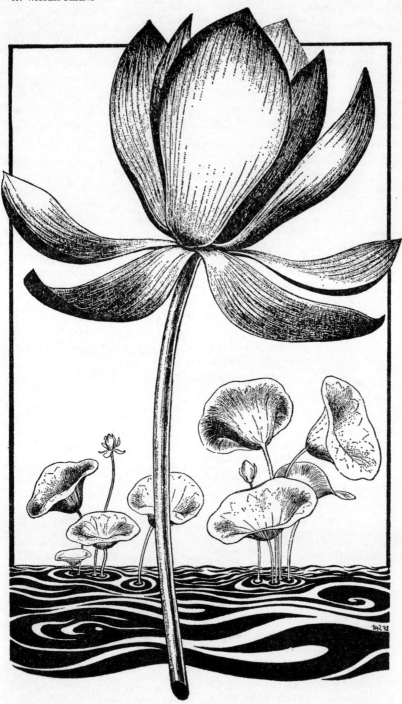

lotus

vessel born on the back of an elephant which, in its turn, is standing on the back of a tortoise (swimming in the primeval deep, no doubt). But for the Buddhists, who have spread their iconography throughout the East, it is pre-eminently the Buddha's throne, symbol of wisdom and enlightenment. He is frequently pictured seated thus in the yogic *padma-asana* (lotus seat), legs crossed so that the feet are raised on top of the opposite thighs. And when a Tibetan intones the Sanskrit mantra *Om mani padme hom* ('Om – the sacred syllable – hail to the jewel in the lotus') he is saluting the emanation of wisdom from the Buddha, and the possibilities of Buddhahood in himself.

So exalted an opinion seems to be held about the lotus that it is a wonder anyone should descend to eat it. And yet there is hardly a part that is not used in one way or another. In India the stamens are used for soothing bleeding piles; in Kashmir the leaves, commonly half a metre across, serve as disposable platters. Before they are fully developed the leaf-shoots are gathered and cooked. The seeds, which are equally sacred symbols of fertility, are dark brown and eaten raw or cooked; they taste like delicately flavoured nuts. The red-brown starchy tuberous root, or underwater stem, may be eaten raw, reduced to a flour generally used for making soups, and roasted, boiled or fried.

The *Indian, Hindu* or *sacred lotus* (N. nucifera), known simply as *pink water lily* in Australia, grows from Persia to Japan, in the Pacific area and adjacent northern part of Australia. All parts are eaten. Its extraordinarily hardy seeds have been known to germinate as long as 200 years after they have left the parent plant. The *Chinese water lily* (N. speciosa) is also known as *Egyptian* or *sacred bean*. It no longer grows in the Nile, however, although lotuses were once abundant there and are depicted on almost every ancient monument. The rhizomes are generally boiled or preserved in sugar, and are the source of the starch known as *lotus meal*. Starch and flour are also obtained from the seed kernels, and the young leaves are boiled as a vegetable. In Indo-China the stamens are used for flavouring tea. The *water chinquapin* (N. lutea) is the American equivalent, found in warm areas in the north and in the Caribbean. It is also known as *American lotus, yellow water-bean, water-nut,* and *duck acorn*. The entire plant is edible, especially the large starchy rhizomes, and was once much used by the Red Indians.

4. Nuphars *(genus Nuphar)*

This species seems to have been eaten chiefly by the North American Indians, especially the *spatter-dock cow lily* (N. advena), otherwise known as *stripe-flowered* or *three-coloured water lily*. The roots were stored as a winter food, and the seeds were eaten in various ways. The Klamantl Indians of California either grill the roots or convert them into flour for bread. In the north-west only the nutlets of the *edible-fruited water lily* (N. multisepalam) may be eaten. They are very attractive to waterfowl. In Europe the Finns have made use of both the seeds and roots of the *yellow water lily* (N. luteum), known to Americans as the *European cowlily*. From the flowers the Turks used to make a calming infusion.

This water lily is very common in Britain and goes under a variety of names: *Bees rest, blobs, bright flag, fairy boats, frog lily, patty pans, spatter-dock*. A number refer to the enlarged seed-pods: *brandy bottle, butter churn, butter pumps*. Since there is a very close relationship between this and the white water lily (Nymphaea alba) they hold many names in common: *can-dock*, with variations such as *water can, can* or *cambie leaves; clot*, with a whole variety of spellings, representing a very old medieval name; *flatter dock; nemphar* or *nenuphar* or *ninnyvers* (variations on the Latin and French names); *drowning lilies*. They are also distinguished by similar sounding names, the yellow variety being called *queen of the river*, and the white *queen of still waters*.

5. Nymphaeas *(genus Nymphaea)*

The root of the *white water lily* (N. alba), mentioned above, is edible and was eaten by the Tartars. There are several English names for this too: *water bell, nymph, socks*, or *rose; lady of the lake; swan amongst the flowers; virgin lily; sweet* or *sweet-scented water lily*. In Cornwall it bears the Cornish name of *alau* which might be allied to the Sanskrit *alu* for any edible root (this lies behind the Latin generic term *allium* for the edible-rooted onion family, q.v.). According to Pliny and later writers the roots were used as an antidote for aphrodisiacs and were supposed to have a relaxing quality. As a matter of fact they contain tannin, an element also found in tea, and are rather stimulating than otherwise.

Many other species have been eaten at one time or another. The *white Egyptian lotus* (N. lotus) was another plant highly honoured by that people; in Gabon it is still a fisherman's fetish. The ancient Egyptians used to make bread from its seeds, and its roots may be eaten like a potato. The *blue Egyptian water lily* (N. caerulea) is another found elsewhere in Africa. Senegal fishermen roast and eat the seeds and also use the roots. The *blue Indian lotus* (N. stellata) grows in tropical Africa and Asia, but its rootstocks and seeds are generally eaten only in times of famine.

Both root and seed of the *red Indian water lily* (N. magnifica, rubra or edulis) may be eaten, as may those of the *bulb-leaf water lily* (N. micrantha). From the roots of the *Rudge water lily* (N. rudgeana) a nourishing and fattening starch is made. The South American *dot water lily* (N. ampla) also has edible roots, while the natives of Transvaal eat the seeds of the *South African water lily* (N. calliantha). The *Australian water lily* (N. gigantea), known simply as *giant blue water lily* in the eastern part of the continent where it grows, is also to be found in the New Guinea area. Its hollow-leaved stalks have proved very useful as breathing-tubes to wily tribesmen escaping their enemies by submerging themselves. The tuberous roots, seeds and stems are eaten.

6. Royal water lily *(Victoria regia or amazonica)*
The *royal* or *Queen Victoria water lily* was named in honour of the long-lived British empress after first being brought to notice by English explorers during her reign. It is to be found on the Amazon in Brazil and is remarkable for its leaves, two metres across, with prickly undersides. These are termed *water platters*, and the seeds, which are used as food by the natives, *water maize*.

Water plants

1. Water hawthorn *(genus Aponogeton)*
The water hawthorn is an aquatic plant which grows in warm areas of Africa and Asia; it has been taken elsewhere and naturalised. For instance the *Cape water hawthorn* (A. diastychum), also known as *Cape asparagus* and *hawthorn-scented pondweed*, which was taken to France in the nineteenth century. The flower-bearing stalks furnish a spinach or asparagus substitute, and the roots are edible though not very digestible. On the other hand Indians are very fond of the roots of A. monostachyum, known to them as *nama dumpa*, which are said to be as good as potatoes. The Bengali A. undulatum may be used in the same way, as can A. microphyllum; the roots of A. crispum, which grows both in India and Australia, are said to be excellent. In Madagascar they eat the roots of the *lace water hawthorn* (A. fenestralis), which grows in the mountain streams. These are called *ouvirandra*, deriving from *ovy* (tuber) + *randrana* (branch).

2. Waterleaf *(genus Hydrophyllum)*
Members of the waterleaf family are eaten as a vegetable in northeastern America, principally H. appendiculatum, which was once much used by the early settlers. The young tender shoots of the *Shawanese salad* (H. virginicum) were, as the name suggests, eaten raw by the Indians. An Asian relative, Hydrolea zeylanica, is cultivated, and the young leaves eaten with rice in Indonesia.

3. Floating heart *(genus Nymphoides)*
Members of this family look very like water lilies and are eaten in much the

same way. The Japanese N. peltatum is to be found growing on ponds and provides a mucilaginous salad, as does the *Chinese marsh flower* (N. cristatum), which is a swamp dweller. The Queensland aborigines roast the small round tubers of the *Australian floating heart* (N. crenatum).

4. Water dropwort (*genus Oenanthe*)

The Greek-derived generic name of this umbelliferous plant means 'wine-flower', and alludes to the smell of the flower. An Asian variety (O. stolonifera) grows wild in marshy places in the Far East and India; it is also cultivated and eaten like spinach. The leaves and stems of the *Pacific water dropwort* (O. sarmentosa) or *water parsley* have a taste rather like celery and are eaten by the Indians of Oregon (U.S.). The black tubers, which are also eaten, have a sweet cream-like taste. But care should be taken over this plant since several species are poisonous; in England these are usually distinguished in their popular names by having the name of hemlock somewhere in them (e.g. *water hemlock*, O. crocata), although the roots of one variety (O. pimpinelloides) used to be eaten by children under the name *earth* or *pig-nuts*.

5. Pickerell weed (*genus Pontedoria*)

A pickerell is a young pike, and members of this family are popularly supposed to be a preferred food of these and other fish. The starchy seeds of the *moose-head* (P. cordata) are used for bread by North American Indians and some of the rural whites, and the plant serves as a vegetable. Members of the related *water hyacinth* family (genus Eichhornia or Piaropus) are said to make a very nutritive vegetable, eaten in North America, its continent of origin. The plant has now spread to Asia and Australia.

6. Arrowleaf (*genus Sagittaria*)

These plants, also known as *arrowheads*, grow in ponds and lakes in the temperate regions of Europe and Asia, and have become naturalised in North America. In warm regions they are cultivated as a food, principally in the Far East, where the *common* or *Old World arrowleaf* (S. sagittifolia or chinensis) is eaten. This is also known as *water archer* (cf. the French *fléchier*). Water birds are very fond of the leaves, and the roots go by the names *swan's* or *marsh potatoes*. They are tuberous and about the size of a hen's egg; some of their pungent bitterness disappears upon cooking. It is this species which is also eaten by the Kalmuks and other Central Asian peoples. Other species are eaten in North and Central America, principally the *wapato* or *duck potato* (S. variabilis), which is a favourite among the Indians.

The fresh roots of the *water plantain* (Alisma plantago), related to the above, are bitter and poisonous, but become harmless after careful

14. WATER PLANTS II

arrowleaf   fringed water lily   water plantain

preparation. These too are eaten by the Kalmuks. In Scotland the plant is known as *devil's spoon*, after the shape of its leaf. Other names include *mad dog weed* and *great thrum-wort*. Other relations include the *yellow velvet-leaf* (Limnocharis emarginata), whose young leaves and blossoms are eaten with rice in Indonesia, and the *water gladiole* (Butomus umbellatus), also known as *lily grass* and *flowering rush*. It is commonly to be found by ponds, rivers and boggy ground in northern Europe and America. The roots are eaten like a turnip or ground and used in bread.

Yams *(genus Dioscorea)*

There are more than 200 species of yam to be found growing in Africa, Asia, the Pacific and the Americas. It is one of those vegetables, growing wild in both the Old and New Worlds, that is of doubtful origin, and was first discovered by Europeans in the Americas. Those most commonly obtainable look like an enormous potato; they are, however, more mealy.

The *Chinese yam* (D. batatas) was among the many tubers considered as a possible cultivated crop to be substituted for the potato during the blights of the nineteenth century. It was first brought to France from Shanghai in 1850, but was soon abandoned because the great depth to which it burrows into the ground made it extremely difficult to gather. It is for this reason that the Chinese and Japanese are, on the whole, prejudiced against yams; they play no great part in their diets, although yams have been cultivated in China from very ancient times. They are fast losing popularity in Africa also.

In Fiji and other Pacific areas the yam still constitutes the preferred food. It is of such importance that almost the whole of the Fijian eleven-month calendar is centred upon its growth-cycle, and most of the names of the months refer to it in one way or another. It is grown in earth that is hard and unprepared in the belief that it is a sporting sort of vegetable that likes to feel resistance before it will show its strength (and therefore grow large).

In the D'Entrecasteaux Islands, off the east coast of New Guinea, there lives a gloomy and suspicious people who believe that yams travel underground from garden to garden. They therefore spend a good deal of time trying to entice their neighbours' yams into their own plots by magic; and yet are righteously indignant if someone else's superior magic (or husbandry) produces a crop better than their own. In the neighbouring Tobriand Islands there is a much happier and more open race. These make a parade of their wealth by constructing fairly open yam-houses in order that all can see the quality of their produce, putting the roots of best quality well to the fore, of course. Particularly fine yams, however, are displayed outside the stores, often framed and decorated with paint.

The way in which the European name was derived makes a curious story. It begins in the Guinean verb to eat, *nyami*. The tale goes that some slaves brought back from Africa to Spain were observed to dig up the taro (*see under* Colocasia) and eat it with evident enjoyment. On being asked what this was they typically underrated the intelligence of their captors and remarked that it was something to eat (*nyami*). Under this name were christened all roots that were found pleasing to 'natives' and, the yam eventually taking pride of place, the name stuck with it. In Spanish it is *ñame*; *inhame* in Portuguese; *igname* in French. In the languages of the New Hebrides, however, there are forty-four separate names for the different kinds grown. To the Indians it is *aloo*, deriving from the Sanskrit *âlu*, a name covering any edible and nutritious root.[1]

All varieties of yam are not equally tasty; indeed, some are not even edible. The West African *poison yam* (D. toxicaria) is toxic, as is the *devil yam* (D. daemona) of India and Indonesia, from the boiled roots of which the primitive peoples of Java make a poison with which they tip their arrows. Below is a list of the more interesting and principal edible species.

D. aculeata, the Fijian *kaavi*, comes originally from Bengal and is to be found wild and cultivated in India, Malaysia and the Pacific.

D. alata, *ten months yam*, is the Fijian *uvi* of South Pacific origin. It is very widely cultivated and the only yam available in some parts of India. The root-stocks, which may be baked, boiled, roasted and fried, or used raw in salads, weigh over 100 pounds; it also bears aerial tubers. Of its two main varieties the white is preferred to the red. Other names for it include *greater*, *water* or *winged yam*.

D. altissima, *Dunguey yam*, is of West Indian origin. Its coarse and very irregular root has been used for food in the American tropics for centuries.

D. atropurpurea, *Malacca yam*, has a large violet-skinned root, and is to be found in much of Asia, especially the tropical south-east.

D. batatas, the *Chinese yam* or *potato* mentioned in the introduction to this section, grows wild as far west as Indonesia and is widely cultivated. An edible starch obtained from the roots is known as *Guiana arrowroot*.

D. bemandry, the Malagasy *bemandry*, is used for food in Madagascar and tropical East Africa.

D. bulbifera, the *air potato* or *potato yam*, originally from tropical Asia, where it grows wild, is also cultivated in the southern U.S., Caribbean and tropical South America, although it lags behind many in popularity. As well as the underground tubers, which usually grow in pairs, it bears aerial tubers on creepers which often climb as high as thirty feet.

D. cayenensis, *alloto* or *yellow Guinea yam*, is of West African origin and

[1] Cf. Latin *allium* (onion), and French *ail* (garlic).

now heavily grown in the U.S. and Caribbean. Since it takes about a year to mature, but may be harvested at any time after that, it is known as *twelve-month yam*.

D. dumetorum is of East African origin and a very important food to some tribes in the area. Of its three varieties *kiwa* is golden-yellow, *vigongo* yellow, and *sinquekano* white and mealy.

D. eburnea grows wild, and is cultivated, in Indo-China.

D. esculenta, the *fancy* or *lesser yam*, is possibly of African origin but is found in tropical Asia from India eastwards, and in the Pacific. It is one of the best varieties, bearing several tubers per plant, and has thus gained the name *potato yam*. These tubers are snow white, mealy and rather sweet.

D. fargesii, *Farge's yam*, grows in the mountainous region of Southern China.

D. fasciculata grows in Bengal, cultivated both for the tuber and the edible starch extracted therefrom.

D. glandulosa is a yam from the Brazilian region. Also used in some areas is D. dodecaneura.

D. globosa, *globe yam*, is the most popular species cultivated in India.

D. hastata, the *shield yam* of South American origin, bearing good quality tubers.

D. hastifolia is from Western Australia, where it grows wild and was once cultivated by the aborigines, to whom it was an important source of food.

D. japonica, the *Japanese yam*.

D. latifolia, *akam yam*, originally from tropical West Africa, is now widely cultivated.

D. lutea, *yellow yam*, of tropical American origin. The tubers serve as a vegetable, or as meal after roasting and grinding. This may also be used as a coffee substitute.

D. luzonensis, *Luzon potato*, the tubers from the wild plants of which are cooked like potatoes in the Philippines.

D. maciba, the Malagasy *maciba*.

D. oppositifolia, a rather woody species growing wild all over Asia, occasionally gathered and eaten.

D. papuana, the *New Guinea yam*, cultivated by the natives of the South Pacific.

D. pentaphylla is another hardy yam (like Farge's) from southern China which grows well at high altitudes and has been recorded at over 7500 feet.

D. praehensilis, *white*, *bush* or *forest yam*, is of West African origin, somewhat primitive, and probably the parent of the *white Guinea yam* (D. rotundata). The top of its underground root is often hard and woody from protruding into the air; the whole is thorny and has to be cooked very thoroughly to destroy these fibres. The plant also bears small aerial tubers.

D. purpurea, *purple-fleshed yam*, retains its colour even after cooking. It is the third most popular yam in India.

D. pyrennaica, the *Pyrennean yam*, is a cultivated European species.

D. pyrifolia grows wild in the Malay peninsula and is eaten by the Sakai peoples.

D. rotundifolia, the *negro* or *Guinea yam* of West African origin.

D. rubella is a red-skinned yam much grown in Bengal, and the fourth most popular species in India.

D. sativa, the *common yam*, is that most widely cultivated in tropical and sub-tropical areas. It grows wild in Northern Australia, where it is known as *karro* by the aborigines. Seventeenth-century writers mention its great popularity in the Pacific. Specimens have been known as long as $7\frac{1}{2}$ feet and weighing up to 100 pounds.

D. spinosa, *prickly* or *wild yam*, is gathered wild in South-East Asia and the Pacific.

D. transversa, the Australian *long yam*, is gathered wild by the aborigines and eaten raw or roasted.

D. trifida, the *Indian yam* of tropical American origin and cultivated in the Caribbean region; it also goes by such native names as *yampi*, *cush-cush* and *mapicey*. Its small well-flavoured tubers grow in clusters of twelve or more.

D. trifoliata is of South American origin. Its sweet tubers may be used as a vegetable or for starch.

D. tuberosa, the *cinnamon vine* of tropical America, whose large starchy tubers are eaten among the rural population of Brazil.

Acorns and beechnuts *(genus Quercus)*

The acorn is the nut (or fruit) of the oak, having a bitter-sweet taste and brittle consistency, as far as I can remember from eating it in childhood. Pigs thrive on it. It will be remembered that 'haycorns' were the favourite food of Winnie the Pooh's friend Piglet, but that Tiggers don't like them. They were part of the normal fare of the prehistoric inhabitants of Lombardy and lake-side Switzerland, although later they were regarded as mere famine fare, often ground and added to other substances to make bread. An Elizabethan writing in 1586 records that the poor 'in some shires are inforced to content themselves with rie, or barlie, yea and in time of dearth manie with bread made either of beans, peason, or otes, or al altogither, and some acornes among'. There are also several earlier mentions of them in medieval texts. Higden's *Polychronicon* says that the early Athenians sowed and ate them, and several classical writers say the same of the idyllic Arcadians in the 'Golden Age' of innocence. Sir Philip Sidney makes acorns a principal part of the country fare in his *Arcadia*. One is given to understand by many of these writers that the present ignoring of acorns and 'treading them under heel' is a sign of man's fall from grace rather than his progress into enlightenment. As the old proverb has it, 'Acorns were good till bread was found'.

There are many dialect variants of the name: *akern, acharne, archard atchen, hatchorn, yacon, yeaker,* and others. Country names include *glans* (from the French *gland*), *Jove's nuts* (the oak was Jove's tree in classical times), and the North Scottish *knappers* (from *knap*, a knob). Their use in children's games gives them such names as *cups and saucers* or *ladles, frying pans* and *pipes*.

In fact acorns are still eaten by some poor people in parts of Europe, principally in Spain and Italy where old ways die hard anyway. But they also represented an important source of food for forest Indians of Northern America, especially those of the *white oak* (Q. alba). By them the acorns were crushed and then left in running water in order to leach out the bitterness, after which they were cooked in grease or oil. Another species used in northeastern America was the *basket oak* (Q. primus), also known as *yellow* or *swamp chestnut-oak* (chestnuts and chinquapins are relatives, and characteristics occasionally overlap). In southwestern U.S.A. and Mexico the acorns of the *emory oak* (Q. emoryi) are occasionally found on sale in local markets and are popular with the Indians. These, like several others, are sweet-tasting and require no prior preparation. Californian Indians, finally, make a bread from the acorns of the *roble* (Q. lobata) also known as *Californian* or *Sacramento white oak* and *valley oak*.

262 / Acorns and beechnuts

In Asia the Japanese roast the acorns of the *Japanese oak* (Q. glabra), while in some parts of India those of the *blue Japanese oak* (Q. glauca) are eaten. In Iran and Kurdistan it is the acorns of the *manna oak* (Q. persica) which are used. The best flavoured acorns of Europe come from the Mediterranean *holm oak* (Q. ilex), also referred to as the *holly* or *evergreen oak*. This attains great ages in the warm countries of southern Europe. Pliny mentions several at Rome popularly supposed to be older than the city itself (founded in the eighth century B.C.). It is a particularly handsome tree which can occasionally be found growing in England. The *Barbary* or *belote oak* (Q. ballota) is sometimes counted only as a variety of the ilex.

There are two British oaks whose acorns were used in times of scarcity, and occasionally as a medicine – an old remedy for the stitch recommends stitch-wort and powdered acorns drunk in wine. The *English* or *female oak* (Q. robur) is widespread. In Yorkshire its leaves were referred to as *rump*; in the New Forest the catkins were called *the trail*, the mast and acorns *the turn-out* or *ovest*. It is also called *black oak* to distinguish it from the *English white* or *durmast oak* (Q. sessilifolia). The latter name is considered to derive from the Celtic for an oak, *derw*. Other distinguishing names include *male* or *maiden oak*. Some woodmen claimed to distinguish the two only by the fact that twigs of the one sank when thrown in water, while those of the other floated. In fact the former has long flower-stalks, while the flowers of the latter are practically stalkless. Other names for it include *bay* or *chestnut oak*.

The beech (genus Fagus) is another relative of the oak whose mast was once used by the European poor and by the Indians of North America. In fact that of the *American beech* (F. grandiflora) is still found occasionally for sale in markets. The leaves also serve as a vegetable and as a tobacco substitute. The *European beech* (F. sylvatica) was anciently considered the tree of the graphic arts, probably because runic tablets were made from thin slabs of its wood. It is from its name that we derive the word 'book'. Since it is resistant to the entrance of water it is preferred for making clogs in France. A long-keeping cooking-oil is still expressed from the nuts, which themselves are now considered slightly poisonous. The name 'beech' derives from the Old English *bece*, variant forms of which were *beoce* and *boc*; these last developed into *buck*, a name once used in England (cf. the German *buche* and Swedish *bok*). The name of the town and county of Buckingham is said to refer to the large beech-forests formerly growing in that area.

Almonds and other fruit kernels

1. Almond (*Prunus amygdalus* or *Amygdalus communis*)

The edible part of this member of the peach family is the nut-kernel. It is a native of North Africa and West Asia and is now cultivated in warm areas on most continents. Almonds are much used in Chinese cookery. From this tree came Aaron's rod that budded and bore fruit during a dispute over the high priesthood, and which was later placed in the Ark of the Covenant (see Numbers 17). Because it blossoms before the leaves are fully extended it became a Biblical symbol of a sudden or hasty event, and is thus used in the prophecy of Jeremiah (1:11–12): 'The word of the Lord came unto me saying, "Jeremiah, what seest thou?" And I said, "I see a rod of an almond tree". Then said the Lord unto me, "Thou hast well seen, for I will hasten my word to perform it".'

The kernels of the peach and the bitter almond are considered too bitter to be edible and, moreover, contain prussic acid, recognisable by the characteristic smell of almonds. Freed from this however, flavouring extracts are made from the latter. The derivation of our word 'almond' is from the Latin *amygdala* which later developed into *amendla*, and then *amandola* in the Middle Ages. This became *amande* in Old French, and also *almande*, the initial 'a' being confused with the Arabic suffix 'al' present in the Spanish *almendra*. Thus it entered English; the tree is referred to as an *almandre* in the late fourteenth-century translation of the *Roman de la Rose* attributed to Chaucer.

2. Other fruit kernels

The kernels of certain other fruits are put to some curious uses. The *Bancoul nut* (Aleurites triloba), alternatively known as *Indian*, *Tahiti* or *country walnut*, is the hard stone of a large spherical fleshy fruit to be found growing in tropical Asia, Australia and the Pacific. South-Sea Islanders string the stones on a reed or leaf-strip and use them as torches which, since they are composed of about 50 per cent oil, will burn for hours. The kernels, which are very nutritious, are popular with the Australian aborigines, although they may cause internal troubles if the fruit is not fully ripe. They are said to taste rather like walnuts. Because of the use to which they are put they are perhaps best known as *candlenuts*.

The fruit of the *quandong* or *Australian peach-tree* (Santalum acuminatum) also has edible kernels which can be used in the same way as candlenuts, as may those of the *bitter quandong* (S. murrayanum), known to the aborigines as *ming*. The roots of the latter are also nutritious and may be roasted. Several other Asian species of this tree provide the fragrant sandalwood which goes into the making of religious ornaments, incense and other pious uses.

The kernels of the *sebestan plum* (Cordia dichotoma), which grows in India and Australia, are edible and taste like hazels. In India the tender young fruits are eaten as a vegetable or used in pickles; when mature they are mucilaginous and sour, and are then used to make expectorants. The pit of another edible plum-like fruit, of Far Eastern origin and now introduced into the U.S., is known as the *pili nut* (Canarium ovatum). This is thick-shelled and fatty and may be eaten roasted or raw, although it is said to bring on diarrhoea in the latter case. It is sometimes used as a source for oil, as is its edible relative the *Java almond* or *Chinese olive* (C. commune), growing mainly in South-East Asia. Other members of the family are used in the same area for incense.

The *New Zealand laurel* (Corynocarpus laevigata) bears an orange-coloured damson-shaped fruit whose small kernel is known to the Maoris as *karaka* and serves them as a staple food. The fruits are, however, extremely poisonous and their effect may permanently twist the limbs of their victim out of shape. The remedy is to bury the victim up to his neck in a pit, having first bound the arms and legs and put a gag between his teeth. The fruits have to be prepared by being baked in an earthen oven for several hours. They are then placed in baskets and soaked in running water for a day or two, after which they are thoroughly cleaned of flesh and skin and stored for use in feasts or as ceremonial gifts.

The trees grow in groves, entry to which is restricted by native custom, since they are regarded as holy. Chaplets of their leaves are worn when visits are paid to the graves of ancestors. According to Maori myth these trees were brought with them when they sailed for New Zealand from the semi-Paradise of Hawaiki. However, although the Maoris really came from the Western Pacific, the tree is not known further west than the New Hebrides; there are, however, somewhat similar trees in the presumed place of their origin, and it must have been the *kanaka's* likeness to these which taught them its use and originated the legend.

Arrowroot, salep and orchids

The true arrowroot is a starchy flour extracted from the roots of *reed arrowroot* (Maranta arundinacea) and put to a variety of uses. The plant is of South American and West Indian origin but has now been introduced into Asia, Africa and the Pacific. There are red and white varieties of which the former is the most esteemed. The roots, which were once believed to cure arrow injuries, may also be eaten boiled and otherwise. Brazilian Indians of the Matte Grosso region roast the root-stocks of a related herb, Saranthe marcgravii, over an open fire and eat them without further preparation.

Yet another relative, genus Calathea, bears tubercules which are used in the Caribbean and South American area. *Sweet corn root* (C. allouia) was the principal food of the West Indian aboriginals before the coming of the whites. From it an arrowroot is extracted, generally known as *leren* or *topee tambo*, apparently a creole contraction of *topinambour blanc de Martinique*. Alternatively the roots may be boiled for three-quarters of an hour and eaten cold as a sweet. The flower-clusters of two Central American species, C. macrosepala and C. violacea, are cooked by the Indians, and are sometimes to be found for sale in the markets.

Toleman or *Purple arrowroot* is extracted from the tubers of several species of the *flowering reed* or *Indian shot* (genus Canna), in particular C. edulis, of West Indian origin but now grown elsewhere in former British possessions such as Australia, where it is known as *Queensland arrowroot*. In Trinidad the starch is known as *tulema*, a contraction of *tous-les-mois*, its name in the French West Indies and elsewhere in the creole-speaking area. About 1836 an arrowroot of this name was exported to Britain from St Kitts, an extract from the related *scarlet canna* (C. coccinea). When young the small tubers of these may be eaten, as may those of several other species: *brick canna* (C. discolor), a less productive species from Trinidad; *Inca* or *Peruvian arrowroot* (C. languinosa); *Andean canna* (C. paniculata); *broad leaved canna* (C. latifolia); and the rhizomous *iris canna* (C. iridiflora). *Mexican canna* (C. glauca), also to be found in Brazil and the Caribbean, is another bearing edible rhizomes; those of the tropical African C. bidentata, however, are only used in times of scarcity. Its leaves are often used for wrapping other foods.

Asian Indians derive similar substances from the roots of the *curcumas* (genus Curcuma, the name deriving from the Arabic *kurkum*), members of the ginger family to which turmeric (*see under* Curry flavourings) also belongs. *East Indian arrowroot* is extracted from C. angustifolia, cultivated in the Himalayan region, and also known as *Tibur* or *Travancore starch*. Another source is C. leucorrhiza. *False arrowroot* (C. pierreana) is of Burman and Malayan origin and cultivated in Indo-China; C. xanthorrhiza is the Indonesian cultivated species. *Indian* or *South-sea arrowroot* is the product of the *pi* or *Tahiti salep plant* (Tacca pinnatifida), whose tubers contain a poisonous substance which disappears upon cooking. There are several other species in this area, and in Tropical Asia and Africa, among them *Hawaii arrowroot* (T. hawaiiensis), whose tubers are a source of food to the islanders, and the product of which was once of commercial importance.

Salep is another starchy meal derived from various roots and bulbs, generally those of the orchis family in the Middle East and Europe. It is mucilaginous, swelling in water, and has a slightly salty taste. As a food it looks rather like sago or tapioca; a nutritious jelly suitable for convalescents is also made from it. It is chiefly used in Southern Europe, North Africa, Turkey and Persia. In the East it is considered an aphrodisiac in common with the products of many other orchids, which are also used medicinally. The tubers are dug up as soon as the flower stalks have decayed, when they are considered in their best condition; the skin is rubbed off and they are then dried ready for use.

The name 'salep' is derived through the Turkish from an Arabic word for fox's testicles. It refers to the appearance of the roots, which have given the plants such English names as *culliens* (from the Italian *coglione*); *ballocks*; *dog's cods*; *dog's*, *goat's*, *fool* or *fox stones*. The very name 'orchis' is derived from the Greek for testicles. Other names applied generally to the plants are *fly-flowers, crowdy Kates, frogwort, standelwort,* (from the German *standelcruyt* or *wurtz*) or *standergrass*, and *satyrion*. The last is a book-name awarded either from a belief that it was the root of these which incited satyrs to their excesses, or from a legend of a satyr named Orchis who died and became this plant (cf. hyacinth, narcissus, etc.).

There are several of the orchis family (genus Orchis) and its relations which are used in the making of salep, of which the rarer varieties include the *ape orchis* (O. simia), *military orchis* (O. militaris) and *tawny orchis* (O. fusca), together with the related *black-spider* or *late spider orchis* (Ophrys arachnites), also known as *cobweb*, and *bee orchis* (Ophrys apifera), so called because 'it is in form and colour so like a bee that anyone unacquainted therewith would take it for a living bee sucking of the flower'. From this likeness it also achieves the names *bee* or *honey flower, humble bee* and *dumble dor*. The Mediterranean *false orchis* (Platanthera bifolia) is yet another from which the product can be obtained.

More commonly used are those better known in Britain also. There is the *broad-leaved* or *marsh orchis* (O. latifolia), also known as *pink marsh orchis*, which has flat- and bent-lipped varieties. The flattened hand-like appearance of the roots gives it such names as *Mary's hand orchis, Christ's* or *Satan's hand*, and the Scottish *Deil's foot*. In Scotland the plant was used in a love charm. There, according to *Mactaggart's Scottish Gallovidian Dictionary*, 'there are few districts in Scotland that have not their own name to this plant; in Annadale, and by the border, it is *meadow rocket*; in the west and greater part of Ireland, *Mount Caper*'. In addition it is known by the name *pull-dailies* around Edinburgh, perhaps a corruption of *bull-dairy*, also used locally, and *baldberry, balderry*, elsewhere in the area. *Cock's kames* is yet another Scottish name for the plant, which goes under the Gaelic name of *dodjell reapan*. In respect to the aphrodisiac quality of the roots, which has given the name *lovers wanton* in the Aberdeen area, it is said that 'rustics believe that if you take the proper half of the root of an orchis and get anyone of the opposite sex to eat it, it will produce a powerful affection for you, while the other half will produce as strong an aversion'. Perhaps the names *male* and *female satyrion* refer to the respective parts of the root; certainly such names as *Adam and Eve* apply, 'the tuber which sinks being Adam and that which swims being Eve', as well as *Cain and Abel*, 'Cain being the heavy one'. These names too are much used in Scotland. Others include *dandy gusset, lamb's horns* and *red lead* (also used for others which follow).

Two other orchids include the *lady* or *old woman orchis* (O. purpurea), names much used in Kent 'from the fancied resemblance of the flower to a lady dressed in a poke-bonnet with a best bib on and wide sleeves'; and *adder's* or *ballock grass* (O. maculata), also called *hen's kames* in Scotland, which is often confused with the *man orchis* (O. mascula). The *fool, salep* or *green winged orchis* (O. morio) has such variations on names which have already appeared as *fool's ballocks* or *stones*, and *bull's bags* (or *segg* in the North and Scotland, where 'to segg' means to castrate), and *ram's horns*. Other names include *bleeding willow, king-finger* or *fisher, nuns, parson's nose, single castle* and the Scottish *puddock's spindles*.

Lastly there is the very common *male* or *man orchis* (O. mascula), also known as *early purple* or *spotted orchis* and *cuckoo orchis*, as well as *wild bog* or *purple hyacinth*. This is popularly called *Gethsemane* because it 'is said to have been growing at the foot of the cross, and to have received some drops of blood on its leaves; hence the dark stains by which they have ever since been marked'. This accounts for other names such as *bloody* or *red butchers, butcher boys, bloody fingers* or *bones, bloody man's fingers* or *hands* and *poor man's blood*. A bluish-grey cast to the roots also gives it names like *blue butcher, granfer greygles* (grey gules) and variants like *grammer, granfer, granfy*, or *grandfather griggles, grigg*, or *gregors*. Whole families of names centre upon bird similies: *cock-flowers*; *crow feet* or *toes* (with many Scottish variants); *cuckoo buds*; *ducks and drakes*; *scab gowks*; *gandergoose* or *gosses, gandigosling, giddy gander, goosie-gander, goslings*, and such corrupted forms as *gossips* and *gussets*. The roots' hand-like appearance gives such names as *dead-men's fingers, hands* or *thumb*; *lords and ladies fingers*; *cling fingers*. To Shakespeare the flowers were *long purples*. Other names include *priest's pintle*, corrupted into *spreespinkle* or *spreeap-rinkle*; *adder's flowers, mouths* or *tongue*: *frog's mouths* or *distaff*; *soldier's caps* or *jackets*; *Devil and angels*; *locks and keys*; *jolly soliders*; *naked nannies*; *spotted dog*; *underground shepherd*; *kettle case* or *pad*; *single ghost* or *guss*; *ring finger*; *Johnny-cocks*; *skeat-legs*; *standel-welks* (cf. the German *stendelwurz*); *clothes pegs*; *fried candlesticks*, The Irish name *mogra-myra* refers to its use as an aphrodisiac.

Orchids are used in various other ways. In Malaysia, where the *king plant* (genus Anoectochilus) is especially abundant, one often finds various species on sale as pot-herbs in the market. Elsewhere in South-East Asia Dendrobium salaccense is used for flavouring rice. The Japanese used the flowers of Cymbidium virescens salted as a beverage, or preserved in plum vinegar; the aborigines of North Queensland consume the pseudobulbs of its relative, C. canaliculatum. In Tasmania the tubers of the *potato orchid* (Gastrodia sesamoides) are eaten under the name *native potatoes*; those of the New Zealand G. cunninghamii are eaten by the Maoris, who also make similar use of the *adder's mouth orchid* (Microtis porrifolia). The Indians of northwestern America eat the bulbs of the *Calypso orchid* (Calypso bulbosa); elsewhere the pseudobulbs of the *tree orchid* (Epidendron cochleatum) of Central and tropical South America and the West Indies are used to provide an edible mucilage like that of gumbo. The *Canyon fringe-orchid* (Habenaria sparciflora), another member of the orchis family growing in the southwestern U.S. and adjacent Mexican territories, is used as a food by the Indians in time of want.

Arums (genus Arum)

The arums belong to the large Araceae family, most of which are poisonous, although the poisonous properties can be dissipated in those species which are considered worth the trouble. Taro and its relations (*see under* Colocasia) belong to this family, as does the *sweet flag* (Acorus calamus), introduced from the U.S. into Britain, whose bitter root-stocks have been used in medicine both commercially and by country-folk. So also does the *dumb cane* (Caladium sequinum) of the West Indies and South America, which makes the tongue swell causing temporary dumbness.

There are also those curious plants whose seed-head is a tubular conglomeration of red berries; it looks vaguely like a large grape-hyacinth. The best known among these in Britain is the *wild arum* or *cuckoo pint* (A. maculatum) from whose poisonous root a starchy substance known as Portland sago or arrowroot can be obtained. The roots have first to be dried and boiled. I suspect that very little if any is produced now but the name *sago-plant* remains. According to Dioscorides the roots were once eaten boiled or roasted. The leaves were pickled or dried and then used as a pot-herb; they were also used for keeping cheese fresh. In Elizabethan times the starch from the roots was considered the best for the wide ruffs and collars then in fashion, and thus it received the name *starchwort*.

Rampe, which once had the sense of wantoning, refers to its supposed aphrodisiac powers. Other names include *Aaron*, a corruption of arum, *nightingales*, *wild lily*, *lily grass*, *pig lily*, and the wildly improbable *Kitty-come-down-the-lane-jump-up-and-kiss-me*.

But by far the greater number of its names, many of which it shares with some of the orchis family (*see under* Arrowroot), refer to the flower, the small berries of which range from dark red to paler and almost near white shades on the same plant. From this it gets such names as *white and red wild arum* and various fanciful pairings: *Adam and Eve, Bobbin and Joan, cows and calves, bulls and cows* or *wheys, stallions and mares, lords and ladies, kings and queens, parson and clerk, Devil's ladies and gentlemen* or *men and women* (the Devil's because they are poisonous), *devils and angels, soldiers and angels* or *sailors*. Their poisonous properties are advertised in the names *poison berry, adder berries,* and *snake's victuals*. Besides the name cuckoo pint the plant is also called *cuckoo babies, cock* or *flower,* and *great* or *small dragon*. Yet other references to the berries, some salacious, are *wake Robin, bobbin Joan, dog bobbins, parson's pillicods, hobble gobbles, fairies fingers* and *silly lovers*. Before the berries appear the flower has a different aspect: a yellow-white hood (known as a spathe) tinged with green encloses the fleshy purple-brown spike on which the berries will eventually grow. This gives rise to such imaginative names as *schoolmaster, priesties, priest's hood* or *pintle, dragon's tassel, parson-in-his-smock, babe-in-a-cradle, Jack-in-a-box,* and *parson, priest, preacher, man, daddy,* or *lamb-in-a-pulpit*.

The root of the *Italian arum* (A. italiacum) may be treated similarly to those of the wild arums in order to produce an edible starch, and the turnip-like root of the European *friar's cowl* (A. arisarum) can also be rendered palatable after long boiling. In America there is another plant very similar to the wild arum which the Indians once used as a food. This is the *Indian turnip* (Arisoema triphyllum) or *bog onion, brown dragon, Jack-in-the-pulpit,* also referred to as *starchwort*. Formerly of Asian origin, its acrid tuberous rhizome produces a starchy arrowroot which has medicinal properties after long boiling to dispel its poisons. The corms of the *Japanese dragonroot* (A. japonicum) are used by the Ainu, and the plant parts are baked in hot ashes.

The related *giant arum* or *devil's tongue* (Hydrosme or Amorphophallus rivieri), known to the Japanese as *koniaku*, is often cultivated for food on the mountainsides. Some of the tubercules, the larger of which look rather like turnips, are so small that they are left in the ground, thus ensuring a crop of new plants for the next year. Their taste and smell is strong and disagreeable, but this is lost after they have been soaked in whitewash (i.e. milk of lime), crushed and cooked. The resulting flour is used for making a kind of pasta and other dishes. In Indo-China the related A. hermandii is occasionally cultivated for its corms, and in India the *white-spotted giant arum* (A. campanulatus), whose tuber is known as *telinga potato*.

The *green arrow arum* (Peltandra virginica) is another American relative growing eastwards from the marshy land of the Missouri. This was once an important food to the Indians, whose native name gives us the alternative *Virginian tuckahoe*. The leaves and fruits are edible and the roots were eaten as a vegetable or dried and ground for bread. Likewise the roots of the *golden club* (Orontium aquaticum), and aquatic plant of eastern North America, after repeated boiling and washing. The seeds may also be ground for flour. Lastly the *bog* or *water arum* (Calla palustris), common in the bogs of northeastern America, Asia and Europe, whose roots may also be used for flour. This plant is known as *wild calla, water* or *female dragons* and *faverole*. It also goes by the name *arum lily* but must not be confused with the true calla or arum lily (genus Zantedeschia) of the florists, although the two are related.

Barley (*genus Hordeum*)

Barley is one of the five most important cereal crops in the world, its grains being used for human food, in malting and for livestock. It is also one of the most ancient in use, going back to the earliest evidences of agriculture. It was in use in Egypt in 5000 B.C.; 3500 B.C. in Mesopotamia; 3000 B.C. in northwestern Europe; and 2000 B.C. in China. For the Hebrews, Greeks and Romans (who represented the goddess Ceres with barley plaited into her crown) it was the chief bread-flour crop. In Europe it kept this paramount position until the sixteenth century. Barley was first brought to Britain in the Bronze Age (about 500 B.C.) and barley bread was still used everywhere in the Isle of Man right into the nineteenth century.

The origins of the plant are probably in South-East Asia and the Ethiopian highlands, where the grains are used for paying land-rent. The low gluten content rules it out from making a good rising loaf, but it is still much used in making porridges and unleavened bread in North Africa and parts of Asia. Pearl barley is excellent in soups. At one time the grain, or barley corn, was used as a unit of measure; there were three or four to the inch.

Barley's use for the making of various alcoholic drinks is well known. Indeed, beer was named after it; one of the plant's older names was *bere*. It is also used in the making of whisky, as a very old folk-ballad of which there are many versions, testifies. It was this which was adapted by Burns into his poem 'John Barleycorn':

There were three kings into the east,
Three kings both great and high,
An' they hae sworn a solemn oath
John Barleycorn should die.

They took a plough and plough'd him down,
Puts clods upon his head.
And they hae sworn a solemn oath
John Barleycorn was dead.

But the cheerful spring came kindly on,
And the showers began to fall;
John Barleycorn got up again
And sore surpris'd them all.

The sultry suns of summer came
And he grew thick and strong,
His head weel arm'd wi' pointed spears
That no one should him wrong.

The sober autumn enter'd mild
When he grew wan and pale.
His bending joints and drooping head
Show'd he began to fail.

His colour sicken'd more and more.
He faded into age;
And then his enemies began
To show their deadly rage.

They've ta'en a weapon long and sharp
And cut him by the knee,
They tied him fast upon a cart
Like a rogue for forgerie.

They laid him down upon his back
And cudgell'd him full sore;
They hung him up before the storm
And turn'd him o'er and o'er.

They filled up a darksome pit
With water to the brim;
They heaved in John Barleycorn,
There let him sink or swim.

They laid him out upon the floor
To work him further woe,
And still as signs of life appear'd
They toss'd him to and fro.

They wasted o'er a scorching flame
The marrow of his bones,
But a miller used him worst of all
For he crushed him between two stones.

> *And they hae ta'en his very heart's blood*
> *And drank it round and round,*
> *And still the more and more they drank*
> *Their joy did more abound.*

Nevertheless, as a proverb says, 'Sir John Barleycorn is the strongest knight'. A Scottish proverb states that 'it is ill prizing green barley', a typically cautious approach meaning much the same as 'don't count your chickens before they're hatched'. Another proverb is said spitefully of those who take after their parents in bad qualities: 'He gave no green barley for it'. The English 'as long in coming as Cotswold barley' is rather more charitable, having the meaning of 'slow but sure' since barley matures more slowly in colder uplands but is the most forward developer after this.

Several country proverbs contrast barley with wheat. Thus 'barley makes the heap but wheat the cheap', dating back to at least the seventeenth century, notes that a good wheat-crop brings down the price of all other cereals. The grains flourish under different weather conditions: 'Sow barley in dree [dry weather] and wheat in pul [wet]'; 'good elm, good barley, but good oak, good wheat', a kind of weather indication of the same order as

> *If the oak be out afore the ash*
> *We're in for a splash,*
> *And if the ash be out afore the oak*
> *We're in for a soak,*

and

> *Bad for barley and good for the corn*
> *When the cuckoo comes to an empty thorn.*

On the other hand they say in Somerset that

> *Wait [wheat] and barley'll strut in June*
> *'Nif they baint no higher'n a spoon.*

Many other proverbial country rhymes deal with the preparation for and planting of the crop:

> *Dry your barley land in October*
> *Or you'll always be sober.*

> *Upon St David's Day [March 1]*
> *Put oats and barley in the clay.*

> *When Westridge wood is motly*
> *Then 'tis time to sow barley.*

> *When the sloetree is as white as a sheet*
> *Sow your barley whether it be dry or wet.*

*When the oak puts on his gosling gray
'Tis time to sow barley night and day.*

*When the elmen leaf is as big as a mouse's ear
Then to sow barley never fear;
When the elmen leaf is as big as an ox's eye
Then say I, Hie boys, hie!*

The several varieties of barley are usually divided according to the number of grains along the ear: the two-rowed *coffee* or *Peruvian barley* (H. distichon); four-rowed *spring* or *common barley* (H. vulgare); and the six-rowed variety (H. hexastichon) which was probably the most popular in ancient times. There is also *fan, spratt* or *brattledore barley* (H. zeocriton), also known as *fulham* or *Putney barley*, popular in Germany and sometimes known as *German rice*; and *beardless barley* (H. trifurcatum) alternatively known as *Egyptian barley*, grown in North Africa and Asia Minor.

Our name derives from the Old English *baerlic*; two older names which have become all but obsolete since the nineteenth century are *bere* and *bigg*. The latter was mostly used in the north, deriving from the Old Norse *bygg* of the Danish invaders; *byug* is the modern Swedish. It is thought that the Old English name is connected with *beow*, the verb to grow, ultimately connected with the Primitive Germanic *bheu*, to grow or be, and the Greek φυειν and Sanskrit *bhu*, to be.

Brazil nut and relations (*Bertholettia excelsa*)

Most of the world supply of this nut comes from the wild trees of Brazil, particularly along the Amazon basin where it is an important food of the Indians, who also use it for oil. Attempts to grow it elsewhere, notably in Florida in the U.S., have failed on the whole. The nuts are enclosed in a large woody fruit the size of a man's head, which generally contains about two dozen of them. Collectors of these and similar nuts have to wear head-padding because of the danger of the enormous fruits falling and stunning them. The nuts were first introduced to Europe by Dutch traders in 1633. Other names includes *castanea* or *castana, paranut, creamnut, butternut* and *niggertoe*, the last alluding to its dark wrinkled toe-like appearance.

The fruits which contain the Brazil nut are sometimes known as *monkey pots* since monkeys wait for the rats to gnaw them open and then thrust in their hands after the nuts, but are unable to withdraw them. The real *monkey pot trees* are relatives (genus Lecythis) whose fruits have removable plug-like tops, and in which are to be found similar nuts. Monkeys are said to thrust their heads into these and become trapped. The most used is the *sapucaia* or *paradise nut* (L. usitata or zabucajo), whose nuts taste like almonds and are more wholesome than brazils. The flesh from the fruit may also be cooked as a vegetable. This and the Brazilian *sabucoja branca* (L. lanceolata) have been introduced into tropical Africa, the latter being grown in Madagascar. The nuts are not much used commercially, however. Other tropical American species include the *South American monkey pot* (L. ollaria) and the *Trinidad monkey pot* (L. laevifolia).

Another relative is the *cannonball tree* (L. or Couroupita guianensis) whose woody fruits are known as *monkey* or *wild apricots* and are as big as a child's head. The pulp is pleasantly acid but smells horrible as it putrifies, and the outer casing is used in tropical America as the bottle-gourd and others (q.v.) are used in tropical Africa. The edible seeds are sometimes known as *Andos almonds*.

Souari (or swarri) *nuts* (genus Caryoca) closely resemble brazils but are larger and richer tasting. These share similar names with the foregoing, such as *butter*, *paradise* and *Guiana nuts*. The thick fruit-shells resemble rusty cannonballs and weigh up to 25 pounds; they are almost impossible to crack. Inside they contain four kidney-shaped nuts the size of an egg. Because of the bulk of the trees they are very difficult to collect. Those most favoured are the *Guiana butter-nut* (C. tormentosum) and the *souwarro* (C. nuciferum), whose fruits are also known as *pekea nut* and *Peruvian almond*.

Buckwheat (*genus Fagopyrum*)

There are about eight species of this family, all native to the temperate climates of Asia, and probably of Manchurian origin. Not all, however, are used for food. The most important is *brank buckwheat* (F. esculentum) chiefly grown now in Central Asia and parts of India and Persia. The seeds are eaten by the poorer classes, and by Hindus on feast-days, but are generally regarded as bitter and heating. The first Chinese mention of its cultivation is in the tenth to eleventh centuries. It was brought to Europe via Russia by the Tartars and is first mentioned as cultivated in Germany in 1436. During the sixteenth century it spread over Europe and achieved some importance on poorer soils, as in Brittany, for example. At that time it was generally known as *Saracen wheat*, a name that remains in the French (*blé*) *sarrasin*. The English name derives from the German *buchweizen* (beechwheat), alluding to a fancied resemblance between the triangular seed and that of the beechnut. The Italians make the same allusion in their name *faggina* (*faggio* is a beechtree and *faggiola* beechmast). *Beechwheat* was a name by which this plant was known in some areas of Britain; it was also called *crap* or *crop, French wheat* and *bullimong*, the last word being an Essex term for a mixture of buckwheat, oats, pease and vetches.

Tartary buckwheat (F. tartaricum) grows wild in Siberia and is cultivated in the Himalayas and the U.S.S.R. It is less sensitive to the cold but yields an inferior seed. It entered Eastern Europe later than brank and bears such names as *tattar, tatarka, tatrika*, in various Slavic languages. *Notch-seeded buckwheat* (F. emarginatum) is another species which is grown in the highlands of northeastern India and China.

Cashew (*Anacardium occidentale*)

This tropical American tree, whose relatives include the mango and pistachio, is now naturalised in all warm countries. In fact it was the first fruit tree from America to be grown in Asia after the Portuguese discovered it in the late fifteenth century and took it to East Africa and India. It is now in cultivation in all these areas and the nut it produces is much exported. In very warm regions it fruits all the year round, but bears its heaviest crops after a dry period.

The nuts' receptacle is surrounded by a watery fruit, known as an 'apple', which has an attractive smell and is popular with birds, fruit-bats, climbing creatures and children. However, it often causes tickling in the throat and one has to be careful about extracting the inner receptacle in which the nuts are found. There is a nasty twist about getting at these, as there is about many things to be found growing in South America, particularly the Amazon area. The plant is also a relative of poison sumac and poison ivy; similarly toxic is the brown oil to be found beneath the receptacle. This has first to be burned away, and if the smoke so much as comes in contact with the eyes or the mucous membrane it can cause extreme pain and sores. Having extracted the nuts they must then be boiled or roasted (lime is also said to help neutralise the poison) before they are safely edible. They have, however, a fine taste whether eaten 'raw' or cooked. An alcoholic drink (known as *kaju*) is made from the outer apple in Mozambique, and in the Malaysian region the young leaves are eaten cooked with rice.

The name is derived from the Portugese *acaju*, taken from the original Tupian name. In Brazil it is *cajus*, and *gajus* in Malaya. There the name has been corrupted by contact with a local word for fruit, *jambu*, into *janggus*, *janggar* and *kanjus*. Other names include *heart-nut*, *mahoganny apple* and *bean of Malacca*.

Cassava (*Manihot utilissima or esculenta*)

Cassava is the mealy flour extracted from the starchy nutritious root of the *bitter cassava* or *tapioca plant*, also known as *Brazilian arrowroot*. Since the tubers do not keep for very long they have to be left in the earth until required or sliced and dried in the sun. Parboiling the 'chips' helps preserve them for several months. Since the roots contain hydrocyanic acid they have to be treated with heat before they are safe to eat. The flour is made by grinding the chips. In South America a coarse meal (*farinha*) is prepared by grating the tuber and then squeezing it in order to express its juices and toasting the compressed pulp over a low fire. The expressed latex and juices may be concentrated by boiling into *casureep*, an ingredient of a West Indian dish known as pepper-pot. Tapioca is obtained by gently heating the washed and cleaned starch extract from the roots on hot iron plates, partly cooking so that it agglutinises into small round pellets. The West African *fufu* paste, which is obtainable dried from some shops, is made by peeling and cutting up the already boiled roots and pounding them in a mortar. In Africa, also, the leaves are used as a pot-herb and in the making of beer.

Cassava was cultivated in tropical America long before the arrival of Europeans, approximately 4000 years ago in Peru and 2000 years ago in Mexico. By the Europeans it was taken to West Africa, whence it spread over the tropical parts of the continent. Today Africa produces more than all the rest of the world; it is replacing yam and sweet potato in many places. The plant was also taken to Asia, where it is used less. The name derives via French and Spanish from Taino, the extinct language of the Haitian Indians, who called it *casavi* and *cuzavi*.

Other members of the spurge family, to which cassava belongs, are used for much the same purpose. In particular the roots of *sweet cassava* (M. palmata aipi) whose non-poisonous roots may be eaten like potatoes. The meal derived from them is known as *manioc*, after the Tupian *mandioca* of the Amazon delta (cf. the Guarani *mandio*). The leaves and young shoots of this plant are highly nutritious as a cooked vegetable. Various South American Indian tribes use the root for the making of a sacred beer. The Jibaro Indians of Eastern Ecquador, for instance, make it for warriors to drink when celebrating the taking of an enemy's head. The roots are

15. NUTS

cashew
Moreton Bay chestnut
Moreton Bay chestnut
pine nut   cashew

Queensland
Brazil nut
Brazil nut
Queensland nut

chewed, mixed well with saliva, and then spat into a dish and left to ferment. This serves as the basis for the beer.

Chestnuts

1. Sweet chestnuts *(genus Castanea)*

The sweet chestnut grows wild in the northern temperate zone, especially in Europe and Asia. In the U.S. the trees were at one time very numerous, but the greater part were killed off by disease during the first third of this century. The tree can grow to a great age. One at the foot of Mt Etna was estimated to be about 2000 years old before it was killed by an erruption. The name comes into English from the Old French *chastaigne*, deriving ultimately from the Latin and Greek. In Middle English the names were *chesten* and *chastaine*.

At one time the chestnut was a staple for many of the peasant class in Europe, and still is for some people in Turkey and elsewhere. In Italy it is popular as the basis of peasant dishes such as *polenta*. On the other hand there is a cautionary tale to be found in the fourteenth-century *Tsurezure-gusa* (Essays in Idle Hours) by the Japanese monk Kenko:

> The daughter of a certain man from the province of Inaba who had become a monk was renowned for her beauty and much sought after by suitors: but since the lady would eat nothing but chestnuts, refusing rice or any other grain, her father would not consent to her marrying, arguing that so strange a person was not destined for it.

Perhaps the father was partly motivated by the fact that in Japan the chestnut suggests haughtiness, although in the same country the dried meal signifies success. To dream of eating the nuts however, means a difficult business situation is in the offing; to dream of cooking them signifies exploitation. In the language of flowers the blooms are an appeal for justice. Elsewhere they symbolise luxury, and chastity in the Christian tradition. The nuts are used in a charm to keep off rheumatism. Perhaps it is because of their peasant associations that the French idiom *travailler pour des marrons* (meaning to work hard for little reward) seems to hold them in so little honour.

The best known tree is the *wild chestnut* (C. sativa), which is very common in southern Europe, and also spreads into South-West Asia and North Africa. The Japanese C. crenata, Chinese C. mollissima and American C. dentata chestnuts have smaller fruits, which are ground down and used for meal by the local inhabitants. In America the *dwarf chestnut* (C. pumila) is also to be found. The fruits of this tree are sometimes called *chinquapins*, a name deriving from one of the Indian languages, but the name is more properly applied to the edible fruits of related trees (genus Castanopsis) such as the *golden* or *Western chinquapin* (C. chrysophylla) of the American north-west.

Several other edible species of chinquapin are to be found in Asia The nuts of C boisii are sold in the markets of Indo-China, as are those of the *Philippine chinquapin* (C. philippensis) in the Philippines, and the *Sumatra chinquapin* (C. sumatrana) in Malyasia and Indonesia. The nuts may be boiled, roasted, or parched and used for meal. The peasants of China and Tibet make a pasta from the nuts of the *Tibetan chinquapin* (C. tibetiana), and from C. solerophylla which grows in eastern and central China. In Japan the nuts of the *evergreen chinquapin* (C. cusipidata) are usually roasted.

2. Horse chestnuts (*genus Aesculus*)

The *horse chestnut* (A. hippocastanum) was first brought to England from Constantinople in the sixteenth century, but the origin of the tree itself is doubtful. Guesses range from Greece to South-West Asia. It is no relation to any of the other chestnuts, and its fruits are bitter-tasting. This is due to the tannin they contain. It may, however, be dispersed by soaking the ground meal in running water, after which it may be mixed with wheat or rye flour and used for bread making, as once often happened in times of shortage. Alternatively they may be boiled in large quantities of water, drained and boiled again.

Its relatives, however, especially the American and Far Eastern species, are more palatable and are usually ground into bread-flour. These include the *Japanese horse chestnut* (A. turbinata), the *Californian horse chestnut* (A. californica), the *yellow, big* or *sweet buckeye* (A. octandra) and the *bottlebrush buckeye* (A. parviflora). The *Ohio horse chestnut* (A. glabra) is said to have been used although the fruits are suspected of being poisonous.

The name *buckeye* is suggested by the white markings on the top of the rich brown-coloured nuts. A use has, of course, been found for the English nuts by children, who bore through them, thread them on strings and play 'conkers' with them, seeing who can split his opponent's nut first. The fruits have, therefore been called *conquerors* (*see* Plantains for a description of a similar game named thus) and *oblionkers*. The latter stems from a little chant made while playing one's first shot: 'Oblionker, my fust conker'. Another game was to try and persuade one's companion to allow one to slash them across the knuckles with a leaf-stem from the tree, thus giving rise to the name *knuckle-bleeders* for this part. The white burgeoning flowers are known as *lambs*, and an old name for the tree was *bongay*.

3. Water chestnuts (*genus Trapa*)

The nut-like fruits of this water-plant are edible and chestnut flavoured. Those of the European *water chestnut* (T. natans) were said to have been a staple of the ancient Thracians. Like chestnuts, they may be eaten raw, boiled, or cooked in ashes. A name of Venetian origin for them is *Jesuit's nut*. Another, *water caltrop*, refers back to the name of an ancient snare for the feet. Other species are to be found in Asia, such as the *sighara nut* (T. bicornis), whose name derives from the Hindi. The two-horned shape of the fruit is referred to in the Latin name. It grows in still waters all over China and is reputedly able to keep them fresh; it often supports people where the rice crop is insufficient to do so. The rather similar T. bispinosa, of mid-Asian origin, grows as far south as Ceylon and southern Africa, and as far east as Japan. It spreads fast and gives an abundant crop. The fruits are often converted into a kind of flour but may also be boiled.

4. Others

Several other fruits go by the name of chestnuts. These include the *Fiji* or *Tahiti chestnut* (Inocarpus edulis) of the Pacific whose fruits are large and fleshy and are usually boiled or roasted, but are not considered suitable for weak stomachs. Another Australian tree which bears large pods with chestnut-like seeds is the *Moreton Bay chestnut* (Castanospermum australe) or *Australian bean tree*, mostly limited to eastern Australia and the New Hebrides. The raw 'nuts' are indigestible and harmful. First they must be steeped in water for 8–10 days; the natives who gather them dry them in the sun, roast them over a hot stone and pound them into meal.

Sea chestnuts are the bean-like fruit of a plant known in Queensland as the *match-box bean* (Entada phaseoloides) which also goes by such West Indian names as *cacoon, coccoon, scimitar pod plant* and *filbert tree*. It is in reality a vine frequently to be found climbing over the trees in the Australian mangrove swamps and elsewhere in tropical Asia, the Pacific and America. Its curved pods are 3–4 feet long and are 3–4 inches wide; they are constricted between the seeds, which are hard, wide, woody and highly polished. They have been used for making fancy matchboxes, snuff boxes, scent bottles and other bric-a-brac. They and their stems may also be crushed and used as a soap substitute. The stems, are water conservers; in times of drought one can cut them and drink the liquid, which is palaable and free from harmful substances. But the aborigines and some peoples in Indonesia have first to rid the bitter tasting seeds of their poisonous principle before they can use them for food, usually by washing, cooking and grinding them into flour.

The *China chestnut* (Sterculia monosperma, formerly nobilis), also known in Cantonese as *pheng phok* (alternatively *phang phor*), grows from China into South-East Asia and is eaten by the Chinese there. It bears 1–4 pods at the end of a stalk; these are green and ripen to scarlet, eventually splitting down one side to reveal glossy black chestnut-sized seeds. They are eaten after boiling and removing the three outer skins, and have a mealy consistency and taste like the European chestnut.

Groundnuts

Best known of these is the *monkey* or *peanut*, the fruit of a Brazilian herb, Arachis hypogaea, which pushes its nuts into the ground, where they ripen. Seeds have been found in old Peruvian tombs, thus testifying to their long use among the original inhabitants of the Americas. The plant was carried to Africa by the slave-traders, who used the nuts to feed their cargo on the outward trip. It now grows in all tropical countries. The shoots and leaves are also edible. Oil is obtained from the seeds and a kind of salty dry butter from the nuts, roasted, ground into flour and moistened. Other names for them include *goobers* (in the U.S.), *katjang* (the Malay and Javan for a bean), *pindars, mandubi, underground kidney beans, earth almond, Manilla* and *grass nut*.

The *Congo, Madagascar* or *Bambarra groundnut* (Voandzeia subterranea), also known as *underground bean,* is a cultivated plant which Negro slaves carried with them to America. Its nuts are a common article of food in tropical Africa, Madagascar and some parts of Asia, and may be eaten roasted or boiled. In Brazil it is called *Angola mandubi*; in Madagascar, *Malagasy pistachio*. In Natal it is known as *hluba-bean* from the native name *iguihluba*. The similar *Kersting's groundnut* (Kerstingiella geocarpe) is grown in Togo, where it is called *kandela*. Lastly there is the leguminous *wild bean vine* (Amphicarpa monoica), also known as *hog* or *ground peanut*, whose seeds are eaten by the Indians of the U.S.

Hazel nut   (*genus Corylus*)

Hazel nuts were once much used by ancient peoples, including the North American Indians, and are very popular now among certain vegetarians. The tree grows widely in the north temperate zone and is the subject of much curious lore. It is the emblem of justice, reconciliation and love, although a girl who gives the nuts to a lover does so as a sign of discouragement. This is contradictory, since the hazel is associated with fertility and erotic fulfilment. To court under a hazel tree was the best opportunity to have one's love returned, even by those who had shown no love elsewhere. As an English proverb has it, 'the more hazels, the more bastard children'. Idioms like the German '*in die Haseln gehen*' and the French '*aller aux noisettes*' mean 'to make love'. Sterile women were beaten with hazel twigs to make them fertile and the nuts were traditionally given to a bridal couple.

In southern Europe it seems to have been customary for unmarried girls to dance under the trees in order to attract lovers. There is a song put in their mouths while doing this by the thirteenth-century Portuguese poet Joan Zorro:

Let us dance now, come, O fair ones,
Under these flowering hazel trees –
And whoever is fair as we are fair,
   If she falls in love,
Under these flowering hazel trees
   She'll dance with us.

Let us dance now, come, O prized ones,
Under these fruit-laiden hazel trees –
And whoever is prized, etc.

Yeats refers to a Celtic tradition connected with a hazel wand in his 'Song of the Wandering Aengus'. The singer cuts and peels the wand one evening and uses it to fish in a stream. The fish he catches changes into a shining girl 'with apple-blossom in her hair', calls him by name and then fades from sight, since when he has been hunting for her. Burns notes a divinatory use for the nuts in his poem 'Halloween': 'Burning the nuts is a famous charm. Young people name the lad and lass to each particular nut as they lay them on the fire and, accordingly as they burn quietly together or start from beside one another, the course and issue of the courtship will be'.

In medieval times hazel twigs were used in courts of law to discover thieves and murderers, and in Prussia the clothing of a man suspected as a thief was beaten with them. If he was guilty he then fell ill. A breast-band of hazel was often worn by horses to ward off evil spirits and the mariner who put them in his cap was supposed to be able to weather any storm. It was these twigs which were used by diviners for the discovery of water and (especially) hidden treasure. They were also credited with making their bearer invisible. Some held, however, that only those cut on St John's Eve or Night (Midsummer) had magic power.

By the early Irish the tree was especially revered and the penalty for anyone caught cutting one down was death. It governed the ninth Celtic month (6 Aug. to 2 Sept.), and its nuts were a symbol of concentrated wisdom. The nine hazels of poetic art might well be an equivalent of the Greek Muses. But, like the Tree of Knowledge in Eden, the hazel also had its sinister side; there is a good and an evil knowledge. It is therefore pictured as dripping poisonous milk, a roost for vultures and ravens. It has this aspect in Norse myth as the tree of Thor. The Anglo-Saxons would seem to have had similar sorts of belief, for their name (*hasel*) derives from the verb *haten*, to call, order or name, in allusion to the use of a hazel rod by magicians, masters and herdsmen. Names like *hall nut* and its Cornish varient *hale nut* (cf. the verb 'to hail') would seem to be connected with this.

At one time the hazel was the commonest nut-tree in Britain (perhaps the only one aside from trees like the beech) and its fruit was known simply as 'the nut'. Thus it appears in the proverb 'many nits, many pits' (many hazels, many graves). Another proverb informs us 'if it rains not on St John's Day [24 June] hazel-nuts will prosper'.

In the fragment entitled 'Nutting' Wordsworth describes vividly a childhood expedition gathering wild nuts.

> *... It seems a day*
> *(I speak of one from many singled out)*
> *One of those heavenly days that cannot die;*
> *When, in the eagerness of boyish hope,*
> *I left our cottage threshold, sallying forth*
> *With a huge wallet o'er my shoulder slung,*
> *A nutting-crook in hand; and turned my steps*
> *Tow'rd some far distant wood, a figure quaint,*
> *Tricked out in proud disguise of cast-off weeds*
> *Which for that service had been husbanded,*
> *By exhortation of my frugal dame –*
> *Motley accoutrement, of power to smile*
> *At thorns, and brakes, and brambles – and in truth*
> *More ragged than need was! O'er pathless rocks,*
> *Through beds of matted fern and tangled thickets,*
> *Forcing my way, I came to one dear nook*
> *Unvisited, where not a broken bough*
> *Drooped with its withered leaves, ungracious sign*
> *Of devastation; but the hazels rose*
> *Tall and erect, with tempting clusters hung,*
> *A virgin scene. – A little while I stood,*
> *Breathing with such suppression of the heart*
> *As joy delights in; and with wise restraint*
> *Voluptuous, fearless of a rival, eyed*
> *The banquet ... Then up I rose,*
> *And dragged to earth both branch and bough, with crash*
> *And merciless ravage; and the shady nook*
> *Of hazels, and the green and mossy bower,*
> *Deformed and sullied, patiently gave up*
> *Their quiet beings; and unless I now*
> *Confound my present feelings with the past,*

Ere from the multilated bower I turned
Exulting, rich beyond the wealth of kings,
I felt a sense of pain when I beheld
The silent trees, and saw the intruding sky.

A variety of our wild tree, C. avellana var. grandis, is cultivated for the nuts, commonly known as *cobs* or *filberts*, of which the Kentish cob is the most favoured in England. Turkey, Spain and Italy are also leading cultivators. In the U.S. breeding for nut-production has been going forward since 1919. The following are a selection of trees, growing in various parts of the world, with edible nuts:

Giant Hazel (C. maxima), Europe.
Constantinople or *Turkish Hazel* (C. colurna), Turkey, southern U.S.S.R., Himalayas.
Himalaya Hazel (C. ferox), Himalayas.
Siberian Hazel (C. heterophylla), Siberia, China, Japan.
Chinese Hazel (C. chinensis), Central and western China.
Japanese Hazel (C. sieboldiana), Japan.
Californian Hazel (C. californica), western U.S.
Beaked Hazel (C. cornuta), northeastern America.
American Wild Hazel (C. americana), northeastern America; the cultivated *Mildred Hazel* (var. avellana) is grown all over the U.S.

There are a variety of country and dialect names applied to the nut. *Cob* comes from a dialect word meaning a large man and referred to as the master nut in games played by children. One game resembles the game of 'conkers', another that of marbles. *Filbert*, with such variations as *fillnut* and *fill* or *full-beard*, is a corruption of *Philibert nut*, referring to the king of Normandy whose saint-day falls when the nuts are ripening. But the beard connection refers to the hairy case in which the nut is to be found and may be compared with the German *bartnusz* (bearded nut). In Scotland they call two nuts growing in one husk a *St John's nut* and once regarded it as a protection against witchcraft and a cure for toothache; a triple nut is called *St Mary's nut*. Here, too, the ripe nuts are known as *leemers*, from the Norwegian dialect word *lemo* or *lima*, to dismember; this may be compared with our *crack-nut*. Others names are *wood-nut*, *hedge nut*, and *cobble-de-cut nut*, a variation upon *cob*.

Lichens

Lichens have been used for making bread intermittently through the centuries and in many different areas, especially those lichens which break down into sugary substances. The most famous is the *manna lichen*

(Lecanora esculenta) of Africa and Western Asia, still used by Tartars and other desert tribes. It is easily torn up by the wind and, after being carried some distance, settles in a globular 'manna-rain'. It has been surmised that this was the manna that the Israelites were 'miraculously' provided with during their wilderness journey for the making of bread:

> ... and in the morning the dew lay round about the host. And when the dew that lay was gone up, behold upon the face of the wilderness there lay a small round thing, as small as the hoar frost on the ground ... And Moses said unto them, This is the bread which the Lord hath given you to eat.
>
> (Exodus 16:13–15.)

And the manna was as coriander seed, and the colour thereof of bdellium.[1] And the people went about, and gathered it, and ground it in mills, or beat it in a mortar, and baked it in pans, and made cakes of it: and the taste of it was as the taste of fresh oil.

(Numbers 11:7–8)

According to the Exodus account the taste was like wafers made with honey, and later tradition had it that its taste was like the gatherer's favourite food. Certainly this substance fits the given facts better than any of the other mannas (such as the exudations referred to as bdellium). There are also some European species of this lichen, but they are used for violet dyes.

In the land of Egypt, from whence the Israelites were fleeing, baskets of the lichen *stag-horn oak moss* (Evernia prunastri) were placed in the royal tombs. No one is really sure why, but it has been suggested that it was for making bread, since it contains an edible farinaceous substance. The very similar *European oak-moss* (E. furfuracea) is to be found climbing over trees and wooden artifacts in mountainous districts over much of the northern hemisphere. This has been used for making perfume since the sixteenth century.

Horsehair lichen (Alectoria jujuba or jubata) is related to the above and is to be found in Europe and North America growing on the twigs of trees and on rocks. This is boiled with camass root (*see under* Lilies) or fermented and baked by the Indians of the Pacific coast of America. Another species (A. fremontii) is used by them as a famine food. The Wisconsin Indians add *dark crottle* (Parmelia physodes), otherwise known as *puffed shield lichen*, to their soups. It is to be found growing in the sun throughout the temperate zone and is blown about by the wind in much the same way as manna lichen. Yet another lichen, Sticta glomulifera, is cooked by other Indian tribes.

---

[1] Possibly the gummy exudation of various desert bushes and trees, also known as manna.

In Arctic North America two bitter lichens which go collectively under the name *tripe-de-roche* or *rock-tripe* (genus Umbilicaria and genus Gyrophora) are edible and have been occasionally used. An Eastern relative of the latter, G. escuilenta, known by the Japanese as *iwa-take* (rock mushroom) is greatly esteemed in Japan and China but, because it is difficult to collect off rocks and cliffs, is highly priced. Another lichen of the Arctic north is known as *Iceland moss* (Cetraria islandica) and is gathered commercially in Scandinavian countries. It is mostly used as cattle-feed but, especially in Iceland, it is also used for the making of bread.

Maize (Zea mays)

Anyone who has ever eaten cornflakes for breakfast has had contact with a product of this plant of American origin, also known as *Indian* or *sweet corn*. It was cultivated in Mexico 2,500 years ago and primitive forms of it have been found in Peru. At one time it used to play a great part in the native American religions. Sacrifices of maize-bread were offered by the Virgins of the Sun at Cuzco. The old Mexican name for maize (*cintli*) is incorporated into the name of Cintleutl, the Mexican goddess equivalent to Ceres. Grains of maize have also been discovered in primitive burial mounds. Columbus found it growing in Cuba during his first voyage; later it was discovered in cultivation from Canada to Chile. Columbus brought it back to Europe and it was taken to Africa by the Portuguese, and from thence into Asia. Now Mexicans make *tortillas* and Indians *chapaties* out of flour from the same plant. In China there is a picture of it dated 1573, thus illustrating that it had toured the earth in less than a century.

There is a fascinating North American Indian legend which describes the beginning of maize-growing which runs thus:

> A long time ago, when Indians were first made, there lived one alone far from any others ... [who] grew tired of digging roots, lost his appetite, and for many days lay dreaming in the sunshine. When he awoke he saw something standing near him, of which he was very frightened at first. But when it spoke his heart was glad, for it was a beautiful woman with long light hair very unlike any Indian. He asked her to come to him but she could not ... and at last told him that if he would do just as she said he would always have her with him. She lead him to where there was some very dry grass, told him to get two very dry sticks, rub them together quickly and hold them in the grass. Soon a spark flew out and

the grass caught it ... Then she said "When the sun sets, take me by the hair and drag me over the burned ground" ... [Then] she told him that wherever he dragged her something like grass would spring up and he would see her hair coming from between the leaves; then the seeds would be ready for his use ... and to this day, when they see the silk on the corn stalk, the Indians know that she has not forgotten them'.

Another story as to its origin appears in the Hiawatha legend. The hero was undergoing a seven-day fast, customary before initiation into full manhood during which the young warrior concentrates on the object he desires most. Hiawatha broods upon the needs of his people, especially as regards their lack of food during the winter. On the fourth day he is accosted by the green-clad, yellow-plumed Mondamin, who informs him that he is impressed by this selfless wish and promises that Hiawatha's prayers will be answered on condition that they wrestle together for three evenings in succession and that Hiawatha triumphs over him. (One is reminded of Jacob's wrestling with an angel at Bethel.) Weakened by fasting though he is, the hero agrees and subsequently resists all temptation to eat in order to sustain himself, finally killing the strange warrior. Hiawatha then buries Mondamin and watches over his grave until the strange new plant, maize, springs from it. This, of course, is the answer to his people's winter plight. Later on his wife Minihaha, practises what must have been an ancient rite, going out at night with only her hair as covering to draw a magic circle about the fields and bless the grain. And Hiawatha, in order to scare off marauding ravens, snares one alive and ties it by its foot to the pole of his wigwam as a warning.

Maize is now a staple food for some people in Pakistan, but is, in fact, inferior to other cereals as a human food. In the U.S. 45 per cent of present production goes into the feeding of pigs, and a further 45 per cent is fed to other animals. The grains can be eaten directly off the cob, or cooked in butter. The name 'maize' is derived from the Spanish *maiz*, in its turn derived from the Cuban Indian *maisi* or *majisi* (cf. the Guyanan Arawak *marisi* and the Carib *marichi*). A South African name is *mealie, milje* in Afrikaans, deriving from the Portuguese *milho* (millet). A whole variety of European names rather confused the experts as to maize's actual continent of origin when they first came to look into such things: *Turkish wheat* (France, Germany); *Roman corn* (Alsace-Lorraine); *Sicilian corn* (Tuscany); *Indian corn* (Sicily, Germany); *Barbary* or *Guinea corn* (Provence); *Spanish corn* (Pyrennees); *Egyptian corn* (Turkey); *Syrian dourra* (Egypt).

Millet, sorghum and other grasses

A surprisingly large number of different plants seem to go under the name of millet, of which the chief is the so-called *French, Indian* or *white millet* (Panicum miliaceum), also known as *hog* or *broom corn millet*. In addition it goes by such names as the English *hirse*, and *proso*, an old Slavic name persisting in Russian and Polish. The name millet is a diminutive form of the Old English *mil* deriving from the Latin *milium*.

Millet seems to have originated in Egypt and Arabia, spreading from there to South Europe and Asia, in both which areas it was cultivated in prehistoric times. Remains of the seed have been found in the Stone-Age lake-dwellings of Savoy and Switzerland. Various other species of the witch grass family, to which it belongs, are also cultivated, such as *kutki* or *little millet* (P. miliare), and *brown-top millet* (P. fasciculatum). In North Africa, especially in the Sahara region, P. turgidum is used by various tribes, while in southwestern U.S. and Mexico the Hopi Indians mix the seeds of the *vine mesquite* (P. obtusum) with corn for grinding.

Botanists have been hard at work redefining the families of the millets, so that several once regarded as forming part of genus Panicum are now isolated from it. Among these is *pearl millet* (Pennisetum typhoideum, formerly P. spicatum) which rejoices in such alternative appellations as *pull-paddy* and *bulrush, spiked* or *cat-tail millet*. It is extensively cultivated in India, where it is known as *bajra*, and is also grown in Australia. The Indian *koda millet* (Paspalum scrobiculatum), called *ditch-millet* in New Zealand, yields a grain of inferior quality with traces of poison.

Another grain now fitted out with a new family name is *barnyard millet* (Echinochloa crus-galli var. frumentaceum), alternatively *Japanese millet, Deccan* or *cockspur grass*. This is grown in the warm dry parts of the East, especially in some area of Bengal, where it is known as *sanwa* and *kheri*, and in Japan and Australia. *Awnless barnyard grass* (E. colonum) grows in the same area and is also known as *sanwa*, or alternatively as *jungle* or *millet rice*. The poor also use the seeds of the *hedgehog grass* (E. stagnina), which is known as *burgu*, and in tropical Africa the natives cultivate the nutritious *Guinea grass* (E. maximum). The *Australian* or *native millet* used by the aborigines is known as *umbrella grass* (E. decompositum).

Finger millet (Eleusine coracana) is of South Asian origin, now to be found further east in China and Japan, and also in Egypt and Ethiopia; some grows in southern Europe, but there it is of little use for food. It goes under a variety of Asian names including *natchnee, raggee* (deriving from the Hindi *ragi*, Sanskrit *rajika*), and *coracan* (from the Sinhalese *kourakhan*).

290 / Millet, sorghum and other grasses

The grain of other inferior species is gathered by the poor in south India; among these is *dog's tail grass* (E. indica), *wire*, *crab* or *goose grass*, which is also to be found in the U.S. The young seedlings of this species are eaten with rice in Indonesia. In North Africa the seeds of E. aegiptica, of Egyptian origin, are used in various countries, while in some parts of Ethiopia those of E. tocussa are used.

Some members of the *crab-grass* family (genus Digitaria) also go by the name *finger millet*. Among these are the grasses known as *fonio* or *fundi* (D. exilis and D. iburna) to the Hausas of Northern Nigeria and elsewhere, for whom they serve as a cereal crop. *Red finger grass* (D. or Syntherisma sanguinalis) is a troublesome weed of European origin which has been introduced into North America. The seeds of this were once used in porridges in Bohemia. *Teff* (Eragrostis abyssinica), a relation belonging to the lovegrass family, is an important food-plant for the poor in Ethiopia, who use its grain, especially in porridges.

Lastly there is the misnamed *Italian millet* (Setaria italica), in reality of Far Eastern origin and transmitted to Europe via the Russian steppes. We first hear of it in China in 2700 B.C., and it is now grown throughout the dry parts of the East, and in Africa, for food; in Europe and the U.S. its use is mainly for animal forage. It also goes by such names as *Indian* or *foxtail millet*, *Chinese corn* and *Hungarian* or *Bengal grass*, and the Indian names such as *korra* and *kargi* derive from the Sanskrit *kungu*. There are a number of varieties named after the area in which they are most popular, such as the Turkistan, Siberian and Kursk millets.

The sorghums (genus Sorghum) are also classed as millets. The chief of these is *great* or *Turkish millet* (S. durra or vulgare), known in Egypt and elsewhere by its Arabic name of *dourra*. The maize-like grain is the leading cereal in Africa, its continent of origin, and very important in India, northern China and the U.S.; it is to be found cultivated over wide regions of most continents. From Africa it entered prehistoric Egypt, eventually reaching India, and then China by the fourth century. It was taken early by the Spaniards to Mexico. Nowadays it is common for the flour to be mixed with that of wheat in the countries in which it is used. In South Africa the cooked products of its flour are referred to as *mealies*, while a fermented drink made from it is called *tialva*; a similar Brazilian fermentation is called *merisa*.

There are many varieties of this sorghum growing in various areas, the colour of whose grains range over black, brown, orange, red, yellow and white. *Kaffir corn* or *African millet* is that grown in South Africa; *milo maize* is another variety which resembles it. The North African dourra is called *Egyptian*, *Guinea* or *rice corn*, and a Near Eastern variety is known as *Jerusalem corn*.

Sugar sorghum (S. saccharatum), *sweet reed*, *Chinese sugar cane* or *imphee cane*, is taller than great millet, but its grain is inferior, although it is sometimes used for food – once much more than it is now. This also originated in Africa, entered Egypt in Pharaonic times, and from thence was taken to the East via Arabia. There are records of its growing in Assyria in 700 B.C. It is now especially important in North India, Africa and the U.S., and cultivated chiefly as an animal fodder or for the sugary syrup obtained from its canes. In China this is made into a spirit. Another name, *broom corn*, refers to the fact that its fibrous branches are used in the making of brooms and that at one time it was cultivated in some areas solely for this purpose.

Several other related and unrelated grasses are, or have been, used by various peoples, although very often only in times of famine. The chief among these is *Canary grass* (Phalaris canariensis), originally from the Canary Islands and taken from there to Spain and southern France, from whence it has been widely distributed. It is cultivated now chiefly for bird seed (often known as *canary seed*) but, as the name *long millet* suggests, it is still used occasionally as a cereal. In the first stages of its growth it is hardly distinguishable from wheat or oats. When mature the grass stands from three to four feet high and bears an egg-shaped head containing about 100 seeds.

Oats (*genus Avena*)

Oats seem to be of West European origin, probably developed from two wild grasses, the *common wild oat* (A. fatua) and *wild red oat* (A. sterilis), *animal* or *fly oat*. Other varieties besides the *common cultivated oat* (A. sativa) are grown in the cool temperate regions all over the world. These include the *short oat* (A. brevis), which grows in high regions where no other cereal can survive; *naked* or *hull-less oat* (A. nuda), also called *pill corn*, in the highlands of East and Central Asia; and *bristle oat* (A. strigosa) in Wales and the islands of Scotland. Many of these are grown mostly as livestock feed, especially the last, which is considered to make a highly palatable hay-crop; they are also grown for human consumption in some areas. Two other species are *Turkish* or *Hungarian oat* (A. orientalis), mostly grown on the Eurasian steppes, which entered Europe from the East at the end of the eighteenth century, and *red oat* (A. byzantina), grown in the warmer areas of the world. The wild oat mentioned above (A. fatua) was once used for flour by several North American tribes of Indians. In Britain it also goes under the name *flaver, drake* or *droke, sowlers, Kentish longtails, poor oats* and *uncorn* (in Scotland).

Oats seem to have been adopted as a cereal later than most, but remains have been found in the Bronze-Age Swiss lake settlements, and in first century B.C. Germany, although they are probably of much older cultivation there. The grain came to Britain in the fifth century B.C. A Celtic group who moved through Macedonia into Asia Minor where they became known as Galatians, seem to have taken it with them. Galen mentions its use there in the second century, principally to feed horses, although people used it in times of scarcity. This is reminiscent of Dr Johnson's jibe in his dictionary that oats were commonly cultivated to feed horses in England and people in Scotland.

Although oats are mentioned by Greek authors, and in T'ang times (seventh to tenth centuries) in China, it might only be in reference to wild uncultivated forms. Only about 5 per cent of the total yield is now grown for human consumption. The grains, known as groats, are generally crushed to make oatmeal, oat flour and oat flakes; the latter are used in the making of porridge. Formerly, however, these products were of prime importance to people who could not grow wheat or afford wheaten flour. An early nineteenth-century agricultural survey of Britain disclosed that oatmeal and oatcake were almost the only cereal foods used by people in Scotland and northern England, and the commonest bread cereal in Wales.

In the language of plants an offering of oats means appreciation of someone's music. It is an allusion to the shepherd's pipe, the popular 'oaten straw' of pastorals. Various metaphors refer to the use of oats as horse food. Thus to 'feel one's oats' means one is inclined to be frisky; conversely 'to be off one's oats' means the opposite. 'Sowing wild oats', the pastime of riotous and lecherous youth in all ages, refers to the wild grass which spreads very easily. Geese, too, are supposed to be fond of the cereal, and a proverb points out 'it is ill buying oats from a goose' (goods from an expert). A country rhyme encourages one to diligence in planting.

> *Who in Janiveer sows oats*
> *Gets gold and groats;*
> *Who sows in May*
> *Gets little that way.*

Burns notes a divinatory use for the grain in his poem 'Halloween': Girls 'go to the barn-yard and pull each, at three several times, a stalk of oats. If the third stalk wants the *top-pickle*, that is the grain at the top of the stalk, the party in question will come to the marriage bed anything but a maid'.

The name derives from the Old English *ate* and is thought to be connected with the verb *etan*, to eat (cf. *edere* in Latin, *ad* in Sanskrit). The Scots *ait* is thus nearer in sound to the original word. But an alternative name, *haver*, which persisted in northern dialects where planted oats were said to be *haved*, is the most closely linked with the generality of European

16. GRAIN

two-rowed barley   Japanese millet   oats   rice

names, including the Latin *avena* (the name is unchanged in Italian and Spanish, *avoine* in French). The Old Saxon *havoro* develops into Dutch and East Frisian *haver* and 'Platt-deutsch' *hawer*; Old High German *habaro* into German *haber*, *hafer*, and Walachian *hafar*. Old Norse *hafre* is unchanged in Swedish, *havre* in Danish. Lastly, Old Slav *ovisu*, *ovesu*, *ovsa* becomes *ovesu* in Russian, and is connected with the Lithuanian *awisa*, Lettish *ausas* and Ossetic (the language of a Scytho-Iranian people in the Caucasian region) *abis*.

Pasta

'Do you like macaroni?' asks R. at the beginning of Somerset Maugham's *The Hairless Mexican*. 'What do you mean by macaroni?' answered Ashenden. 'It is like asking me if I like poetry. I like Keats and Wordsworth and Verlaine and Goethe. When you say macaroni, do you mean spaghetti, tagliatelli, vermicelli, fettucini, tufali, farfalli, or just macaroni?' 'Macaroni', replied R., a man of few words. 'I like all simple things ... but the only one I can eat day in and day out, not only without disgust but with the eagerness of an appetite unimpaired by excess, is macaroni'.

So, it would appear, can many Italians, especially the poor of Sicily, for whom it is a staple. The word sometimes serves to cover all kinds of pasta of which there are at least 600 and for many of which there are no standard names but a different one in every Italian dialect. Pasta is made from a glutinous paste or semolina, the product of a hard-grained wheat unsuitable for the making of bread and growing originally in the poor soil of Italy and southern Russia. This with various additives, such as egg in some cases, is dried in the forms of tubes for some kinds and flat ribbons for others. At one time all had to be made by hand. Boccaccio mentions manufacturers in the *Decameron*, but there could not have been many of them. The big step forward in its popularity came with the introduction of machines, after which production flourished. The first big attempts were made in Naples and adjacent territories in about 1800; it was soon taken up elsewhere.

Pastas are generally classified into fifty main divisions with infinite subdivision, some of them only local. The most common are the tubular pastas, with *macaroni* (derived from the dialect word *macare*, to crush or bruise) as chief. Macaroni proper is the thick tube about a yard long which has to be broken up to cook. A very large variety is known as *ziti*, and a smaller as *macaronelli*. *Spaghetti* (little cords) looks like its description and may be cooked whole, sliding it into hot water, where it bends and coils round the saucepan. *Vermicelli* (little worms) is a thinner variety than spaghetti but cooked like it; this is dried in bundles on a tray during manufacture. It is very good in soups but is too slimy and soft to act as a staple and, whatever other books may say, most definitely is not an acceptable substitute for Chinese noodles. Smaller still is *capellini* (little hairs).

Many of the smooth ribbon pastas are made with the addition of eggs. Among these are to be found *fettuce* (ribbons) and *fettucine*, *linguine* (little tongues), *canelone* (canals or gutters), and the Genoese *terenette*, which are matchstick-shaped. There are also the thick and thin Italian noodles, *tagliatelle* and *tagliatellini* (derived from *tagliare*, to cut). *Lasagne* consists of sheets measuring about 5 inches by $1\frac{1}{2}$, cut with a wavy pattern round the outside; *ricce* is another variety of this. Long lengths may also be platted together, and these are called *fussili*.

There are many kinds which come in short lengths and in various shapes. Short-length types include *ditale* (thimbles) and the ridged, *rigatoni*. Those cut obliquely are *penne* (feathers) and *maltagliati* (badly cut). Among the stamped shapes are *conchiglie* (shells), *anelli* (rings), *stelli* (stars), *rombi* (square shaped), *crocette* (crosses, also called kisses), *fiocci* (bows), *farfalle* (butterflies), and the letters of the alphabet referred to in various children's songs about alphabet soup from the U.S. *Gnocci* (lumps, knobs) are made from semolina and look rather like shells; they are tinned in cheese or tomato sauce in Switzerland and exported.

A curious kind of noodle to which spinach has been added goes under the name of *spinache*. The Chinese also make a version of this. Noodles have been used for a very long time in China and are made from various combinations of wheat, soya beans, other flours and eggs, and include such varieties as flat, narrow, tubular and transparent. The twelfth-century Venetian traveller Marco Polo is popularly supposed to have introduced them into Europe.

Pine nuts (*genus Pinus*)

Pine nuts or *pignons* are the seeds from the cones of various pines, those best known in Europe being the almond-flavoured seeds of the Mediterranean *stone pine* (P. pinea). These have been valued as an article of food and as a dessert since Roman times. They do not ripen until the fourth year, and are best kept in the cone until required since their abundant oil soon turns rancid. Very similar seeds are borne by the *Swiss stone pine* (P. cembra), also known as *Russian cedar*, which abounds on the continental ranges such as the Alps and Carpathians. Lamp oil is expressed from the seeds, which have also been used as a food from ancient times. The *Siberian pine* (var. sibirica) is a variety from the Siberian mountains.

The gigantic *Californian pine* (P. lambertiana), also known as *great, sugar* or *shake pine*, is one of the largest known pines and grows in the southwestern U.S. The local Indians gather the large nut-like seeds and pound and bake them. The name sugar pine does not refer to the taste of the seeds but to the fact that the burning resin gives off a smell like burnt sugar. The best-known American pine, however, is the *pinyon* (P. cembroides) or *American nut pine*, which grows in the U.S. and Mexico and bears very large and tasty seeds. It is these which are generally marketed, tinned or otherwise, under the name of pine nuts. There are several varieties of pinyon: *Colorado pine* (var. edulis); *single leaf* or *Californian nut pine* (var. monophylla); *Parry nut pine* (var. parryana).

Many more pines grow in the Californian area the seeds of which are used for food by the Indians. The most important to them is the *grey-leaf* or *bull pine* (P. sabiniana), also named *Digger pine* after one of the local Indian tribes. The *Torrey pine* (P. torreyana) has large seeds which are eaten raw or roasted, and the *Coulter* or *big cone pine* (P. coulteri) is used in the south of the state. The *white bark pine* (P. albicaulis) grows over a somewhat wider area and has large sweet seeds. In Asia the *Korean pine* (P. koraiensis) is used for food in Korea and the Kamchatka peninsula. The *Nepal pine* (P. gerardiana) is a Himalyan species, also known as *chilghoza* or *neoza*, whose seeds are eaten in India and Afghanistan.

A very distant relative, the *Australian* or *bunya-bunya pine* (Araucaria bidwillii), grows in Queensland. The bunya nuts grow in large cones and are eaten raw when young by the aborigines, or roasted among ashes when mature. They are also used as a means of barter. Every third year in this area, when the crop is particularly plentiful, the natives gather in large numbers, as many as 700–1000, for bunya picnics which may last for as long as a fortnight. It has been observed that they come away from these gatherings looking particularly well-nourished. Most other species of this pine are to be found in the Australasian area but we know a south American relative, the *Chile pine tree* (A. imbricata), as the *monkey puzzle*. The seeds of this too, are eaten roasted; in Chile they go under the name *piñones* and are an article of commerce.

Queensland nut (*genus Macadamia*)

This spherical nut is encased in a brown shell and grows in a cone-shaped cluster. Originally from Eastern Australia, it is now grown in the Pacific area and in the U.S., and considered one of the finest in the world. Before the coming of the whites it was a very important part of the aboriginal diet and is still very popular either raw or roasted. The tree principally used is M. integrifolia, since it bears the fruits with every rainfall, although in Australia such a phenomenon can manifest itself but rarely. The other species which is also cultivated, M. tetraphylla, fruits only once a year. M. ternifolia bears smaller fruits which, since they contain traces of prussic acid, are considered harmful by the aborigines. Those in Northern Queensland, however, do eat the rather similar fruit of M. whelani after first washing and cooking it to disperse the acid traces. The nuts go by several other names including *Australian* or *bush nut* and *Australian hazel*. *Bopple nut* is said to contain a corruption of Bauple Mountain in Queensland, although the name *bopple-popple* would seem to refer rather to the long round clusters.

Rice *(Oryza sativa)*

Rice, one of the principal Asian staple foods, was developed from a wild species growing from China to Bengal. In China it has been in use since early neolithic times (*c.* 20,000 B.C.) although it is not mentioned in their writings until 3000 B.C. About 200 years after this the Emperor Chingnong instituted a new-year ceremony in which he sowed this, the chief crop of the country, while lesser princes planted the less important staples. Many superstitions have grown up around the valuable grain. To upset a rice-bowl is an omen of ill-fortune, and deliberately to overturn someone else's is a terrible insult. Children are discouraged from wasting any by being told that for every grain they leave a pock-mark will appear on the face of their future bride or groom. The first European to encounter it would seem to have been Alexander the Great, who found it growing in the valley of the Euphrates and in India. It was never grown in Europe, however, until the Arabs brought it to Spain. The Italians grew it in Pisa in 1468, and in the seventeenth century it was taken to America.

During his first exile in 1081 the Chinese poet Su Tung-p'o wrote a series of poems on the hardships he was undergoing and his need to grow his own food. After he had been granted land he set to work, describing the growing of rice in one poem thus:

> *I planted rice before Spring Festival*
> *And already I'm counting joys,*
> *Rainy skies darken the Spring pond;*
> *I chat with friends by green-bladed paddies.*

Transplanting takes till the first of summer,
Delight growing with wind-blown stalks.
The moon looks down on dew-wet leaves
Strung one by one with hanging pearls.
Fall comes, and frosty ears grow heavy,
Topple, and lean propped on each-other.
From banks and dikes I hear only
A sound of locusts like wind and rain.
Rice, newly hulled, goes to the steamer,
Grains of jade that light up the basket.
A long time I've eaten only government fare,
Old rusty rice no better than mud.
Now to taste something new!
I've already promised my mouth and belly.

The point about government fare in the penultimate couplet is that government officials used to be paid in rice, as were the samurai in medieval Japan. In China, three grains of rice figured in the earth-worship dance, representing the rising, rousing and increasing of the sap. The grain, since it is so important a staple, has always served as an emblem of happiness, nourishment and fecundity. Because of this last connection, of course, it used to be thrown at weddings. In Malaya they also have some curious beliefs concerning rice, such as that it has a soul similar to humans and is presided over by the great Rice Mother; flowering rice, therefore, is treated as considerately as is a pregnant woman and no sudden or loud noises are allowed in its vicinity for fear that it should take fright and miscarry. A Japanese belief is that lightning causes rice to head, which might be compared with the similar Somerset lore that thunder frightens up the beans. One might also compare with this the belief that wheat does not ripen properly until there has been sheet lightening and should not be gathered until then.

About 90–95 per cent of the world production of rice still grows in Asia, although it is now grown in Egypt, Italy, Spain, Brazil and the U.S. There is also a species, O. glaberrhima, native to West Africa. A wild aquatic rice, Zizenia aquatica, gathered by the Indians of the North American midwest, is also to be found in the Far East. The plants grow rooted underwater and the Indians paddle their canoes among them, flailing the grains, which are easily dislodged, into the boat bottoms. They are then kept (ideally) underwater and then dried and parched. Names for the plant include *water oats* and *Indian, Canada* or *tuscarora rice*. Another plant used for grain by the Indians of the western U.S. is the related *mountain rice* or *Indian rice-grass* (Oryzopis hymenoides).

The name 'rice' derives from the Greek through the Latin *oryzon* (cf. the Sanskrit *vrihi* and Arabic *rouz, arousa*) into Italian *riso*, and French *ris*. There are aquatic varieties which grow in flooded paddy-fields and others which grow on ordinary soil. Of that grown 90 per cent is of the hard-grained type, but there is one with a glutinous soft grain, hardly exported at all, which is highly prized in Asia and used for ceremonial dishes. From Italy comes a fat-grained variety which cooks up as large and fluffy; there is a red variety of this which comes from Piedmont. Other rices obtainable in the West come from Madagascar, Spain or California. Lastly one should mention the brown unhusked rice, which is simply the grain in its original state before polishing, and which takes much longer to cook – anything up to three-quarters of an hour. It is this which is used by those who follow the Japanese macrobiotic diet. According to them it is not at all fattening, contains valuable protein, and is good for varicose veins and other disorders.

Rye *(Secale cereale)*

Rye seems to have developed from the wild species of South-East Asia. It was cultivated very early in western Asia and in mountainous regions of northern Europe. The Romans did not come in contact with it until they had dealings with the German and Slavic tribes whose staple it was. At the collapse of their empire rye displaced wheat even in the Mediterranean regions. It was probably brought to Britain by the Anglo-Saxons. In seventh century Kent the month of August was once known as *Rugern* (the rye-harvest month). Now the grain has declined in importance, only making up about 1 per cent of the world cereal crop, although it is still grown intensively in the poorer soils and high altitudes of northern Europe and other cool temperate regions. One old proverb expresses it thus, 'Good rye thrives high'. It is used for whisky and gin distilling, and in France for the making of various *eaux-de-vie*. The flour is used to make the dark heavy 'black bread' which, despite the fact that many people affect to dislike it, stays fresh longer than wheaten bread, has more taste and is very filling.

In pagan times images of the gods used to be made of rye-dough and placed on the altars. The practice is still remembered in the saying 'he looks like an image of rye-dough'. A variety of country proverbs are concerned with it. 'Sow wheat in dirt and rye in dust' runs one, alluding to the favoured weather conditions for the grains:

> *March dry, good rye;*
> *April wet, good wheat.*

A cryptic Worcestershire rhyme informs us:

Sell wheat and buy rye
Say the bells of Tenbury.

December cold with snow [is] good for rye.

In July some reap rye;
In August, if one will not, the other must.

There is, however, many a slip 'twixt the cup and the lip, for 'Peter and Paul will rot the roots of the rye' This refers us back to two bits of weather lore:

If St Paul's [25 Jan.] be fine and clear
It doth betide a happy year,

but 'if it rains on St Peter's Day [29 June] the bakers will have to carry double flour and single water; if dry, they will carry single flour and double water'.

Because it was once of over-riding importance to the various tribes of northern Europe there is a remarkable consonance in their names for rye. The Old English was *rughe* (cf. the Welsh *rhyg*), and Old Norse was *rugr*, giving *rugur* in Icelandic and Faeroese, *rug* in Danish and Norwegian, *rugh*, *rogh* and *rygh* in Medieval Swedish, developing into the modern *råg*. Old Prussian was *rugis*, giving the Lithuanian *rugys*, Lettish *rudsis*, Esthonian *rukis* and Finnish *rukin*. Old Slav *ruji* and *roji* become *rozhi* in Russian, *rez* in *Polish*, and so on. The Dutch is *rogge* and the German *roggen*.

Various more or less wild grasses in the same family are cultivated in some areas. *Mountain rye* (S. montanum) is grown as a fodder crop and for grain in the northern Mediterranean region. S. sylvestre, a native of Europe and tropical Asia, is grown in the mountainous regions of China and Tibet. The seeds of *giant wild rye* (Elymus condensatus), also known as *lyme grass*, and in British Columbia as *bunch grass*, are used for flour by the Indians of northwestern America.

Sago palms and cycads

Sago is an easily digested and very nutritious starch obtained from the stems of a variety of palms flourishing in wet tropical areas, principally the *smooth sago palm* (Metroxylon sagu or laeve) and the *spiny sago palm* (M. rumphii) of South-East Asia. These flower at fifteen years of age when they have stored up a vast amount of starch in the pith of their stems. In order that this starch is not channelled into the flower the palm is cut down just before infloresence and the pith removed. This is then grated to a powder, and kneaded with water over a strainer through which the starch passes, leaving behind the woody fibre. The starch may then be dried into sago meal by the natives, who eat it in pottage or make it into biscuits. But for export the flour is mixed into a paste with water and rubbed through sieves into various sized grains (pearl sago, bullet sago, etc.). A large proportion of European imports comes from Borneo, and the word 'sago' derives from the Malay.

There are many other palms, several cultivated for quite different reasons, from which sago may be obtained. The *Gombuti sugar palm* (Arenga pinnata) grows throughout South-East Asia and is cultivated for the sap which flows from wounds made at the lower-leaf level; this gives a brownish sugar on evaporation and wine by fermentation. The candied unripe fruits are considered good for the stomach, but when ripe their juice causes intolerable itching if applied to the skin. The inhabitants of the Moluccas used to steep these in water for the production of 'devil water', which they would pour over their enemies from the palisades in time of war.

The *fish-tail palm* (Caryota urens), *Kittul palm* or *bastard sago-palm*, is also a native of South-East Asia and of India, and is occasionally cultivated in the warm areas of Europe. This too is grown principally for the sap obtainable from the stem and roots, as several of its other names suggest. *Jaggery palm* incorporates a common name for palm sugar, deriving from the Indo-Portuguese *jagara* or *jagre* (cf. *sharkare*, Kanarese; *carkara*, Sanskrit; *shakkar*, Urdu). *East Indian wine* or *toddy palm* incorporates the term used of palm wine, deriving from the Hindi *tari*, an adjective based on the word for a palm. The very young unfolding leaves are edible, as are those of the *tufted fish-tail palm* (C. mitus), also used for sago. The fibre from the leaf-bases is used by primitive peoples in South-East Asia as wadding behind the darts in their blowpipes. The *Buri palm* (Corypha utan), is another useful sago-producing tree from the same area. Its buds serve as a salad or vegetable, and both sugar and palm wine (*tuba*) are obtained from it.

Palm wine and sago are also obtainable from the date palms (genus Phoenix), especially the East Indian *shrubby date palm* (P. farinifera) and the *Senegal date palm* (P. reclinata) of tropical West Africa. This is also true of the raffia palms (genus Raphia). Raffia, whose name derives from Malagasy, is obtained from its fronds. These palms are to be found spanning the equatorial belt around the world. They include the *Piassava wine palm* or *bamboo palm* (R. vinifera), which is the most widely spread, the East African and Madagascan R. ruffia and R. flabelliformis, and the Japanese *jupati palm* (R. taedigera).

Several palms native to America provide similar substances, as for instance the *cabbage* or *royal palm* (Oreodoxa oleracea) of the West

Indies which gives sago and palm wine, and whose terminal bud may be eaten like a cabbage (*see under* Palms). The South American *miriti fibre palm* (Mauritia flexuosa), known as *ita* and *chiqui-chiqui* in various Indian languages, provides a fibre from its leaves, wine and sago (*ipuruma* in Guarani) from its trunk, and its nuts may be pounded and made into bread. Lastly there is the *peach palm* (Guilielma gasipaes, formerly Bactris utilis), also known as *pewa* and *pejibaje*, a spiny palm of Central and South America, which also produces both palm wine and sago. Its fruits grow in date-like clusters which may weigh up to 25 pounds; they are known as *peach nuts*. The mealy flesh tastes rather like chestnuts after they have been boiled in salty water.

Cycads or fern palms, so called because many species look more like ferns than palms, are practically living fossils, having their ancient origin in common with the Chinese *ginkgo* (Ginkgo biloba). They are some of the most primitive seed-bearing plants, already present in Mesozoic times and having near ancestors on the Palaeozoic, that era when land plants were just beginning to develop and amphibians were emerging from the waters. Various genera are to be found in the tropics of most continents, from the pith of most of which a sago (or arrowroot as some call it) can be extracted.

The fern palm proper, from whose generic name (genus Cycas) the cycads get their name, grows in the tropics of Asia and Australasia. There are about sixteen species in all. Those used principally for sago-extraction are the *cycad sago palm* (C. circinalis) of Australasia and the East Indies, and the *Japanese sago palm* (C. revoluta). The young leaves of these and various others are occasionally used as a vegetable, and the seeds of others, including the *Gaveu nut-tree* (C. media), are also ground into a flour. In Northern Australia these seeds, which may be as large as a hen's egg, are a staple of the aborigines. They are, however, in common with the 'nuts' of many other cycads, highly poisonous in a raw state. To dispel the poisonous principle, the aborigines grind the seeds and place the resulting flour in running water for 24 hours or more before cooking.

In Western Australia the aborigines make use of the seeds of the *giant fern palm* (Macrozamia reidlei), having dispelled the poisons by long washing and cooking. To the east they use those of M. miquelii and M. spiralis, referring to them as *burrawang nuts*. There are about fourteen species of this cycad, all confined to Australia. There are also two species of genus Bowenia which the aborigines use in various ways, chiefly for starch-extraction. But the yam-like rhizomes of the *tufted fish-tail palm* (B. spectabilis) are also eaten, especially in Queensland.

In South Africa there are twelve species of *Kaffir bread* (genus Encephalartos); the stems are particularly rich in starch, and the natives make use of the seeds as well. Two of these, E. caffer and E. altensteinii, are now grown in warm European areas. In the very ancient past it was this palm which was to be found all over Europe when it formed part of a tropical zone. Another, the *chestnut dion* (Dioon edule), from whose stem sago can be obtained, is of Mexican origin and can also be found growing in Europe now. In its native country it is known as *chamal* and *palma de la Virgin*. The seeds are also boiled and roasted.

Another of the cycads, genus Zamia, grows in the tropics of all the continents but is of most importance in the Americas. The seeds of the *Ecquador zamia* (Z. lindenii) are used for food by the Indians, while in the West Indies a starchy substance known as wild sago or arrowroot is obtained from the stems of two species both known as the *Jamaica sago-tree*: *scurfy zamia* (Z. furfuracea) and *coontie* (Z. integrifolia). This last name derives from *kunti*, the name the Seminole Indians of Florida give to the flour they obtain from both stem and parsnip-shaped rhizomes of various species. These include the *comfort root* (Z. floridana) and the *St John's coontie* (Z. pumila), the last of which is also to be found growing all over the West Indies. The seeds are also used. It is necessary in all cases to dissipate the plant's poisonous principles first. The Indians do this by pounding the rhizomes or seeds to a pulp and washing it thoroughly in a straining cloth through which the starch is allowed to pass. During the American Civil War the Federal Armies lost a number of men who did not take this precaution, merely eating the roots in a crude unprepared state. The flour is now extracted in factories and sold under the name of Florida arrowroot.

The *ginkgo* (G. biloba), as has already been mentioned, is the nearest allied to the cycads, and dates back to Palaeozoic times. Fossil remains of it have been discovered in the British Isles and North America. It is now largely confined to China and Japan where, thanks to its being regarded as a sacred tree and planted outside temples, it has been preserved. There it is also regarded as the legendary 'tree of milk' and a protector of nursing mothers. It was re-introduced to England, after so many millions of years, as an ornamental tree in 1754, and from there it was taken to America. Because its leaves look like the maidenhair fern it has been christened *maidenhair tree*. Its fruits are the size of a small plum and have an evil-smelling pulp. But the kernels are much eaten in the East and are said to taste like maize. Like those of the cycads they are poisonous raw and cause dermatitis. They are prepared by fermenting off the fleshy covering and then boiling or roasting the seed.

The majority of tree-ferns proper (Cyatheaceae) are now to be found in the Australasian area, although they were once much wider spread. Fossil remains of them have been found in Europe dating from before the ice-ages, in Eocene times. From the stem-pith of many of these a sago can be obtained, as for instance from the so-called *sago fern* (Cyathea medullaris), alternatively known as the *black-stemmed* or *grey tree-fern*, of the

Pacific, New Zealand and Australia (especially the south east). The New Zealand Maoris also use thus the *silvery tree-fern* (C. dealbata), while in the Pacific a mucilaginous substance is obtained from the base of the fronds of C. viellardii.

Other members of the family are also used. The aborigines eat the pulp from the upper part of the *Tasmanian* or *woolly tree-fern* (Dicksonia antarctica) roasted in ashes. There and elsewhere in Australia can be found the *Australian grove-fern* or *hill tree-fern* (Alsophila australis), occasionally to be found growing up to snow level. Sago can be obtained from this and from the *Norfolk Island tree-fern* (A. excelsa) which is apparently threatened with extinction because of its use for this purpose. A South American relative, A. rufa, is occasionally used in times of emergency.

A number of ordinary ferns can also serve, like the Hawaian Sadlera cyathecides of the polypody family, whose starchy pith is cooked and eaten by the natives. The *ashleaf fern* (Marattia fraxinea) of Australia and New Zealand reaches tree proportions and is used for sago by the aborigines. Those of Queensland also find a rich supply in the *turnip* or *oriental vessel fern* (Angiopteris erecta). Lastly one might mention the totally unrelated *macambira* (Bromelia laciniosa), a Brazilian member of the pineapple family, from the inner sheath of which a starchy mass can be obtained. This is sun-dried and made into an ill-tasting bread by the very poor and hungry. In Central America the very young buds of a large relative, the *curra* (B. karatas), are sold in the markets as a kind of palm cabbage substitute.

Walnut *(genus Juglans)*

The walnut now grows in Asia, Europe, and America (whither it was taken by the early settlers). The Romans considered the tree to be of Persian origin, but since it is now so widely spread it is very difficult to tell whether this is the truth. The tree was cultivated by Romans very early in the time of their kings, and it was their custom to throw the nuts at weddings in the same manner as others throw rice. The Romans are thought to have brought the tree to Britain, but there is no definite mention of it until 1562. The name is derived from the Old English *wealh hnutu*, which can mean either Welsh or foreign nut. We do find such names as *Welsh* and *French nut* later applied to it, as well as *bannut* in the West of England. The Welsh name is *cnau ffreinig*.

The wood of the walnut can be very useful, and some species are grown only for this, but of those grown for their fruit the *Persian walnut* (J.

regia), known in the U.S. as the *English walnut*, is the best known. In Canada and the U.S. some prefer the *butternut* (J. cinerea), also known as *long* or *white walnut*. The following are a selection of those others grown chiefly for the fruit:

Cathay walnut (J. cathayensis), China.
Manchu walnut (J. mandshurica), North China.
Japanese walnut (J. sieboldiana), Japan.
Argentine walnut (J. australis), Argentine.
Bolivian black walnut (J. boliviana), South America.
Texas walnut (J. rupestris), North Mexico, southwest U.S.
Californian black walnut (J. californica), south California.
Arizona black walnut (J. major), southwestern U.S.

Walnut blooms are the birthday flower for 15 March and symbolise intellect, longevity, presentiment and stratagem. To dream of them, however, presages danger and misfortune. The nuts were a fertility symbol for both the Greeks and the Romans, which accounts for their presence at classical weddings. In Roumania, however, a bride who does not wish to bear children immediately places as many roasted walnuts in her bodice as the years she wishes to remain childless. Another link between the tree and women occurs in the well-known and very popular proverb:

> *A woman, a dog [spaniel or ass] and a walnut tree,*
> *The more you beat them, the better they be.*

Wheat (*genus Triticum*)

Wheat is now the principal cereal for flours and bread. It is of Neolithic origin, probably in the Mediterranean area, although others claim it for South-West Asia or Ethiopia. It was one of the grains sown in the annual ceremony instituted by the emperor Ching-nong (or Shennung) in the twenty-eighth century B.C. (*see under* Rice), the others being rice, sorghum, Italian millet and soya beans. It was probably brought to Britain about 2000 B.C. by New Stone Age immigrants from Europe. Pytheas, a traveller from the Greek colony at Marseilles (Marsilea), visited Britain in 330 B.C. and noted that wheat was extensively grown in the south-east, where it was threshed under huge barns. The Romans encouraged further advances in cultivation during their occupation and by 360 A.D. the grain was being exported to the armies on the Rhine.

Wheat was of paramount importance to Rome, and it has been suggested that many of her initial overseas conquests were undertaken in order to obtain areas most favourable to its growth, in particular Sicily and Sardinia, Carthaginian Africa and Egypt. It has been estimated that under optimum conditions Sicily was able to export three million bushels, Africa ten and Egypt twenty millions in a year. Merely by occupying Egypt and cutting off the supply to Rome Vespasian eventually gained the emperorship in the 'year of four emperors' (69–70) after the death of Nero. Eventually wheat was also imported from Spain (whose main exports were minerals) and the Crimean region. After the division of the Empire and the rapid extinction of the West, most of these exports were re-routed to Constantinople.

The invasion of Germanic peoples largely displaced wheat-growing for a long time in Europe. In Britain the Anglo-Saxons showed their preference for rye, but the two crops were often mixed and ground together, the resulting flour being referred to as *maslin*. As time progressed wheaten bread regained its popularity but was largely restricted to the rich; the poor had to make do with a mixture of various grains, leguminous plants and even acorns (*see under* Acorns) to make bread-flour. It was not until the nineteenth century that wheaten bread became commonly accessible to the poor.

The English name derives from Old English *hwæte*, and may be compared with Old Saxon *hweti*, becoming *weit* in Dutch; Old High German *weizzi* becomes *weizen* in German, and Old Norse *hveiti* becomes *vete* in Swedish and *hvede* in Danish. All derive from the word for white, thus distinguishing the grains from the other darker cereals with which the Germanic races were more familiar.

A large variety of wheats have been grown, although many have now declined in popularity and are restricted to cattle-feed. They may broadly be divided into three groups.

(*a*) *Einkorn*. *One grained wheat* (T. monococcum), also known as *French*, *German* or *little spelt*, *St Peter's corn*, and by the German name of *einkorn*, thrives in the poorest and most stony soils. Although it is not very productive it yields an excellent meal. It is now mostly restricted to animal fodder and is rarely cultivated, although some is grown in Spain, France and Eastern Europe, Asia Minor and Morocco.

(*b*) *Emmer*. The chief of these is the two-grained *starch wheat* (T. dicoccum) known by the German name of *emmer*. This was once the most used in ancient times amongst the Neolithic peoples of central Europe, and in early Egypt (*c*. 6000 B.C.) it was the only type under cultivation. Now it is only grown in a limited way in India, Iran, the U.S.S.R., Ethiopia, Morocco and the Basque areas. *Durum*, *hard-grained* or *flint wheat* (T. durum) is generally that used for pasta (*see* Pasta); it is grown in the Mediterranean area, the U.S.S.R., central Asia, India, South

Africa and the U.S.A. *Abyssinian wheat* (T. abyssinicum), grown in Egypt and Ethiopia, resembles it very closely.

Other species include *Persian wheat* (T. persicum), grown in the Caucasus, and *Polish wheat* (T. polonicum), also called *giant rye*, which is mostly grown in Spain, not Poland. *Humpy-grained, English* or *Egyptian wheat* (T. turgidum), also known as *pollard, rivet* or *cone wheat*, is much grown on the north Mediterranean shore. It is starchy and best used for biscuits, etc.

(c) *Spelt. Spelt* or *starch corn* (T. spelta), also known by its German name of *dinkel*, has long lax-bearded or beardless ears. It is a hardy wheat grown in mountainous European regions and was not known to the Romans until about A.D. 400. Its name, deriving from a German root, then entered Latin and thus Italian, as *spelta* (cf. the Spanish *espelta*, Medieval French *espeltre, espiautre*, becoming the modern *épeautre*). Its naked grains are often harvested prematurely in German regions; they are then parched and added to soups as *grunkern*.

The commonest species grown now, and the widest distributed, are *common* or *bread wheat* (T. vulgare), *winter* or *lammas wheat* (T. hybernum) and *spring* or *summer wheat* (T. aestivum). These may have been the first types to have been brought to Europe; now they are especially important in the U.S. and Canada, the U.S.S.R., Argentina, India and Australia. *Club* or *square-eared wheat* (T. compactum) is closely related to these and often mixed with them in the same field, although pure crops can be found in Central Asia and China, and in the north-west Pacific states of the U.S. Other species include *shot* or Indian *dwarf wheat* (T. sphaerococcum), drought-resistant and grown in North India, whose grains are round rather than elongated; and *macha wheat* (T. macha), also from Asia.

Other varieties have been produced in an effort to stem world hunger and arrest crop diseases (e.g. rust). One of the newest ways is to expose the seeds to radioactivity until a more favourable variety turns up. (More monstrosities than useful products occur, of course, but these can be destroyed.) A rust-free variety was thus produced and efforts were made to induce Indian farmers to take it up. Unfortunately the variety was beardless and the farmers refused it because, as they claimed, a beard inhibits birds from attacking the crops. Formerly this would have been an *impasse* and much ingenuity and time would have been spent in re-education and propaganda. In this case, however, the scientists simply went back to irradiating until a viable rust-free and bearded variety turned up.

Because of its paramount importance wheat is the symbol of prosperity as the birthday plant for 3 October, emblematic of abundance and life. In heraldry it stands for the earth's bounty. Wheat in a sheaf, however, as well as being the emblem of agriculture and fertility, stands for autumn, harvest and death. It is connected with Thanksgiving Day in both its light and dark aspects (life continues because of the sacrifice of a few), and is an attribute of the leaden Saturn. A white ear of wheat symbolises Christ, a green ear the Egyptian equivalent, Horus. Both these are connected, as has often been pointed out, with the corn fertility gods of ancient myth who died (and were resurrected) yearly for the benefit of mankind. It is the rites connected with these that underlie the John Barleycorn ballads (*see under* Barley) as well.

Bearded wheat stands for faithfulness, rejuvenating fire and the vital (including the sexual) urge. The fertility god looked after the continuance of the race both from within itself, by the begetting of children, and from without, in assuring that both parents and children had enough to eat. In connection with this it is interesting to note that the word corn (connected with such words as 'to grow' and the Latin equivalent, *crescere*) stems from the same primitive Indo-European root as gives us the word 'horn' (*corne* in French), itself another symbol of the sexual urge. One wonders from what deep wells of the Jungian racial unconscious sprang the proverb 'corn and horn go together'. Of course its overt meaning is that the price of grain and cattle is interconnected, as is evidenced by another proverb, 'Up corn, down horn'.

The reason for the phenomenon of the see-sawing prices is to do with the price of cattle-feed which, if it was prohibitive in the bad old days when there was little enough to feed a farmer during hard winters, let alone his cattle, meant a large influx of stock onto the market:

Corn in good years is hay
In ill years straw is corn.

Weather lore and prophecies as to how the crops will do combine in other country proverbs concerning this subject:

When the cuckoo comes to the bare thorn
Sell your cow and buy [or plant] your corn;
But when she comes to the full bit
Sell your corn and buy your sheep.

A similar formula from the Welsh marches informs one

If the cuckoo sings when the hedge is brown
Sell thy horse and buy thy corn;
If the cuckoo sings when the hedge is green
Keep thy horse and sell thy corn.

Another Welsh proverb, from Pembroke, tells us that

> *A fog from the sea*
> *Brings honey to the bee.*
> *A fog from the hills*
> *Brings corn to the mills.*

Other predictions include 'a good bark year makes a good wheat year'; 'a good nut year, a good corn year'; and 'a good wheat year, a fine plum year'.

The proverbs follow the crop's progress through the year. 'A wet March makes a sad harvest' for it inhibits its growth. 'Good wheat in March should cover a sitting hare.' But

> *A peck of March dust and a shower in May*
> *Makes the corn green and the fields gay;*
>
> *A dry March and a wet May*
> *Fill barns and bays with corn and hay.*

'May makes or mars the wheat'. 'A cold May is good for corn and hay' (of bad for the latter in some versions).

> *A cold May and windy*
> *Makes a fat barn and findy.*
>
> *A cold May is kindly*
> *And fills the barn finely.*
>
> *A dripping [or wet and windy] May*
> *Fills the barns with corn and hay.*
>
> *A misty May and heat in June*
> *Bring cheap meal and harvest soon.*

It is not until June that one can be really sure of success:

> *Look at your corn in May*
> *And you'll come weeping away;*
> *Look at the same in June*
> *And you'll come home in another tune.*
>
> *Calm weather in June*
> *Sets corn in tune.*

And just to top off its progress nicely:

> *A shower in July when corn begins to fill*
> *Is worth a plow and oxen and all belongs theretill.*
>
> *No tempest, good July,*
> *Lest corn come off bluely [with mildew].*

One must make up one's mind which variety to grow for 'wheat will not have two praises', one cannot grow both winter and summer crops. Having once decided upon this question,

> Plough deep whilst sluggards sleep
> And you shall have corn to sell and to keep.

At the beginning wheat likes wet weather;

> Sow your wheat all in a flood
> And it will grow up like a wood,

for 'wheat always lies best in wet sheets' or, if you have planted the other variety, 'the corn hides itself in the snow like an old man in furs'. Long maturing underground actually improves it since, nicely punning,

> When wheat lies long in bed
> It rises with a heavy head.

Finally, in Cornwall, all good things come at once:

> When the corn is on the shock
> Then the fish are on the rock.

The Chinese poet Su Tung-p'o, in the series of poems already mentioned (*see under* Rice), also passes on a few hints on the raising of a crop:

> A good farmer hates to wear out the land;
> I'm lucky this plot was ten years fallow.
> Its too soon to count on mulberries;
> My best bet is a crop of wheat.
> I planted seed and within the month
> Dirt on the rows was showing green.
> An old farmer warned me,
> Don't let the seedlings shoot up too fast –
> If you want plenty of dumpling flour,
> Turn a cow or a sheep in here to graze.
> Good advice. I bowed my thanks:
> I won't forget you when my belly's full.

In Arabic tradition Adam took wheat with him from Eden as the chief of all the foods he had known there. It was certainly important to the Jews who evolved much ceremony around its ingathering. Both at the beginning and the end of harvest:

> ... when ye be come into the land which I give unto you, and shall reap the harvest thereof, then ye shall bring a sheaf of the firstfruits of your harvest unto the priest: And he shall wave the sheaf before the Lord, to be accepted for you: on the morrow after the sabbath the priest shall wave it. And ye shall offer that day when ye wave the sheaf an he lamb

without blemish of the first year for a burnt offering unto the Lord. And the meat offering thereof shall be two tenth deals of fine flour mingled with oil, an offering made by fire unto the Lord for a sweet savour: and the drink offering thereof shall be of wine, the fourth part of an hin.[1] And ye shall eat neither bread, nor parched corn, nor green ears, until the selfsame day that ye have brought an offering unto your God: It shall be a statute for ever throughout your generations in all your dwellings.

And ye shall count unto you from the morrow after the sabbath, from the day that ye brought the sheaf of the wave offering: seven sabbaths shall be complete: even unto the morrow after the seventh sabbath shall ye number fifty days, and ye shall offer a new meat offering unto the Lord. Ye shall bring out of your habitations two wave loaves of two tenth deals: they shall be of fine flour; they shall be baken with leaven; they are the firstfruits unto the Lord. And ye shall offer with the bread seven lambs without blemish of the first year, and one young bullock, and two rams; they shall be for a burnt offering unto the Lord, with their meat offering, and their drink offerings, even an offering made by fire, of sweet savour unto the Lord. Then ye shall sacrifice one kid of the goats for a sin offering, and two lambs of the first year for a sacrifice of peace offerings. And the priest shall wave them with the bread of the firstfruits for a wave offering before the Lord, with the two lambs: they shall be holy to the Lord for the priest. And ye shall proclaim on the selfsame day, that it may be an holy convocation unto you: ye shall do no servile work therein: it shall be a statute for ever in all your dwellings throughout your generations.

And when ye reap the harvest of your land, thou shalt not make clean riddance of the corners of thy field when thou reapest, neither shalt thou gather any gleaning of thy harvest: thou shalt leave them unto the poor, and to the stranger: I am the Lord your God.

(Leviticus 23:10–22)

Many a moral lesson is to be drawn from wheat. It is a teacher of faith, for 'who sows his corn in the field trusts in God' or, as the Italians have it, 'when the corn is in the field, tis God's and the saints'. It counsels patience in adversity: 'Corn is cleansed with the wind and the soul with chastenings'. Patience in action also: 'Corn is not to be gathered in the blade but in the ear', and the man who 'eats his corn in the blade' is too precipitate. He would do well to consider that 'much corn lies in the chaff [or under the straw] unseen'. A philosophical view of life, where there is no pleasure without pain, is promoted: 'in much corn is some cockle' and 'corn and cockle grow in the same field'. Finally, a man who was 'corn-fed' was pampered in the times when wheaten bread was not for the likes of the poor.

[1] About 2½ pints.

SECTION 3

Various herbs, spices and condiments

Balm

1. Bee balm *(genus Melissa)*
The best-known and most-used of this genus is *Garden* or *bee balm* (M. officinalis), occasionally found growing wild in the south of England, whose leaves may be used as a flavouring or in salads. A sweetened infusion of its leaves is considered pleasant by some. From classical times onward it has been thought to have curative powers, especially as a poultice for wounds, and as a tonic for digestion and the circulation. Hence the growth of phrases such as 'a balm for a troubled mind' referring to these powers. The name is a contraction of *balsam*, with which this member of the mint family has no connection.

2. Bergamot *(genus Monarda)*
Another member of the mint family, also called *bee balm*, is probably better known as *bergamot*. A belief that its leaves and flowers are especially beloved of bees has given rise to the former name. The Latin generic name derives from the Spanish doctor Nicholas Monardes, who first discovered the herbs in North America and named them bergamot because of the likeness of their smell to the bergamot tree, Citrus bergamia, of the orange family, from which bergamot oil is extracted.

The colour of the flowers differs according to the species: white, blue, pink, mauve and scarlet, the most popular being the variety known as Cambridge Scarlet. Many in this family are known somewhat disparagingly as horse mints, but some are esteemed by various peoples in North America. *Pony bee balm* (M. pectinata), from the western U.S., is used by the Indians of New Mexico. The best known is the *mountain mint* (M. didyma), also known as *Oswego tea* from its use by that tribe. It is a vivid scarlet plant, growing as high as two feet by the side of woodland streams and pools. The nectar is seated at the base of a long tube in the flower and can only be reached by long-tongued insects such as bumblebees and certain butterflies; hummingbirds are also attracted to it.

Basil *(genus Ocimum)*

Basil belongs to the mint family, but its taste is more evasive than mint. The three species most used are *common basil* (O. minimum), *sweet basil* (O. suave), and *bush basil* (O. basilicum). The names are variously juggled about and applied to each other, but those given above appear to be the most authoritative. The herbs are now fairly widespread. Bush basil, for instance, has been introduced into Queensland (Australia), and is very common in New Guinea, where the natives gather sprigs of it to wear in their arm-bands. Yet another species, *purple-stalked basil* (O. sanctum), also known as *holy* or *monk's basil* and *mosquito plant*, is very common in India, where it is sometimes chewed as a betel substitute, and a decoction for coughs is made from it. It is used as a pot-herb in Australia and the aborigines regard it as a blood-purifier and an antiscorbutic. *Hoary basil* (O. canum) is also used as a pot-herb in India and is to be found on sale in the markets. All of them blend particularly well with tomatoes and may be used otherwise as pot-herbs. The name is said to derive from the Greek word for royal, βασιλικος.

The herb is regarded ambiguously by various nations. It comes originally from India, where it is known as *tulasi* and is sacred to Vishnu and his avatar Krishna, god of love, saviour and Christ-figure. Their devotees thread its fruits together for rosaries. The connection with Krishna makes it a herb connected with love, of course. But both in India and Egypt the dead were buried with it as a propitious plant. In fact a bush was grown in every village home and dying men were placed under it, both because this was good for the departing life and because it was considered unlucky for death to occur in the house.

Perhaps it was because of its connections with death that it has come to be regarded as a somewhat gruesome plant in the West since classical times. It was said to be a breeder of poison and 'everybody knew' that it spontaneously engendered scorpions. It was, too, a causer of insanity and said to feed on the brains of men – many of these traditions are beautifully combined in Keat's poem 'Isabella' (or 'The Pot of Basil'), where the heroine puts the head of her murdered lover into a pot and grows a basil-bush on top of it; when it is taken from her she pines away, goes mad and dies. The plant was first grown in England in Tudor times after which, as is well known, Englishmen took to collecting colonies and walking about in the mid-day sun!

Bay (*Laurus nobilis*)

The leaf of the bay tree, a native of the Mediterranean countries and of India, is much used as a flavouring agent in cookery. Men have long known about it, and have attributed to the tree many remarkable and powerful properties. It was hung up in medieval churches as a sign of welcome to the fairies and elves, and a lover crushing the leaves in his hand could tell the truth about his love from their crackling. Since it is of the laurel family, we find it going under such names as *bay laurel*, *sweet laurel*, and *Roman* or *royal laurel*.

It was bay-leaves which made the triumphant laurel crown for Roman victors, for its evergreen properties made it a symbol of the god-like, the everlasting. It is in this connection that it is found in the beautiful simile of the Psalms (37:35):

> *I have seen the wicked in great power*
> *And spreading himself like a green bay tree.*

It is also with reference to the laurel crown that the poet-laureate gets his name. The laureate Dryden was satirised under the name Bayes in the anonymous *The Rehearsal* of the seventeenth century.

Botanically the sweet bay, or laurel, is the true laurel. It used to be grown in Britain during the period of phenomenally fine weather which was at its peak about 1250–1350 and declined with the waning of the Middle Ages. A blight was destroying the trees in Wales during the reign of Richard II, a fact referred to by Shakespeare (*Richard II*, II. iv). In medieval England the bay was referred to as laurel, or variations upon the French *laurier* such as 'laurer'. The word 'bay' was then confined to its black berries; it derives from the French *baie* (a berry), a corruption of the Latin *baca* (cf. the Provencal *baga*). It was not extended to cover the tree as a whole until the early sixteenth century. One of its first appearances in this sense is in Coverdale's translation of the verse from Psalm 37 already quoted.

What we now call the laurel should properly be known as the *cherry laurel* (Cerasus laurocerasus). It is a shrub of Near Eastern origin which was first brought to England in Elizabethan times, and its leaves and berries are generally accounted poisonous, but I was once given a subtly flavoured home-made blancmange by an old lady in Oxfordshire who had boiled a laurel leaf in the milk and am still here to tell the tale. It would be advisable to wash the leaf well first as the laurel is a dirty tree.

In the Greek myth the maiden Daphne, fleeing from the lecherous advances of Apollo, was changed into a bay tree; in contrition he, as god of medicine, took the tree as symbol of his protection against evil. It was for this reason that the bay, and another evergreen, myrtle, were used at funeral ceremonies. Milton thus addresses them in his elegy for Lycidas:

> *Yet once more, O ye laurels, and once more*
> *Ye myrtles brown, with ivy never sere,*
> *I come to pluck your berries harsh and crude,*
> *And with forced fingers rude*
> *Shatter your leaves before the mellowing year.*

Belief in the bay's protective powers persisted into the Middle Ages, when it was said to be a good defence against witchcraft and 'all the evils old Saturn can do to the body'. The seventeenth-century Nicholas Culpepper reiterates the old lesson: 'Neither witch nor devil, thunder nor lightning, will hurt a man where a bay tree is'.

Burnet (*genus Sanguisorba*)

Burnet, a member of the rose family, is like borage in taste, and is used similarly, both in the kitchen, in salads, and as an additive to drinks. It tastes best when the young leaves are used. The name is derived from their browny colour, coming from the French diminutive for brown, *brunette*, a word we use for chestnut-haired girls. (The word was once applied to a kind of brown cloth as well.)

In England the name burnet, as such, seems to be mentioned first in the twelfth century, although the herb was known to the Anglo-Saxons, who used it as an ingredient in their famous green-salve, among other things. Earlier it was called *pimpernel*, deriving from the Latin *bipinella, bipinulla*, meaning little two-wings, a reference to the shape and setting of the leaves. It is not to be confused with the true pimpernels, members of the primrose family, of which *scarlet pimpernel* (Anagallis arvensis) is probably the best known.

From the fifteenth to seventeenth centuries burnet was very popular and, in addition to its culinary uses, was used for healing wounds and as a protection against the plague. In Tudor times it was also taken for gout and rheumatism, and is still used now as an astringent medicine. It is mentioned in a poem by Ronsard where he pictures himself setting out with his valet to gather a pastoral salad:

> *Tu t'en iras, Jamyn, d'une autre part*
> *Chercher soigneux la boursette toffue,*
> *La pasquerette à la feuille menue,*
> *La pimprenelle heureuse pour le sang*
> *Et pour la ratte, et pour le mal de flanc;*
> *Je cuillerai, compagne de la mousse,*
> *La responsette à la racine douce,*
> *Et le bouton des nouveaux groiseliers*
> *Qui le printemps annoncent les premiers.*

A rough translation reads:

> *You'll go off, Jamie, to one side*
> *Looking for the corn-salad tufts*

And marguerite with slender leaves,
Burnet, a medicine for the blood,
For rheumatism and the spleen;
While I, companion of the moss,
Will pick sweet-rooted rampion
And buttons of the fledgling currants,
First messengers of spring.

There are three members of the burnet family and its relations that may be used. *Great* or *common burnet* (Sanguisorba officinalis), commonly growing in meadows; *lesser* or *salad burnet* (Poterium sanguisorba), which prefers chalk; and the *burnet saxifrage* (Pimpinella saxifraga). The first of these was once known as *bloodwort* from its supposed power to stanch bleeding.

Capers   (*genus Capparis*)

The young berries and buds of this low prickly shrub, a native of the Mediterranean region, are often used in sauces and otherwise, most frequently pickled in European cookery. At one time capers were imported into Britain from Southern Russia, but now they come generally from Spain and northern Africa. That most commonly used is the *prickly caper* (C. spinosa), while in northern Africa the most common is the *Timbuctoo caper* (C. sodala), and below the equator the *South African caper* (C. corymbifera), whose flower heads are pickled in salted vinegar. Among the alternatives which may be used similarly are the buds of the Leventine *bean caper tree* (Zygophyllum fabago). The Coahuilla Indians of Mexico cook the green pods of the related *burro fat* (Isomeris arborea).

Various other relatives are used in a variety of ways as, for example, the species genus Boscia in Africa. The seeds of B. angustifolia are eaten by natives of the Sahara; the leaves and berries of B. senegalensis are added to soups and mixed with cereals, while in South Africa the natives eat the roots of B. caffra in times of scarcity. The leaves of the *garlic weed* (Crataeva macrocarpa) serve as a vegetable and pot-herb in Indo-China, and those of the *African spider herb* (Gynandropsis pentaphylla) in areas as far apart as India and Nigeria. The leaves of the *showy spider herb* (G. speciosa), a tropical American species, are also eaten as a vegetable.

Lastly there is the North American *spiderflower* (Cleome integrifolia) which, to the Pueblo Indians, is a source of black dye. The Tewa Indians of New Mexico, on the other hand, gather the plant in the spring and, having submitted it to various processes in order to remove the alkaline taste, eat it with cornmeal porridge. Another tropical species, the *clammy weed* (C. viscosa), is sold in the herb shops of the Far East as an appetite stimulant, and its pods serve as pickles. The leaves of the related *Indian copper leaf* (Acalypha indica) are reckoned to have purgative properties, and are eaten by the Asian poor.

Capsicums *(Capsicum frutescens)*

Capsicums, or peppers, are the various-sized seed-pods of a solanaceous plant, having nothing in common with the true peppers (Piperacae) except the hot quality of some, but by no means all, of them. The plants grow wild all over South America and are thought to have spread from the Guiana region, since most of their native names can be traced back to a Carib root. They were in cultivation before the arrival of the Spaniards, to whom they were introduced by the Aztecs. They were then brought to Europe in the sixteenth century and quickly spread over the Mediterranean region and from there into the tropics of the world. Three separate varieties were growing in India by 1542, and that country is now the greatest exporter.

Formerly the plants were divided into different species but are now thought to be only varieties of a superspecies which goes under such names as *garden ginger, bush* or *spur pepper, goat* or *bird's eye pepper*. They are thus divided into varieties:

1. Var. abbreviatum. *Wrinkled pepper.*
2. Var. accuminatum. *Long cayenne* or *pod-pepper*. The name *chilli* is derived from the Mexican Nahuatl Indian. This usually goes into the making of various condiments, such as the hot red cayenne pepper. The pods are laid out to dry, then mixed with flour and baked into a very hard biscuit which is, in its turn, ground and sifted. It is reputed to be adulterated with red lead on occasions. Spanish pepper, *pimenton* or *pimiento*, is a hot very pungent spicy product. *Paprika* is a Hungarian pepper which may be either hot or mild. The name derives from the Hungarian for Turkish pepper. *Keonigs paprika* is made from pods ground whole and is exceptionally hot. The milder *rosenpaprika*, more a colouring agent than anything else, is made from selected pods whose stalks, stems and seeded centre has been removed.
3. Var. cerasiforme. *Cherry pepper.* It is coloured red, yellow and purplish, and is very pungent.
4. Var. connoides. *Cone pepper.* Very acrid and used as the constituent of tabasco sauce.
5. Var. fasciculatum. *Red cluster pepper.* Very pungent.

6. Var. longum. *Long pepper*. Many of these are sold shredded and dried, or ground into chilli powder and used as a spice, especially in curry. The fresh fruits may also be used and are not too difficult to obtain.

Sweet, bell or *bull-nosed peppers* (var. grossum) are the very large pods, neither pungent or acrid, which can be eaten sliced as a salad or used in various ways as a vegetable. They may be used either in a green unripe state or when sweeter, red and ripe.

Cardamoms

Cardamoms are the seeds of various members of the ginger family used in medicines and as a peppery spice. True cardamoms belong to genus Elettaria. *Small* or *cluster cardamom* (E. cardamomun), also known as *Siam* or *Malabar cardamom*, is cultivated chiefly in South-East Asia. The leaves of an Indonesian relative, E. speciosa, are eaten with rice, and its flowers and fruits serve as a tamarind substitute (q.v. *the section* 'Others' *under* Beans).

Other tropical Asian cardamoms belong to the *Ammomum* family. In the East Indian area these include *Bengal cardamom* (A. aromaticum), *round cardamom* (A. kepulanga), the Cambodian *Mount Krevanh cardamom* (A. krevanh), and the Indonesian and Malaysian A. cardamomum and A. dealbatum. The whole of the *Java cardamom* (A. maximum) is used as a condiment. Among others widely grown are the *Nepal cardamom* (A. subulatum) and A. thyrsoideum. *Wild* or *bastard Siamese cardamom* (A. xanthioides) is cultivated in the Indo-Burmese region.

The African near-equivalent (genus Aframomum) which includes *alligator* or *Melegueta pepper* (A. melagueta), whose seeds are known as *Guinea grains* or *grains of Paradise*, comes from West Africa. It was after this spice that the Pepper Coast of Guinea was named. It has long been imported from the Ghanaian region, first carried overland by Arab traders, and appears in official English pharmacopoeias from the thirteenth century onwards. Mixed with ginger and cinnamon, it was used to flavour the wine known as Hippocras whose name refers back to 'the father of medicine', Hippocrates. Keats, with his genius for muddling up hard facts, applies to it the name Hippocrene (actually a legendary fountain whose waters impart poetic inspiration) in his *Ode to the Nightingale*:

> *O for a beaker full of the warm South,*
> *Full of the true, the blushful Hippocrene,*
> *With beaded bubbles winking at the brim*
>   *And purple stained mouth,*
> *That I might drink and leave the world unseen.*

Other species from tropical Africa include *East African cardamom* (A. mala), *Cameroon cardamom* (A. hanburyi), and *Madagascar cardamom* (A. angustifolium).

Cinnamon and cassia (*genus Cinnamomum*)

The bark of several species of this member of the laurel family, especially C. zeylanicum and C. loureirii, is used as a spice in cooking. The trees grow in the East and West Indies, India, China, Brazil and Egypt, but the best cinnamon comes from Ceylon, where the tree is so common that it is even used for firewood.

The spice is best bought in sticks of tightly rolled bark,[1] but chips and other broken parts of it are ground into a powder which is useful if you only want to make small and sparing use of it. There is mention of cinnamon in the Bible, and it seems to have been used both as a spice and a scent by the Jews (Solomon's Song 4:14). The name comes to us from the Hebrew *quinnamon* via the Classical languages and French.

Chinese cinnamon or *cassia bark* (C. cassia) is mentioned in China as early as 2700 B.C. The bark is thicker, less aromatic but more pungent, and preferred to cinnamon in southern Europe. Because it is cheaper it is often substituted for the true variety and sold as such. The bark is also used for the extraction of oil. In the Bible it is only mentioned as a perfume (cf. Psalms 45:8); the Hebrew name *ketziah* passed into Greek and Latin and thus comes down to us. Cassia buds, the dried unripe fruits of this tree, are also used as spices, and an oil is expressed from the leaves.

The bark and leaves of *Indian cassia* (C. tamala) are used as spices in India, while the bark of *Padang cassia* (C. burmanii) from Indonesia is widely used in the East and exported to the U.S. Elsewhere inferior substitutes can be obtained, such as the Australian *Oliver's bark* (C. oliveri) and the *massoia bark* (C. massoia) of New Guinea. *White cinnamon* is the inner bark of the totally unrelated *wild* or *West Indian cinnamon tree* (Canella alba).

Various other members of the family, which includes the bay (q.v.), are used for the purpose of flavouring food. The flower-calyces and bark of the *Santa-Fe cinnamon* (Nectandra cinnamonoides) are used as a spice in the Andean areas of Ecuador, while elsewhere in tropical America the clove-scented bark of the *clove cassia* (Dicypellium caryophyllatum), also known

---

[1] The appearance of these gave cinnamon the name *canella* (diminutive of *canna*, a reed) in Medieval Latin. The French *canelle*, derived from this, was also used in Middle English.

as the *clove-bark tree, pepper wood* or *Cayenne rose*, is collected and sold as a flavouring. In Brazil the seeds of the *pichrim bean* (Aydendron firmulum) are used as a condiment. In Asia, from China to Indonesia, the scented young fruits of the *pond spice* (Litsea cubeba or citrata) are used as a flavouring. In tropical West Africa the seeds of the *spicy cedar* (Tylostemon manii) are eaten raw or roasted, used in soups, as a vegetable or with rice, and are to be found on sale in the local markets. In addition the fragrant flowers are used to flavour rice and other foods.

Cloves *(Eugenia aromatica)*

Cloves are made from the dried unopened flower buds of a tropical tree of the myrtle family. The name refers to their nail-like appearance, deriving from the French *clou-de-girofle* (gilly-flower nail). They are used in cooking as a spice, and are crushed for their oil, used in both perfumes and medicines. (The oil is supposed to be good for allaying toothache, but chewing the actual cloves most certainly is not.) After crushing, the spent cloves are parched and shredded into tobacco to make a flavoured cigarette in the East Indian region.

One of the earliest mentions of cloves is in China, 226–220 B.C., when officers of the court were required to hold them in their mouth while addressing the emperor. They came originally from the Moluccas in Indonesia. At first the Portuguese held the monopoly for European supply. The Dutch, however, expelled them from the region in 1605 and decided to corner the trade still further by destroying all the trees in other islands except for those in their plantations around Amboyna. It was not until 1770 that the French, after enormous difficulties, succeeded in smuggling shoots of the clove and nutmeg to the islands of Reunion and Mauritius, opposite Madagascar. Afterwards, the plant was introduced into the Guianas, Brazil, the West Indies and, of course, Zanzibar.

At one time Zanzibar, and the neighbouring island of Pemba, produced 80 per cent of the world supply of cloves, but Indonesia is now fast catching up. Nevertheless the latter is still an importing nation, producing only one third of its requirements both for home use and exported products. At one time Madagascar was the chief source of supply to the U.S. for clove-oil, but the amount of trade has been dropping for some years.

Other plants bear the name of clove. The fruit of the *Ravensara nut tree* (Agathophyllum aromaticum) is known as *Madagascar clove* or *clove-nutmeg* because the kernels are supposed to combine the flavour of both the last-named species. They are used locally in Madagascar and some are exported to France. The *Brazil clove* is the fruit of the *clove bark tree* (Dicypellium caryophyllatum), and is used locally, while an oil is extracted from the bark. Other names for the tree include *Cayenne Rose*, *pepperwood* and *clove cassia*.

Coriander (*Coriandrum sativum*)

Coriander seeds are hot and aromatic when dried. The leaves may be used in salads and soups, if picked young; this is common practice in Egypt and Peru, where they are regarded as a great delicacy. The seeds are also used as a food additive and as an ingredient of Indian curry. Further east we find the Chinese using the plant as well; they once believed that it bestowed immortality. Nowadays this is attributed to the sayings of Mao Tse Tung.

The Romans brought coriander to Britain, and introduced it to much of the rest of Europe as well. It used to be a commercial crop in England at one time, used for gin manufacturing and in veterinary drugs. The name derives from the Greek κοριαννον, itself an untraced borrowing. In an Anglo-Saxon translation of a Roman medical work, and in other places, we learn a marvellous little spell for inducing pregnancy in women. A number of the seeds, it varies from eleven to thirteen with a preference for the odd numbers, are bound to the left thigh near to the matrix. We are assured that it cannot fail to work.

Curry flavourings and spices

Curry powder has no place in the true Indian curry. Afficianados will inform you that the best and only authentic results can come from freshly grinding the ingredients usually to be found in such powders and using them immediately. Many of the spices used appear under their respective headings, but a few of the less familiar, not to mention the *recherché*, appear here.

One of the more familiar spices is *turmeric*, obtained from the aromatic root of a plant belonging to the ginger family, Curcuma longa. It is a brilliant yellow dye used also for flavouring and medicine, and very useful in the colouring of fried rice. In the native medicine of India it is used for skin diseases and sore eyes. The plant has been cultivated and used widely in Southern Asia from remote times. The name comes from the Medieval

17. AROMATICS AND MUSTARD

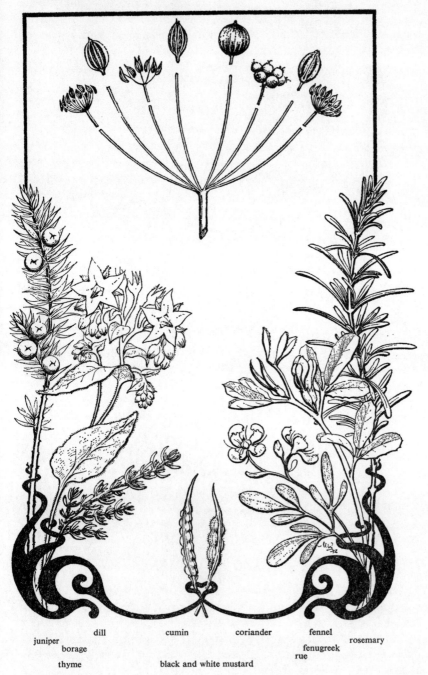

juniper   dill   cumin   coriander   fennel
       borage   fenugreek   rosemary
  thyme   black and white mustard   rue

Latin *terra merita* (deserving earth), *terre mérite* in French, and *tamaret*, *tormerik*, in Middle English. Also used as a spice are the roots of the related *round zeodary* (C. zeodaria) and *long zeodary* (C. zerumbet).

The leaf of the *Jasmin orange-tree* or *mock orange* (Murraya koenigii), a native of Ceylon, is very popular there and known as *curry leaf*. This can be found in Australia, and elsewhere in the tropics. The white fleshy leafstalks of *lemon grass* (Cymbopogon citratus) are used in curries by peasants in the southern Indian area; they are also used in pickles and an oil is obtained from them. The poor also gather the thick succulent stems and leaves of one of the *snap-weeds* (Impatiens flaccida), so called because the seed-pod bursts open at a touch. Its English relative I. noli-me-tangere is known as *quick-in-hand* and *touch-me-not* (*n'y touchez pas* in French) for a similar reason. The edible acid-tasting fruits of the tropical Asian *Indian dillenia* (Dillenia indica), known as *hondapara* to the Sinhalese, are used in curries and also for making jellies and cooling drinks.

The seeds of the pale pink-flowered umbelliferous *cumin* (Cuminum cyminum), which are supposed to have similar digestive properties to dill, are used as a curry ingredient and in many other Near and Far Eastern dishes. The plant was in use in the Mediterranean area and in China from very early times, and is now grown in North Africa, the Near East, India and Mexico. In South America it is one of the most widely used spices. The seeds of the fennel-flower (*see under* Pepper) are sometimes used as a substitute, as are those of the Mediterranean *wild* or *bastard cummin* (Lagoecia cumminoides).

Dill (*Anethum graveolens*)

Both the leaves and seeds of dill may be used as a garnish, and in soups and sauces. it is an ingredient in some vegetable pickles. The name derives from the Germanic languages: In Old English it was called *dili, dille*, and in the old dialects of the continent *dilli* (O.L.G.) and *tilli* (O.H.G.). The alternative name of *anet* is a shortening of the Latin *anethum*. Dill has probably been used since time immemorial, appearing in an Egyptian medical work of about 3000 B.C., and among the Greeks both as a food, a perfume and an incense. It was among the herbs of which the over-scrupulous Pharisees paid a tithe to the temple, together with mint and cummin (Matthew 23:23), but is wrongly translated as anise in the Authorised Version.

Culpepper informs us that 'dill, being boiled and drunk, is good to ease swellings and pains; it also stayeth the belly and stomach from casting [and] it stayeth the hiccough'. Some of these beliefs must have come down from Anglo-Saxon times, when we find dill used in the cure-all magic holy salve and as an ingredient in a charm for the water-elf disease (probably chicken-pox), whose symptoms are colourless fingernails, watery eyes and a downcast appearance. It was also good for the elf-disease, the symptoms of which, yellow eyes and complexion, sound like jaundice. In fact the herb is still put to good use as a stimulant and for stomach upsets. An extremely old recipe for quieting babies with wind, dill water, retains its popularity.

Fennel (*Foeniculum vulgare*)

The umbelliferous fennel, whose seeds and tips taste of aniseed, was originally a maritime plant from Southern Europe but is now to be found quite far north. The Greeks believed it had slimming powers and was able to give a man strength, courage and long life. The Romans employed all parts of it – roots, stems, leaves and seeds – eaten either raw or cooked, and believed it would sharpen the eyesight. For the Anglo-Saxons it was one of the most esteemed of all herbs:

> *Thyme and fennel, a pair great in power,*
> *The Wise Lord, holy in heaven,*
> *Wrought these herbs while he hung on the cross;*
> *He placed and put them in the seven worlds*
> *To aid all, poor and rich.*
> *It stands against pain, resists the venom,*
> *It has power against three and against thirty,*
> *Against a fiend's hand and against sudden trick,*
> *Against witchcraft of vile creatures.*

Mixed with incense, hallowed soap and salt, it was put in a bored hole at the end of a plough for the improvement of the land's yield. The juice was an ingredient in an eye-salve, and the plant of a head-salve. It figures in a potion against insanity, fiendish temptation and typhoid, and in another against elf-sickness. It also appears in the green and holy salves, in a holy drink against the tricks of elves and demons, and in the elf-salve efficacious against the race of elves and spirits that walk in the night, and women who have had sexual intercourse with the devil (witches). In later times it was still believed to be a good defence against witchcraft, as garlic was against werewolves and vampires. At all events it was a most effective flea-repellant. The Italian idiom *dare finocchio* (to give fennel to someone) has the meaning 'to flatter'. Perhaps this is why an English proverb runs 'sowing fennel is sowing sorrow'.

As a vegetable it was widely used by the Franks in the time of Charlemagne. In our times it is only the seeds and tops which are used, although in Italy they eat the blanched shoots of *Italian* or *sweet fennel* (F. dulce), *finocchio* or *finicho*. This, however, both smells and tastes so powerfully of aniseed that some find it too strong for enjoyment, even as an acquired taste. Our name is a corruption of the Latin diminutive for hay, *foeniculum*; in the north there exist nearer equivalents such as *fynkle* or *fenkelle* (cf. the German *fenchel*), deriving from the Late Latin *fanculum*. *Spingel* is another regional name.

The seed is a carminative and, like dill, is an ingredient of gripe-water. When the effects of over-eating began to affect Louis XIV of France he had made up a special liqueur known as *Rossoly*. This consisted of the seeds of aniseed, dill, fennel, coriander and carraway, macerated in spirit for seven weeks; more spirit was then added together with an infusion of camomile and sugar. A general licence to sell this was granted in 1676 and the liqueur had a considerable vogue so long as it was fashionable at court.

The related *hog-fennels* (genus Peucedanum) have a whole variety of uses. *Masterwort* (P. ostruthium), a plant found in both Asia and Europe, has been used for making gin and in some herb cheeses. The roots of the *brimstonewort* (P. palustre), also known as *milk parsley*, serve as a ginger substitute in some Slavic countries. The so-called *East Indian dill* (P. graveolens) is used as a food flavouring in tropical Asia, and in Europe as an aromatic stimulant for cattle; it is also applied as a poultice to their skin when it cracks with the cold.

Ginger *(genus Zingiber)*

The ground roots of several of this family are used as condiments and as a medicinal stimulant. They grow in tropical Asia, Africa, Central and South America, and Australia. Ginger has been used since remote times in India, where the fresh roots are also eaten in salads; from here it was imported to the classical nations. In Melanesia it is used to gain the affection of a woman, and in the Philippines chewing it is a drastic means of driving out the evil spirits. In his *Sexual Life in England*, Ivan Bloch informs us that 'in England ginger was not only taken inwardly as an aphrodisiac, but it was thought that its outward application had a stimulating effect upon the *libido sexualis*. Thus in an erotic pamphlet, *The Amatory Experiences of a Surgeon* (London, 1881), the hero seduces two women whilst he touches them with his hands rubbed in ginger'.

The *common* or *red ginger* (Z. officinale), also known as *Jamaican* or *East*

Indian pepper, is that most commonly used. In Japan the buds of the *mioga* (Z. mioga) may be cooked in several ways, like members of the onion family, after they have first been lightly boiled. The fruits are also eaten, and the roots are a source of Japanese ginger, said to taste like bergamot. In South-East Asia *cassumnar ginger* or *Bengal root* (Z. cassumnar) is cultivated as a condiment, and Z. amaricans can be found on sale in the markets as a condiment and appetiser.

The name derives through the Greek and similar Latin *zingiber* or *gingiber*; in Old French this becomes *gengibre, ginginbre*. The 'b' disappears in both English and the Welsh *sinsir*. Galingale (*see under* Rushes) is a milder substitute. The roots of the *Canada snakewort* (Asarum canadense), known as *colic root* and *wild ginger*, also serve. The whole plant, but especially the root, gives out a strong peppery smell. There are several other American species, some of which might be similarly used, and others are cultivated in the Far East. But the European *hazelwort* or *wild nard* (A. europaeum), known by such corruptions of the Latin name as *asarabaca* and *cabaret*, and found in humid mountain woods and shaded stony ground throughout the northern hemisphere, only serves as a cathartic and emetic, an antidote for drunkenness in Russia, and for snuff.

Horseradish and susumber
(*Cochlearia armoracia or Armoracia lapathifolia*)

The rather hard root of this member of the mustard family is used in medicines or, grated, as an ingredient of the famous sauce. Countrymen regard the root as a remedy for rheumatism. The plant is originally from Eastern Europe, but has spread into Central Asia and Western Europe, and in North America. It is most popular in northern countries and districts, such as in Britain, Germany, and Alsace – where it was once so popular that there were very few meals where it was not present, either raw as an appetiser, or cooked like a vegetable. When horseradish was first used is doubtful, but it is mentioned in several countries during the sixteenth century.

The name is a somewhat disparaging reference to the root's coarseness; it is sometimes written, or understood at least, as *hoarse radish*, because of its extremely hot taste. *Mountain radish* and *horse plant* are other names. The French *grand raifort* (strong root) is echoed in such names as *great raifort* and *reefort*, while the northern names such as *redcole* and *rotcoll* may be a combination of the Old English *rettih* (radish) with the old word for a cabbage, to which family it is related.

18. SPICES I

cinnamon   ginger   vanilla pod   cardamom   turmeric

The roots of the *scurvy grass* (C. officinalis) may be used in a similar way. Also called *scurvy cress* or *weed*, this plant, which grows near the sea, used to be taken on board ship as an antiscorbutic in the old days when fresh vegetables were unobtainable during long voyages and the sailors used to come down with scurvy. The names *spoonwort* and *sea spoons* refer to the shape of its leaves. Other names include *crawlers* and *pinch-herb*.

The roots of the related *toothworts* (genus Dentaria), so named because of their appearance, and also known as *American pepper-root*, serve as a horseradish substitute in North America. Amongst these are the *crinkle-root* (D. diphylla) and the *cutleaf toothwort* (D. laciniata), whose roots are also eaten raw or boiled by the Iroquois Indians. *Wasabi* (Eutrema wasabi) is a Japanese relation cultivated for the sake of its roots, which are used as a condiment.

In Asia the roots of the widely cultivated *horseradish tree* (Moringa pterygosperma) are used as a substitute. The leaves also are used as a vegetable, a seasoning, and in pickles. Its unripe pods, from 2–4 feet long and known as *susumber* or *drum sticks*, are cut up and boiled like beans. They are available from delicatessens in tins and are excellent in curry, but the outsides are extremely hard and woody and impossible to eat; one has to pick them up and suck off the mucilaginous inside and pips, which are lightly hot and taste delicious. The tree is also grown in Mexico under the name *coatli*. In Indian native medicine the flowers and bark are used, and a valuable oil is obtained from the seeds. This is known as Ben oil and is much used by watchmakers, and by perfumers to extract the fragrant principals from various plants.

Marigold (*Calendula officinalis*)

The marigold comes originally from southern Europe. It was once highly regarded for supposed curative properties and appeared in an old preparation for curing the plague. It was also much used in cookery as a pot-herb, as a cheese colourant and a soup flavouring. A seventeenth-century work refers to it as 'the herb general in ale pottage'. The petals are not particularly striking in their taste but look very decorative floating on the top of soups; both they and the leaves may also be used in salads.

The plant belongs to the daisy family, whose peculiarity is that the 'petals', known as florets, are, in botanical fact, each separate little flowers and not strictly petals at all. The centre of the flower is male, the outer florets female, the plant thus being bisexual. As such it is fitting as an emblem of the Virgin Mary, to whom allusion is made in the name, other old versions of which are *Mary Golde* and the slurred *Mary Gowles*. In our less romantic times it rejoices in the name *pot marigold*. Another name is *Mary bud*. The association is born out by other Germanic names such as the Dutch *Marienbloemkijn* and the German *Marienblome*.

In the East the flower is often associated with the Buddha since its saffron colour is that of the robes of monks in Theravadin countries such as Ceylon, Thailand and Burma. It is also used as a votive flower in Hindu worship. In heraldry the marigold stands for devotion and piety. It is also emblem of the sun and has thus gained the name *sunflower* (cf. the French *soleil*), upon which Charles I touchingly observed during his imprisonment in Carisbrooke castle in the couplet

The Marigold observes the sun
More than my subjects me have done.

As the birthday flower for 15 January the marigold is symbolic of cruelty in love, pain, grief and cares, a fact born out by the usual French name for the plant, *souci*. Other religious connections joined with reference to the golden or saffron colour of the flower appear in old names like *holigold* and *Christ's eye*. The latter becomes corrupted into the form *crazy* and is applied to other bright yellow flowers like buttercups and celandines. The colour is also referred to in the names *ruddies, rod's gold, gold* and *goldins* (from the Dutch *guilden*, a guilder or gold florin). More mysterious names include *hardhow* and *jackanapes-on-horseback*.

Marjoram and oregano (*genus Origanum*)

These are members of the mint family. In classical times marjoram was regarded as the herb of happiness. It was once used for brewing ale before the popularity of hops, and is still used as a beer preservative. There was an old belief that sprays of this and wild thyme laid by milk in a dairy would prevent it going sour during thundery weather. It is also considered good for any type of nausea, especially sea-sickness. Taken by English settlers to America, marjoram was there used as an infusion. In his novel *Dog Years* the German novelist Gunter Grass records various beliefs held by the Koshnavians, a Slavo-Teutonic race of the Danzig district. They said it was good for a dog's sense of smell, and had the following proverbs centred about it: 'Marjoram is good for your looks. Marjoram makes money go further. Against devil and hell strew marjoram over the threshold'.

The native English species (O. vulgare), the *grove marjoram* or *joy-of-the-mountain*, was originally known as *organ, organy* or *oregano*, but after the introduction of foreign species it was demoted to *wild* or *bastard marjoram*. The Anglo-Saxons used it in a headache plaster and advised its being eaten for coughs, on the principle of like will cure like, since 'it is of a hot and vehement nature'. The leaves are, indeed, much better than in other species and it would be as well to use them sparingly. In older English documents it was mistakenly called *sweet marjoram*, but this name rightly belongs to O. marjorana, also called *Portugese, North African* or *knotted marjoram*. As a kind of compliment in return this plant has also been called *oregano*. Other species include *French* or *pot marjoram* (O. onites) and the *winter marjoram* from Greece (O. heracloeticum). There is also a genus Marjorana of which the American M. hortensis is used as a flavouring in cooking.

The name is derived via the Old French *majorane* from the Medieval Latin *majorana*. In some of the Romance languages the name has become perverted by being assimilated with the local derivative of the Latin *major* (greater). The Italian *maggiore* appears in *maggiorana*, but *majorana* is also used. The latter is also the Spanish name, but *mayor* has gone to make up *mayorana*. The Latin word entered High German in medieval times, and the variants found therein, *margam, meigramme, meyeron, maigarm, maioran*, have gone towards the formation of the name in other Germanic languages.

Mints *(genus Mentha)*

Mint was used by the Greeks and Romans as a scent; for the Athenians in particular it was the odour of strength, and they used to rub it on their arms. Floors were strewn with the plant during feasts, either because the smell was supposed to cause humans to rejoice and incline them to eating, or because mice were supposed to be frightened by it. The Romans are said to have brought the herb to England, and with it the secret of their invention, mint sauce. Several mints were used in the Anglo-Saxon magic green-salve, and in Tudor times it appears in a wash for heads inclined to sores. Mixed with honey it was dropped into the ear for earache, and was rubbed on rough tongues. The name derives from *minza* (O.H.G.), which became *menthe* in the Wessex dialect of Old English and *mint* in the Midland Mercian dialect.

It might come as a suprise to hear that there are so many species. After an open-air poetry reading in Ireland an old lady, who was handing my wife and me slabs of cake out of her car window, asked us if we would like to come and see her herb garden. Once there we were required to nibble our way along a whole row of different mints. One of these, pineapple mint, I have not since been able to trace, but some of the following were included in our orgy.

Hairy water mint (M. aquatica) is so called because it is to be found growing in damp places. Other names include *capitate* or *marsh whorled-mint, fish mint* (cf. the German *fischmünze*), *lilac flower* and *bat-in-water*.

Corn or *field mint* (M. arvensis), also known as *lamb's tongue, wild penny-royal*, and *field-balm mint*.

Australian peppermint (M. australis) or *river mint*, a native of Australia.

Bergamot mint (M. citrata), also known as *lemon* or *eau-de-Cologne mint* from its sour-sweet smell and taste.

Fringed mint (M. gentilis), *gentle* or *bushy red mint*.

Slender mint (M. gracilis), or *cardiac mint*.

Hairy mint (M. hirsuta), also known as *balm mint, bishop's weed* or *wort*, and *brook* or *horse mint*. This was used among the Anglo-Saxons for bladder trouble.

Country mint (M. javanica) is used in the area of India and Ceylon.

Forest mint (M. laxiflora), indigenous to Australia. Its oil tastes like peppermint.

Peppermint (M. piperita), also called *brandy* and *lamb mint*, of which there are subvarieties known as *black* and *white mint*. This is the official menthol grown commercially and good for colds.

Downy mint (M. pubescens), or *blunt spiked mint*.

Penny royal (M. pulegium), was once the most popular of all the mints and much used for flavourings, as an ingredient in medicines, and as a flea-repellent, from which use it was known as *flea mint* (cf. the French *chasse puce* and German *flohkraut*). The Anglo-Saxons used it for sea-sickness and in an infusion for 'itching of the stomach'. It was especially popular as a meat-stuffing, and was thus known as *pudding grass* or *herb*, and *lurkey-dish*. It was also known as *brother* or *church-wort, hill-wort, organ*, and such corruptions of its name as *lillie-riall* and *puliall*.

Red mint (M. rubra).

Apple mint (M. rotundifolia) is now accounted the best for cooking. It also goes under the name of *wild mint* in some parts, and the village of Mintie in Wiltshire is said to have been named after it since it grows in abundance in the region.

Small or *thyme-leaved mint* (M. serpyllifolia) is only found in Tasmania.

Spear or *spire mint* (M. spicata) also called *common garden mint*, is the one used in some brands of chewing-gum and for other commercial flavourings. There is a subvariety, var. crispa, known as *curly, crisped* or *cross mint*.

Smooth water mint (M. subglabrata).

Horse mint (M. sylvestris) is the wild mint which was included among

19. SPICES II

clove   pepper   chilli   red pepper
     mace   allspice
           nutmeg

the bitter herbs eaten at the Jewish Passover feast; at one time it was extensively cultivated in England and also goes under the names *brook, fish, wood, wind* and *water mint*.

Whorled mint (M. verticillata or sativa).

Lamb or *pea mint* (M. viridis), possibly named after those dishes which it accompanies, is also known as *sage-of-Bethlehem, Our Lady's mint, brown, garden* or *mackerel mint*.

Nutmeg and mace (*genus Myristica*)

Nutmeg is the product of the seed of an evergreen native of the Moluccas Islands, the whole area of which is covered by the trees. Only the inner kernel is used, the outer covering being known as mace. The ready-ground variety is reputed to be far less fragrant than spice freshly ground from the kernel. A species of American roguery has given Connecticut the title of the nutmeg state from an alleged practice there of selling wooden nutmegs as the real thing.

Like other spices from the former Dutch possessions, it was once treated as a monopoly crop by the colonisers, The home government is said to have dispatched orders to the Colonial Governor to reduce the number of nutmeg trees and increase the number of those bearing mace, being ignorant of their common origin. But the tree now grows elsewhere: in Ceylon and the islands of the Indian Ocean, Brazil, the Guianas and the West Indies. Nevertheless, nearly the whole of the world supply still comes from Indonesia. Knowledge of the spice, which is supposed to have medicinal and aphrodisiac qualities, spread only slowly westwards. It was not known in India until about the sixth century, and in Europe until the twelfth century, when it was introduced by the Arabs.

The name nutmeg means musk-nut, *noix mugue* in French and *notamuge* in Middle English. Though others of the genus may be used, the true nutmeg comes from M. fragrans; also used are the seeds of the *Otoba wax tree* (M. otoba) known as *Santa-Fe nutmeg*, which yields white mace, and the *wild nutmeg* (M. laurifolia) of India and Ceylon, which is commercially useless. In Australia a substitute to true nutmeg can be obtained from the *Queensland nutmeg* (M. insipida) and *native nutmeg* (M. muelleri). *Bombay nutmeg* and *mace* (M. malabarica) are practically tasteless adulterants of true nutmeg, deriving from the west coast of India. Also used for the same purpose are the seeds and mace of the *Macassar* or *Papua nutmeg* (M. argentea) of New Guinea.

Other fruits and seeds go under the same name. These include the

Ravensara nut tree, Malagasy or *clove nutmeg* (Agathophyllum aromaticum); *Brazilian nutmeg* (Cryptocarya moschata); *Peruvian nutmeg* or *Chile sassafras* (Laurelia sempervirens); *American, Jamaican, Mexican* or *calabash nutmeg* (Monodora myristica); *Californian* or *stinking nutmeg* (Torreya californica).

Parsley (*genus Petroselinum*)

Curiously enough, parsley is poisonous to birds but is very good for animals and can cure foot-rot in sheep. It contains as much vitamin A as cod-liver oil, and three times as much vitamin C as oranges. In the old days parsley tea was prescribed for rheumatism.

The name derives from the Greek πετροσελινον, which means rock-parsley, so called because it is most frequently found growing throughout Southern Europe in stony places, where it flourishes best. The late Latin version of the name is *petrocilium*, in Old English *petersilie*, and in Old French *peresil*, which becomes the present-day *persil*. The Medieval English *persele* derives from this and was later written *perseley*. There is, therefore, some doubt as to whether the plant was introduced from France in early Tudor times or whether the Romans brought it to Britain with them. Since the Greeks and Romans lumped parsley together with the wild celery-like plants, it is always difficult to tell which they mean when they refer to it.

Usually understood by the name in Britain is *curled parsley* (P. crispum), the cultivated variety of which has slightly larger leaves than the wild European variety. Subvarieties of this are *fern-leafed parsley* (P. crisp. filicinum) and *Italian parsley* (P. crisp. neapolitanum). The related wild parsleys of Britain, which are also related to the celeries, Apis sylvatica and Apis saniculum, were well known to the Anglo-Saxons and used by them in a salve for broken heads. It was probably these alone to which their word for parsley referred.

The plant is supposed to be given to those whose birthday falls on 30 October. For the Hebrews, however, parsley was a symbol of the coming spring and human redemption. Parsley wound around a carrot means fecundity. I imagine that the symbology has Freudian implications, especially as parsley seems to have some connection with lust and lovemaking. 'Fried parsley will bring a man to his saddle and a woman to her grave' runs one proverb, and another declares

> *When the mistress is the master*
> *The parsley grows the faster.*

Children are told that they come out of their mother's parsley-bed as well as that they were found under a gooseberry bush or a cabbage (a male in France) or in a rose-bud (a female). A person said to have been born in a parsley bed is naïve and gullible.

A very popular Korean poem by the sixteenth-century Yang Hui-ch'un has secured the author's fame, although nothing more is known about him:

> *I offer you a bundle of parsley*
> *Fresh, clean,*
> *A bundle of parsley just for you:*
> *No one else, God forbid! just you.*
> *Maybe not too delicious, my dear,*
> *But try again; once more test and taste.*

Parsley also has more sinister connections. It is considered lucky to transplant parsley from the old home to the new because the Devil takes his tithe of it or, according to some sources, leaves only a tithe. One proverb states that 'parsley, before it is born, is seen of the Devil nine times', a Yorkshire version of which runs 'parsley-seed goes nine times to the devil'. Elsewhere it is asserted that it must be sown nine times. 'Welsh parsley' is an old expression for hempen rope; to be threatened with it meant hanging.

Pepper (*genus Piper*)

Pepper was one of the first items of commerce between the East and Europe. Its long history might be instanced by the name itself, which has passed unchanged through the Latin from the Greek πιπερι, possibly deriving from some old Aryan root. The Sanskrit for the *long pepper* (P. longum) is *pippali*. It was well known to the nations of the classical world and certainly much used by the Romans for we find that one of the articles demanded by Alaric for the ransom of Rome in 408 was 3000 pounds of pepper. So highly was the spice regarded in medieval times that peppercorns commanded high prices, and tribute and rent were paid in them. A peppercorn rent most certainly was not inexpensive when the phrase was first invented.

For a long time the Arabs held the monopoly of the pepper trade into Europe, obtaining the spice at first from India, later from Java, and bringing it through Arabia to Alexandria in Egypt, from whence it was shipped to Venice. By 1180 there was a Guild of Pepperers to handle its distribution in England. At about the same time there was a flourishing

Java to China pepper trade. Desire for a more ready access to this and other spices lead to European attempts to find alternative routes to India. Columbus' search for a western route ended in the discovery of America; and Vasco de Gama's search in the opposite direction resulted in his circumnavigating Africa, breaking the Arab monopoly, and setting up the Portuguese colony of Goa in India.

Pepper gives rise to many proverbs. One of Jewish origin states that 'one grain of sharp pepper is better than a basketful of gourds'; a wise man is better than a load of fat-heads, in other words. 'To take pepper in the nose' means to take offence; of a man much given to wrath it is said that 'he is not pepper-proof'. Hot emotions, however, are not necessarily followed by like deeds for 'pepper is hot in the mouth and cold in operation'. One is doing the wrong thing when one 'puts pepper on strawberries' and a pointless thing in 'carrying peppers to Hindustan', no better than 'taking coals to Newcastle'. A Scottish proverb seems to date back to the time when the value of pepper was of much account: 'I wold I had as mickle pepper as he counts himself worthy mice dirt!'.

There are, of course, many more kinds of pepper than the grey stuff poured with abandon over badly cooked small-town lunches before they can be coaxed into tasting like anything at all! That best known to us comes from the fruits of the Asian climbing shrub P. nigrum, originally a native of western India and now introduced into Malaysia, Indonesia, Thailand, the Philippines and the West Indies. White pepper is made from the ripe fruit after removing its outer husk. It is much milder than black pepper, which is made from the same berries dried till dark and hot and then ground. One can also obtain the peppercorns unground and then grind them when needed, the result being an even hotter spice.

Another Asian pepper is derived from the *long peppers* (P. longum and P. officinale) of tropical Asia. The first of these grows wild in northern India; it was known to the Greeks and Romans, who preferred it to black pepper. P. saigonense grows in Vietnam and is only used locally, while P. bantamense is found throughout Indonesia and Malaysia. Its bark is added to water in order to give clothes a pleasant smell after washing. Another useful species is *betel pepper* (P. betle), whose evergreen leaves are used as a masticatory throughout the East. In Indian native medicine the roots and dried flower-spikes are used for dyspepsia. The *kava pepper* (P. methysticum) is the basis of the narcotic *kava* drink of the Polynesians, who chew the roots and spit them into a gourd, adding water later.

Cubebs are the dried unripe fruits of the Indonesian P. cubeba; these were used as a spice in medieval Europe, and later medicinally. *African cubebs* are the product of P. guinense from tropical West Africa. *Ashanti* or *West African pepper* is obtained from these and P. clusii. It was imported into Europe by the merchants of northern France and Flanders as long ago as 1364. *Native pepper* (P. excelsum) grows in Australia, New Zealand and the Pacific islands. In Polynesia the natives eat the flower-clusters uncooked. There are various peppers also to be found in tropical America, among which might be mentioned *holy pepper* (P. sanctum), whose aromatic leaves are used for flavouring soups in Mexico. The fruits of others are used as spices.

Pepper-elder (Peperomia vividispica) is a relative, also to be found in South and Central America, whose young leaves and shoots are used in salads by the natives. Similarly used in the same area are those of the *white alder* (genus Clethra). Some people in North America use the *sweet pepperbush* or *summer-sweet* (C. alnifolia) and *cinnamon* or *painted leaved white alder* (C. acuminata), either for salads or as a pot-herb. In mountainous districts in Japan the young leaves of the *Japanese white alder* (C. barbinervis) are eaten with rice.

The wrinkled black berries of the *West Indian bay berry* (Eugenia pimenta), known as *pimento*, are counted among the peppers. They are popularly supposed to combine the flavours of cinnamon, cloves and nutmeg, and are often called *allspice*. The plant is a member of the myrtle family native to Central America and the West Indies; an infusion of its leaves is known as Trinidad tea. During the eighteenth century the North American settlers used as a substitute for pimento the dried and powdered berries of the *benzoin* (Lindera benzoin), alternatively known as *Benjamin, spice* or *fever bush*, and *wild allspice*.

Among other American peppers are such products of the capsicum family as *paprika* and *Cayenne, Guinea* and *Spanish pepper* treated elsewhere (*see under* Capsicums). The small purple berries of the *Peruvian mastich tree* (Schinus molle) can be made into a spice known as *false pepper* which is sometimes used as an adulterant of the true pepper. The berries and bark are much used in native medicine and its gummy exudation is used as a masticatory. Although it is of Andean origin the tree also goes under the names *Brazilian, Californian* and *Australian pepper tree*. Its fruits, and those of the *Chilean pepper tree* (S. latifolius), are used for making fermented drinks.

There are other members of the pepper family to be found in tropical South America. A less sharp but more flavourful spice can be obtained from the fruits of one of these, genus Xylopia. The wood, bark and berries of the *bitter wood* (X. frutescens) have the taste of orange-pips; so strong is this that the flesh of the wild pigeons feeding on them is said to be permanently flavoured with it. Also used for spices are the seeds of the *pindaiba* (X. sericea), *bugre* (X. grandiflora) and *Brazilian pepper* (X. carminative), known locally as *pao d'Embira* and *pimenta de Macaco*. The West African *habzeli* (X. aromaticum) provides the spice known as *Ethiopian, negro* or *kimba pepper,* and *grains of Selim*. The fruits of the closely related genus Unona are used for similar purposes in tropical Africa and Asia. Members of the *prickly ash* family (genus Zanthoxylum) bear small black seeds also used as a pepper-like spice and tasting faintly of orange peel. The most important among these include *Chinese pepper* (Z. bungei) and *Japanese pepper* (Z. aromaticum). Others are to be found growing from Indo-China to India.

Another Eastern substitute for pepper is found in the hot and aromatic seeds of the *fennel-flower* family (genus Nigella), long used as a condiment and still very popular, especially in curries. *Black cumin* (N. sativa) is used among the Arabs; in Charlemagne's time it was known to Europeans as *gith*, and was later used under the name *allspice*. The seeds are still added to bread in some countries. During medieval times, however, the most popular was *love-in-a-mist* (N. damascena) whose wiry spiky appearance has also given it such names as *old man's beard*, *Devil-in-a-bush* (the Gaelic name is *chase-the-Devil*), *Devil-in-a-mist*, *Love-in-a-puzzle*, and in Cornwall *love entangle* (probably a corruption of *love-in-a-tangle*).[1] Other names include *kiss-me-twice-before-I-rise* and *St Katherine's flower*.

The leaves of the *yarrow* or *milfoil* (Achillea millefolium) are rather similar to the above, as is instanced by such names as *hundred-leaved grass* and *thousand-leaved clover*. The French *millefeuil* has become the English *milfoil*. These are used in soups and salads, especially in northern France and Germany (eg. the German *gründonnerstag suppe*) to provide a characteristic tang. It is also known as *old man's mustard* or *pepper*, and as *Devil's nettle* because of the children's practice of drawing the leaves across the face to leave a slight tingling sensation. It is also used to make an infusion in various herbal remedies. In Sweden the leaves serve for tobacco and to make a kind of beer, and the roots of the allied *sneezewort* (A. ptarmica) have been used as snuff.

The plant is to be found throughout the Northern hemisphere. Its leaves are popularly supposed to be a great healer of open wounds and abcesses, a fact celebrated by such names as *nose-bleed*, *carpenter grass* and *thousand seal*. This, it is said, was discovered by Achilles, after whom the Romans named the plant. As birthday flower for 16 January it is supposed to be a healer of broken hearts. Sprigs of it plucked from a young man's grave while repeating a mystic formula were thought to bring dreams of their lovers to love-sick maidens. It also figured in bridal bouquets under the name *seven-years' love*. Witches used it in spells and love-potions, and it was also worn as a protection against witchcraft.

[1] Cf. the similar ideas behind the German names *braut in haaren, Gretchen im busch, Gretel in der staude, Jungfer im grünen.*

Plantain   (*genus Plantago*)

This herb grows widely over Britain and was once cultivated as a condiment. Its leaves, which have a sour-sharp taste, were used in salads. The name plantain is a corruption of the Latin *plantago*, the root element of which is *planta*, the sole of the foot; it refers to the broad shape of the leaves. Usually understood by the name is the *greater plantain* (P. major), but there are several other kinds. These include the *broad leaved plantain* (P. maxima); *hairy* or *hoary plantain* (P. media); *ribwort* or *long plantain* (P. lanceolata) known in the U.S. as *English plantain*.

The old name used in Anglo-Saxon times for the plantain was *wegbreed*, the plant that grows by the roadside. This meaning persists in the names *wayborn* and *wayberan*, corrupted in the name *warba leaves*. Influenced by the shape of the leaves, it was also corrupted into *waybroad* (cf. the German *wegebreit*). But the most correct survival, *waybred*, soon came to be thought of and spelt *waybread*, something that can be plucked and chewed by the traveller on his journey; under this influence the name proceeds further from its original in the form of *wayside bread*. Less complimentary references to it exist in the names *rabbit's beef* or *pudding*. The rounded shape of the flower-head has earned it the name of *scent-bottle*, *bobbins*, *baskets* and *blackjacks* and the lined leaves are referred to in the names *rabbit's* or *donkey's ears* and *banjo leaves*, with which one may compare *violin strings*, properly applied to P. lanceolata. Such names as *baccy* or *wild tobacco* may simply refer to the flower-head's resemblance to a plug of tobacco, or it might be that the leaves really were used as a substitute. The heads of the hairy plantain, called *Welsh plantain* by some, are further characterised as *chimney sweeps* or *sweep's brush*. *Devil and angels* (P. lanceolata) refers to the dark and light contrasts of its curious head.

Returning to the leaves again, it is their appearance which gives ribwort its name; it is also known as *ribgrass*, which has been corrupted to *ripple*. But the names *fire leaves* or *weed* refer not to their colour but to a country belief about them. In the *Gardener's Chronicle* for 1860 it is accounted for thus:

'In Gloucestershire the name is given to the plantain, more especially P. media ... In Herefordshire [it is] used for Scabiosa succisa ... Both are named fire-leaves on the same principle, for we have seen the farmer of Gloucestershire with a plantain leaf, and he of Herefordshire with a scabious leaf, select specimens and violently twist them to ascertain if any water could be squeezed out of them. If so, this moisture is said to induce fermentation in newly-carried hay sufficient to fire the rick.'

Apparently it is also believed in other parts of the country that a large quantity of ribwort has the effect of causing hay to heat.

The Latin root of the name, with its implications of the leaves resembling footsteps, has given rise to some of the more ingenious names of the greater plantain; explanations given for them often sound even more far-fetched. *Traveller's* or *Englishman's foot* is explained by a belief that it sprang up in the wake of the empire-makers, while *white-man's footprint* is supposed to have been the invention of Red Indians. *Follower-of-the-heel-of-Christ* is said to be based on the legend that it sprang up in the Master's footsteps. I suppose that such a superstition would explain the belief that the leaves, placed in a traveller's stocking, will relieve feet sore with walking, but it seems that it has more to do with sympathetic magic, and the doctrine of signatures that like will cure like. But certainly it would refer us back to the name *traveller's foot*. It is such a belief which is put forward to explain the name *healing blade, herb* or *leaf* (*läkeblad* in Swedish dialect), and those belonging to ribwort, *herb of healing* and *leechwort*. But, as will be seen below, they may equally be reminders of the plantain's enormous reputation in Anglo-Saxon times, and even earlier. A Roman medical work says plantain is good for diseases of the stomach, the bite of a snake or mad dog, the sting of a scorpion, fever, swellings, blisters, ulcers and other kinds of sore and poison. The same belief would appear to have been held in Ireland, where the plant bore the name *slanlas* from the Gaelic *slan*, meaning health or healthy. Incidentally another name, translated out of the Gaelic, *Patrick's* (*begging*) *bowl*, refers to the head.

The peak of the plantain's popularity was in the Middle Ages, especially in Anglo-Saxon Britain, when it was lauded thus in the Nine Herbs Charm:

> *And thou, plantain, mother of herbs,*
> *Open from the east, mighty within,*
> *Over thee chariots creaked, over thee queens rode,*
> *Over thee brides made outcry, over thee bulls gnashed their teeth.*
> *All these thou didst withstand and resist;*
> *So mayest thou withstand poison and infection*
> *And the foe who fares through the land.*

After enumerating the other herbs: mugwort, lamb's cress, cock's spur grass, mayweed, crab-apple, nettle, thyme and fennel, the charm continues

> *Now these nine herbs avail against nine evil spirits,*
> *Against nine poisons and against nine infectious diseases,*
> *Against the red poison, against the running poison,*
> *Against the white poison, against the blue poison,*
> *Against the yellow poison, against the green poison,*
> *Against the black poison, against the grey poison,*
> *Against the brown poison, against the crimson poison,*
> *Against the snake-blister, against water-blister,*
> *Against thorn-blister, against thistle-blister,*

> *Against ice-blister, against poison blister;*
> *If any poison comes flying from the east,*
> *Or any comes from the north.*
> *Or any from the west upon the people.*

With an advertisement such as this it is not surprising to find the plantain figuring widely in the magic and medicine of the time. It is an ingredient for a bonesalve against headache and limb weariness; in a poultice for a broken head, and another for snake-bite. Hairy plantain is specified for another head-salve. In the case of headache, yet again, one is advised to 'dig up plantain without iron before sunrise. Bind the root about the head by means of a wet bandage'. For a quartan fever one should 'drink juice of plantain twice in sweet water before the fever attacks'. Even more mysterious, 'if milk has turned sour, bind together plantain, cockle and cress, lay them in the milk-pail and do not set the vessel down on the earth for seven days'.

In the thirteenth century we find very much the same traditions concerning plantain's efficiency against headache and poison. In his *Book of Secrets* Albertus Magnus testifies thus:

> The root of this hearb is marvellous good against the paine in the head because the Signe of the Ram is supposed to be the house of the planet Mars (whose herb plantaine is), which is the head of the whole world. It is good also against evil customs of a mans stones and rottennesse or filthy biles, because his house is the signe Scorpio, and because a part of it holdeth sparma, that is, the seed which cometh against the stones, whereof of all living theings be engendred and formed.
>
> Also the juyce of it is good to them that be sicke of the perillous Flixe, with excoration or rasping of the bowels, contuinual torments, and some blood issuing forth, and more it purgeth them that do take a drinke thereof, from the sicknesse of the fire of blood or emerhods, and of the disease of the stomacke.

Portentous as the reputation of the plantain may have been, this did not stop children having a marvellous time with it, with ribwort especially, whose names testify to a variety of games into which it entered. *Carl* or *curly-doddie*, a name which is carried by some other plants, refers to the head which was twisted round to the limit of the stalk by Scottish children, who repeated 'a carl-doddie rhyme' or spell as it unwound. One of them runs:

> *Curly-doody, do my biddin'*,
> *Soop my house and shool my widden.*

I think the last line may be translated 'sweep my house and shovel my garden'.

Tinker-taylor grass refers to a game played by girls, doubtless like the one still played with fruit stones now. They would collect as many heads as they could find and then count them over to find the state or profession of their husband. A girl from Cambridgeshire told me that she used it for 'taking photographs', winding the leaf round the blade and trying to shoot the head into the client's face, in which case the photograph was successful.

By far the most popular of the games was one now generally outmoded by the use of the horse chestnut, although it is still played in some parts of the country. A Yorkshire girl told me she played it when she was young. A nineteenth-century description of one form of the game as played in Scotland informs us 'it is customary with children to challenge each other to "try the kemps". A kemp consists of the stalk and the head or spike. Of these an equal number is skilfully selected by the opposed parties; then one is held out to be struck at with one from the opponent's parcel, which is thrown aside if decapitated, but if not is used to give a stroke in return. Thus, with alternate strokes given and received, the boys proceed until all the kemps but one are beheaded, and he who has the entire kemp in possession considers himself the victor'. In some versions of the game the number selected is fifty each, and the winner is declared 'cock' over the others. The stalk is quite tough and it sometimes requires skill and persistence to behead it.

The name *kemp* derives from Old English *cempa* and Danish *kæmpe*, meaning a warrior. This meaning remained right into Middle English times, and the Swedish dialect name *kampar* is still used for the plant, which in this country was usually ribwort. Other names having their origin from the game include *cocks, cocks and hens, fighting cocks; knock heads, head man, heads and tails; soldiers, soldier's tappie, scat Tommy, black-jacks, jack-straws, swords and spears. Conquerors* has developed into the alternative conkers (conquers), the game played with horse chestnuts threaded on a string (q.v. under Chestnuts).

Rosemary (*Rosmarinus officinalis*)

The Latin name for this member of the mint family means sea-dew; it was so called because it was found growing wild on cliffs by the sea. The English name is a corruption growing out of a mistaken belief that it was connected with the Virgin Mary. Some say it was first introduced to England sometime between the fourteenth and sixteenth centuries, but it is often mentioned in Anglo Saxon medico-magical works. Chewing the root was supposed to be good for the toothache, and it is included as a poultice and a drink for a pain in the right side.

Later on it was regarded as a powerful defence against evil, so much so that it was often to be found in churches and was even used as incense, in reference to which practice it was called *incensier* by the French and *weihrauchwurz* by the Germans. It was also used as a Christmas decoration and appeared in festive dishes on this and other occasions, as well as a flavouring in ale and wine. It is one of the ingredients of the eau-de-cologne made in its once native areas of Southern Europe and Asia Minor.

The practice of hanging rosemary in churches is sometimes explained by saying that it is a herb of hospitality, and that at Christmas it is put there to welcome in the elves and fairies. As a herb of affectionate remembrance it is the birthday plant for 17 January. An English proverb seems to link it with parsley (q.v.) as an emblem of petticoat government: 'Except where the missus is master, there rosemary will never blossom'.

Rue (*Ruta graveolens*)

The Greeks wore rue as protection against spells and the evil eye, and amongst the Romans it was regarded as an eye salve. Mithradates VI, king of Pontus in the first century B.C., had it as the main ingredient of a famous poison antidote which was highly regarded until medieval times. He interested himself in such things for reasons of self-defence, and became something of a legend as well as a thorn in the side of the unscrupulous Roman imperialists. In fact so successful was he that when he wished to avoid being taken alive by his finally victorious enemies it was necessary to burn himself alive.

Rue was introduced into Britain by the Romans, who needed it to flavour their wine. Then we find the Anglo-Saxons also recommending it as an eye-salve and bone-salve, and the leaves for stanching nose-bleeds. It also appears in potions for jaundice, typhoid, and demonic possession. After the Norman Conquest it became known as a herb of repentance through being mistakenly connected with the verb 'to rue' which, however, comes from an entirely different root, the Old English *hreowan*. The herb itself was called *rude* by the Anglo Saxons after the Latin *ruta*, deriving originally from the Greek.

The mistaken derivation underlies the mad Ophelia's touching herb-distribution in Shakespeare's *Hamlet*. 'There's rue for you, and here's some for me; we may call it herb of grace o' Sundays.' Other forms of this name appear in *herb-grace, herby grass* and *yerb-a-grass*. The name *Ave-grace* is also connected with this, although it probably looks towards the prayer to the Virgin beginning 'Ave Maria, gratia plena'. The name *herb-of-repentance* is said to go back only to about 1750, about which time grew up the custom of strewing the dock at the Old Bailey with rue because an outbreak of 'jail fever' at Newgate made it necessary to avoid contagion. (Dickens refers to this in *A Tale of Two Cities*, part II, Ch. 2.) *Countryman's treacle*, another old name, looks back at its supposed curative properties. and perhaps to the practice of Mithridates himself. The melancholy A. E, Housman devotes his poem 'Sinner's Rue' to the plant.

> *I walked alone and thinking,*
> *And faint the nightwind blew*
> *And stirred on mounds at crossways.*
> *The flowers of sinner's rue.*
>
> *Where the roads part they bury*
> *Him that his own hand slays,*
> *And so the weed of sorrow*
> *Springs at the four cross ways.*
>
> *By night I plucked it hueless,*
> *When morning broke 'twas blue;*
> *Blue at my breast I fastened*
> *The flower of sinner's rue.*
>
> *It seemed a herb of healing,*
> *A balsam and a sign,*
> *Flower of a heart whose trouble*
> *Must have been worse than mine*
>
> *Dead clay that did me kindness,*
> *I can do none to you*
> *But only wear for breastknot*
> *The flower of sinner's rue.*
>   (*Last Poems*, xxx)

Sage (*genus Salvia*)

Sage is yet another member of the enormous mint family, and is of Mediterranean origin. Bees make an excellent honey from its flowers; the sage honey of Fiume was a prime luxury of the ancient world. The Romans regarded it as a kind of heal-all, about which grew up their proverb 'Why should a man die while sage grows in his garden?' This belief persists in England, where a similar proverb runs:

> *He that will live for aye*
> *Must eat [butter and] sage in May.*

Possibly an underlying reason for this tradition growing up was another belief that sage was practically indestructible. This is given utterance in the proverb

Set sage in May
And it will grow alway.

The Roman tradition persists in Anglo-Saxon times, for we find sage in an all-purpose health-potion which 'avails against sudden illnesses, fever, tertian fever, poison and infection. Whoever applies this remedy may keep himself free from all danger of diseases for twelve months'. It is also to be found in a headache salve, both the Holy and the Green Salves, and in potions for jaundice, typhoid, and demonic possession. The magician Albertus Magnus puts it to a most curious and unorthodox use. 'This hearb being purified under dung of cattell, in a glasen vessel, bringeth forth a certain worme or bird, having a taile after the fashion of a bird ... and if the aforesaid serpent bee burned, and the ashes of it bee put in the fire, anon ther shal be a rainbow, with a horrible thunder.'

More mundanely, sage-seeds crushed in water were used as a poultice to draw out thorns and splinters, and to bring down swellings. And, as with so many herbs before tea came along, the leaves were once used for an infusion but, more importantly, served as an ingredient in brewing ale. This, according to one Tudor commentator, made it 'more heady, fit to please drunkads, who thereby, according to their severall dispositions, become either dead drunke, or foolish drunke, or madde drunke'. In America the *Californian chia* (S. columbariae) provides an important food for the Indians. Its seeds are parched and ground into *pinole* meal and made into dark-coloured loaves, or alternatively, added to wheat and corn flour. The roasted seeds of the *thistle sage* (S. carduacea) are made into a flour by the Californian Indians.

The most common sage in England is *garden sage* (S. officinalis), but amongst other species are *clary sage* (S. sclarea); *wild English clary* (S. verbenaca); *painted red* or *scarlet sage*, (S. splendens), referring to the colour of the leaves; *pineapple sage* (S. rutilans), so called because of the smell of the leaves; and *red-topped sage* (S. horminum). Various popular names attach themselves to some of these, all tending to the same meaning: *God's* or *Christ's eye, occulus Christi, Goody's eye, see bright, clear eye*. The seeds, referred to as eye-seeds, were once used in eye infections. *Salvia*, the Latin generic name, refers to the healing properties of the plant; from the same root we derive our word salvation and the French *salut* (health). The name *sage* itself was borrowed in the medieval period from the French *sauge*.

Savory   (genus *Satureia*)

Savory, whose name derives from the Old English *sæderie*, influenced by the adjective 'savoury', is a member of the mint family. It is a Mediterranean plant which is thought to have been brought to Britain by the Romans. So popular did it become that it was one of the first herbs to be taken by early settlers to America. The leaves have a peppery aromatic taste and may be used as a pot-herb or salad flavouring. They are said to be especially good with broad beans. Most commonly cultivated is the *summer* or *garden savory* (S. hortensis) of southern Europe; *mountain* or *winter savory* (S. montana), from southern Europe and North Africa, is also cultivated as a seasoning. The so-called *Canadian savory* (S. thymbra) is originally from southeastern Europe and the Near East.

Sesame   (*Sesamum indicum*)

Sesame, also known variously as *simsim, gingelly, beniseed* and *til*, is grown in Africa, Asia, South and Central America, for the sake of its seeds, which are used to express an oil used on salads and in cooking. In many ways it is similar to olive oil, to which it is sometimes added as an adulterant; it, in turn, is adulterated with peanut oil. In India it is used as a ghee substitute, and for anointing the body and other healthy purposes. The cake left after the oil has been expressed is used as food by the poor.

The plant was first cultivated in Africa, becoming known to Egypt and the countries of the Near East about 1300 B.C. It was taken early to India and reached China by the first century. Slavers took the plant to America, and it is now cultivated also in the southern U.S. In Africa and Asia the seeds are fried and added to soup, bread and other preparations. In West Africa the young leaves are also used as a pot-herb.

The seeds of two wild species, S. alatum and S. radiatum, are eaten in Africa, as are those of Ceretotheca sesamoides, which is occasionally cultivated. The leaves of the latter are also used as a vegetable, comparably with the East African S. calycinum. *Niger seed* (Guizotia abyssinica), also of African origin, is used in much the same way, and yields a yellow odourless oil with a pleasant nutty taste used in cooking. In India the seeds are fried and eaten, and added to chutneys and condiments. Originally from Ethiopia, it was taken early to India where it is often grown together with finger millet (q.v. *under* Millet); these are the only two countries where it is under extensive cultivation.

Tarragon and family (*genus Artemesia*)

Tarragon (A. dracunculus) is a Eurasian member of the daisy family. Its name derives from the Arabic *tarkhun*, itself a borrowing from the Greek for a dragon (δρακων). The French call it *estragon* and the Italians *draconcello* (little dragon, cf. the Latin *dracunculus*). Its first mention in Europe dates from the twelfth century. It is widely used in salads and pickles, and the leaves serve as an appetiser in Persia. In ancient times the plant was believed to cure the bites of mad dogs and venomous creatures; it was also considered an aphrodisiac and its roots were used to cure toothache. Although one can buy dried tarragon there is really no substitute for the fresh leaves.

False tarragon (A. dracunculoides) is an American relative whose oily seeds are used for food by the desert Indians of the western U.S. and whose leaves are baked between hot stones. The leaves of the *silky wormwood* or *fringed sage brush* (A. frigida) are used for flavouring sweet corn by the Hopi Indians, while other tribes use the seeds of various species in much the same way as sage (q.v.). These include *black sage* or *large sage brush* (A. tridentata), *three-tipped sage brush* (A. tripartita) and *Wright's sage brush* (A. wrightii). Several more are used medicinally both in the U.S. and Europe.

Tarragon has several more or less useful European relations. These include *mugwort* (A. vulgaris), mentioned below; *wormwood* (A. absinthium), from which the faintly poisonous oil that goes into flavouring the drink absinthe, symbol of the evil 1890s, is expressed; and *southernwood* (A. abrotanum), whose name is a contraction of 'southern wormwood', which was once a constituent of beer. About this last, which goes under other names as well, there is a very beautiful poem by Edward Thomas which gives one some idea of what the family is like.

> *Old Man, or Lad's-love, – in the name there's nothing*
> *To one that knows not Lad's-love, or Old Man,*
> *The hoar-green feathery herb, almost a tree,*
> *Growing with rosemary and lavender.*
> *Even to one that knows it well the names*
> *Half decorate, half perplex the thing it is:*
> *At least, what that is clings not to the names*
> *In spite of time. And yet I like the names.*

354 / *Tarragon and family*

> The herb itself I like not, but for certain
> I love it, as some day the child will love it
> Who plucks a feather from the door-side bush
> Whenever she goes in or out of the house.
> Often she waits there, snipping the tips and shrivelling
> The shreds at last on to the path, perhaps
> Thinking, perhaps of nothing, till she sniffs
> Her fingers and runs off. The bush is still
> But half as tall as she, though it is as old,
> So well she clips it. Not a word she says;
> And I can only wonder how much hereafter
> She will remember, with that bitter scent,
> Of garden rows, and ancient damson trees
> Topping a hedge, a bent path to a door,
> A low thick bush beside the door, and me
> Forbidding her to pick.
>   As for myself,
> Where first I met the bitter scent is lost.
> I, too, often shrivel the grey shreds,
> Sniff them and think and sniff again and try
> Once more to think what it is I am remembering,
> Always in vain. I cannot like the scent,
> Yet I would rather give up others more sweet,
> With no meaning, than this bitter one.
> I have mislaid the key. I sniff the spray
> And I think of nothing; I see and I hear nothing;
> Yet seem, too, to be listening, lying in wait
> For what I should, yet never can, remember:
> No garden appears, no path, no hoar-green bush
> Of Lad's-love, or Old Man, no child beside,
> Neither father nor mother, nor any playmate;
> Only an avenue, dark, nameless, without end.

Beside the names *lad's* or *boy's love*, or *savour*, and *old man* (which makes a pair with *old woman*, a name belonging to wormwood), it is also called *maid's love* or *ruin*, and *kiss-me-quick-and-go*. These may apply either to a fancied aphrodisiac effect similar to that imagined of tarragon, or else because an ointment made from its ashes was supposed to stimulate the growth of a beard in young men. The archaic name *averoyne*, which may be connected with the Scottish *apple-ringie*, derives from the Latin *abrotanum* through the French *aurone*.

Mugwort has been used as a culinary herb, and medicinally as a diuretic and antispasmodic. A Scottish proverb vouches for its health-giving powers:

> *If they wad drink nettles in March*
> *And eat muggins in May*
> *Sae mony braw maidens*
> *Wad not go to the clay.*

The young leaves and shoots also serve as a condiment, and in the Far East are used for flavouring festival rice-cakes, and other dishes. In Japan the young plant is boiled in the spring. The name is supposed to derive from the Middle English *mough* or *moughte*, a moth or maggot, and therefore pairs with wormwood (*wyrmwyrt* in Middle English). Because of its name the plant is said to be a good defence against insects. On the other hand some think the name to be a contraction of 'motherwort'; an alternative name for the plant is *maidenwort*.

Mugwort is the birthday flower for 3 November and symbolises happiness. For the Aztecs it was the emblem of the mother goddess Chalchiutlicue. It was once worn as a charm against the ague, and it was believed that to carry it in the hand cured a wanderer of fatigue. Scottish boys once used its hollow stem as a popgun, from which practice come such names for the plant as *bluchtan* and *bowlochs*. *Old Uncle Harry* links up with other names in the family, such as old man and old woman for southernwood and wormwood. It also goes under such curious names as *fellon herb*, *dog's ears*, *sailor's tobacco*, *green ginger*, and *grey* or *white serpent*, *toad* or *weed*. *Petty mugweed* and *broad-leaved wormwood* are more staid designations. Finally, the names *St John's plant* and *John's feastwort*, referring no doubt to the time of its appearance, may link the plant's supposed medicinal virtues to those of balm, if it is indeed mugwort to which a seventeenth-century debunker referred when he wrote 'Balme, with the destitution of God's blessing, doth as much good as a branch of herb-John in our pottage'.

Thyme (*genus Thymus*)

For the Greeks thyme was the herb of courage, elegance and grace; the highest compliment that could be paid a man was to tell him he smelled of thyme. Indeed it was very popular as a bath scent and was used in other ways as a male cosmetic. From the practice of burning it with sacrifices comes its Greek name $\Theta \upsilon \mu o \nu$, from the verb, $\Theta \upsilon \varepsilon \iota \nu$ (to burn sacrifice). The name was written as *thymum* by the Romans, who grew it near bee-hives for the sake of its honey, and also flavoured cheeses and liqueurs with it. The word passed into English via the French *thym*. The plant was highly regarded by the Anglo-Saxons, who coupled its power with that of fennel (q.v.) in the Nine Herbs Charm. Headaches were once treated with thyme vinegar. Pounded and mixed with syrup, the herb was used for sore throats and whooping-cough. Out of it an ointment was made for rheumatic swellings.

Thyme is the birthday plant for 9 June. It was once sacred to both Mars and Venus and thus, in the Middle Ages, was given by ladies to their knights in order to protect them in battle. Young girls used to put it under their pillows on St Agnes' Eve (20 Jan.) in order to discover whom they would marry. Keats, in a poem dealing with this subject, mentions other parts of the ritual. The girl had to fast all day, retiring to bed without a word, looking neither sideways or behind, and sleeping on her back. An alternative tradition made her eat an egg filled with salt; being very thirsty she would dream of drinking from a goblet which, according to situation and circumstance, would signify who the man was to be.

Originally a native of Spain and Italy, *common garden* or *black thyme* (T. vulgaris) is the most common. *Musk* or *lemon scented thyme* (T. citriodorus) is considered the best for cooking; this has a variety called Silver Queen, in reference to the colour of its leaves, and there is also one with golden foliage, T. cit. aureus. *Variegated thyme*, also called *lemon scented thyme* (T. cit. variegatus) by some, is yet another variety. There are other scents as well: carraway (T. herba-barona); orange (T. fragrantissimus); pine (T. azoricas); lavender (T. mastachinus); the American *savory thyme* (T. virginicus).

All over England and Europe is to be found *wild thyme* (T. serpyllum), also called *brotherwort* and *bank* or *horse thyme*. From its commonly growing on high pastures it has gained the names *hillwort*, *shepherd's* or *sheep's thyme*. A number of names cluster round its original Latin designation, Serpyllum montanum, in which the first element is derived from the Greek ἑρπειν, to creep. It is corrupted in various ways into *pennymountain*, *pell-a-mountain*, *serpell* and *serpolet*, or *serpolet oil plant*, the last referring to the fragrant oil extracted from it. The word 'serpent' is derived from the same Latin element, and thyme shares with that reptile the property of swiftly progressing low to the ground, and is thus called *running* or *creeping thyme*. It is also named *womb* or *mother thyme*, *old mother thyme* and *mother-of-thyme*. One explanation of this is that the old name for the womb was *mother* and thyme was supposed to be good for it. But I think there is here and also in the case of creeping or running thyme, a deliberate play on the word 'time', especially as thyme was also spelt this way once. Creeping, running and old mother time are familiar phrases, as is the expression 'the womb of time'.

In an early Elizabethan lyric, which describes the various flowers and herbs to be found in a lover's nosegay, the same pun appears:

Thyme is to try me
As each be tried must.

It is treated much better in one of the versions of a traditional folk-song,

'The Spring of Thyme', in which, also, it appears to stand as a symbol for the girl's virginity:

Twas early in the springtime of the year
When the sun did begin to shine,
Oh I had three branches all for to chose but one,
And the first that I chose was thyme.

Now thyme, it is a precious thing,
It's a root that the sun shines on,
And Time, it will bring everything unto an end,
And so our time goes on.

And while that I had thyme all for my own
It did flourish by night and day,
Till who came along but my jolly sailor-boy
And stole my thyme away.

And now my thyme is perished for me,
And I never shall plant it more,
Since into the place where my thyme did used to spring
Is grown a running rue.

But rue, it is a running root
And it runs all too fast for me;
I'll dig up the bed where thyme of old was laid
And plant there a brave oak-tree.

Stand up, oh stand up my jolly oak,
Stand you up, for you shall not die,
And I'll be so true to the one I love so dear,
As the stars shine bright in the sky.

INDEX OF LATIN NAMES

Abelomoschus (Hibiscus) esculentus, 147–8
Abutilon: A. megapotamicum, 146; *A. muticum*, 146
Acacia, 59; *A. aneura*, 59; *A. augustissima*, 60; *A. bidwilli (pallida)*, 59; *A. cibaria*, 59; *A. dictyophleba (stipuligera)*, 59; *A. greggi*, 60; *A. longifolia* var. *sophorae*, 59; *A. oswaldii*, 59; *A. rivalis*, 59; *A. salicina*, 59
Acalypha indica, 320
Acanthopeltis japonica, 214
Acanthophora specifera, 213
Acanthosicyos horrida, 109
Achillea: A. millefolium, 344; *A. ptarmica*, 344
Achyranthes aspera, 27
Acorus calamus, 144, 268
Acrostichum aureum, 115
Adansonia: A. digitata, 146; *A. gregorii*, 146
Adenophora: A. communis, 195; *A. latifolia*, 195; *A. stylosa*, 195; *A. verticillata*, 195
Aegopodium podagraria, 29
Aerva lanata, 27
Aesculus, 280; *A. californica*, 280; *A. hippocastanum*, 280; *A. glabra*, 280; *A. octandra*, 280; *A. parviflora*, 280; *A. turbinata*, 280
Aframomum: A. angustifolium, 322; *A. hanburyi*, 322; *A. mala*, 322; *A. melagueta*, 321
Agave: A. americanum, 140; *A. deserti*, 140; *A. utahensis*, 140
Agathophyllum aromaticum, 323–4, 340
Agriophyllum globicum, 225
Agrostema githago, 228
Ahnfeltia: A. concinna, 214; *A. plicata*, 214
Alaria esculenta, 207
Alectoria: A. fremontii, 286; *A. jujuba (jubata)*, 286
Aleurites triloba, 263
Alisma plantago, 252–5
Allium, 20, 155–63; *A. akaka*, 158; *A. ampeloprasum*, 162; *A. angulare*, 158; *A. ascalonicum*, 157, 159; *A. canadense*, 162; *A. cepa*, 155–7, 159; *A. cepa aggregatum*, 157; *A. cepa bulbiferum*, 157; *A. cepa canadense*, 157–8; *A. cernuum*, 162; *A. fistulosum*, 157, 159;
A. lebedourianum, 159; *A. leptophyllum*, 158; *A. macleanii*, 158; *A. neapolitanum*, 162; *A. nipponicum*, 158; *A. odorum*, 162–3; *A. obliquum*, 162; *A. porrum*, 158–9; *A. roseum*, 162; *A. rubellum*, 158; *A. sativum*, 159, 160–3; *A. senescens*, 158; *A. schoenoprasum*, 159–60; *A. scorodroprasum*, 162; *A. sphaerocephalum*, 162; *A. splendens*, 158; *A. tricoccum*, 159; *A. tuberosum*, 160; *A. ursinum*, 159, 163; *A. vineale*, 159
Allmania nodiflora, 28
Alpina: A. galangala, 196; *A. officinarum*, 196
Alocasia: A. indica, 98; *A. macrorrhiza*, 98; *A. odora*, 98
Alsodeia physiphora, 244
Alsophila: A. australis, 305; *A. excelsa*, 305; *A. rufa*, 305
Alstromeria: A. edulis, 141; *A. haemantha*, 141; *A. ligtu*, 140–1; *A. revoluta*, 141; *A. versicolor*, 141
Alternanthera: A. amoena, 28; *A. philoxeroides*, 28; *A. sessilis*, 28; *A. triandra*, 28
Althaea officinalis, 145–6
Alyssum maritimum, 102
Amaranthus, 25–8; *A. blitoides*, 26; *A. blitum*, 27; *A. caudatus*, 26; *A. dianthus*, 26; *A. flavus*, 27; *A. frumentaceus*, 27; *A. gangeticus*, 27; *A. graecicans*, 26; *A. grandiflorus*, 27; *A. hybridus*, 26; *A. hypochondriacus*, 27; *A. mangostanus*, 27; *A. melancholicus*, 27; *A. oleraceus*, 27; *A. palmeri*, 27; *A. polystachys*, 27; *A. retroflexus*, 26; *A. spinosus*, 27, 239; *A. torreyi*, 27; *A. tricolor*, 27; *A. tristis*, 27; *A. viridis*, 27
Ammomum: A. aromaticum, 321; *A. cardamomum*, 321; *A. dealbatum*, 321; *A. kepulanga*, 321; *A. krevanh*, 321; *A. subulatum*, 321; *A. thyrsoideum*, 321; *A. xanthoides*, 321
Amorphophallus (Hydrosme): A. campanulatus, 270; *A. hermandii*, 269–70; *A. rivieri*, 269
Amphicarpa monoica, 282
Anacardium occidentale, 275–6
Anagallis arvensis, 318
Anchomanes difformis, 98
Anchusa officinalis, 61

Anemone, 84; *A. flaccida*, 84; *A. narcissiflora*, 84; *A. nemorosa*, 84
Anemonella thalictoides, 84
Anethum graveolens, 327–8
Angelica, 28–9; *A. archangelica*, 28; *A. edulis*, 28; *A. sylvestris*, 28; *A. villosa*, 28
Angiopteris erecta, 305
Anoectochilus, 268
Anthemis, 101
Antigonum leptopus, 233
Anthriscus: A. cerefolium, 88; *A. sylvestris*, 88
Apios tuberosa, 189
Apis: A. saniculum, 340; *A. sylvatica*, 340
Apium: A. australe, 86; *A. graveolens*, 85–6; *A. grav.* var. *dulce*, 86; *A. grav.* var. *rapaceum*, 86
Aponogeton, 251; *A. crispum*, 251; *A. diastychum*, 251; *A. fenestralis*, 251; *A. microphyllum*, 251; *A. monostachyum*, 251; *A. undulatum*, 251
Arabis alpina, 102
Arachis hypogaea, 282
Araucaria: A. bidwilli, 297; *A. imbricata* 297
Archangelica (*Coelopleurum*) *gmelini*, 28
Areca catechu, 168
Arecastrum romanzoffianum, 167
Arenaria, 227; *A. peploides*, 227; *A. serpyllifolia*, 227; *A. trinervis*, 227
Arenga pinnata, 302
Arisoema: A. japonicum, 269; *A. triphyllum*, 269
Armoracia lapathifolia (*Cochlearia armoracia*), 330, 333
Arracacia esculenta, 170
Artemisia, 353–5; *A. abrotanum*, 353–4; *A. absinthium*, 353, *A. dracunuloides*, 353; *A. dracunculus*, 353; *A. frigida*, 353; *A. tridentata*, 353; *A. tripartita*, 353; *A. vulgaris*, 353; *A. wrightii*, 353
Arthrocnenum: A. glaucum, 226; *A. indicum*, 226
Arthrothamnus, 207
Artocarpus, 62–5; *A. altilis*, 62; *A. brasiliensis*, 65; *A. champeden*, 65; *A. dadak*, 65; *A. elastica*, 65; *A. integrifolia*, 65; *A. kemando*, 65; *A. lakoocha*, 65; *A. odoratissima*, 65; *A. rigida*, 65
Arum, 268–70; *A. avisarum*, 269; *A. italiacum*, 269; *A. maculatum*, 268–9
Arundinaria gigantea, 41
Asarum: A. canadense, 330; *A. europaeum*, 330

Asclepias: A. incarnata, 148; *A. speciosa*, 148; *A. syriaca*, 148; *A. tuberosa*, 148
Asparagopsis sanfordiana, 213
Asparagus, 33–4; *A. abyssinicus*, 34; *A. acutifolius*, 34; *A. africanus*, 34; *A. allus*, 34; *A. asparagoides*, 34; *A. falcatus*, 34; *A. horridus*, 34; *A. lucidus*, 34; *A. officinalis*, 34; *A. off. maritima*, 34; *A. pauli-guilielmi*, 34; *A. racemosus*, 34; *A. sarmentosus*, 34; *A. scaber*, 34; *A. tenuifolius*, 34
Asphodelus, 126; *A. fistulosus*, 126; *A. luteus*, 126
Asplenium esculentum, 115
Astragalus: A. aboriginum, 174; *A. canadensis*, 189; *A. caryocarpus*, 174; *A. edulis*, 174; *A. fasciculifolius*, 174; *A. pictus*, 189–90
Atractylis (*Circelium*) *cancellata*, 235
Atriplex, 225–6; *A. canescens*, 226; *A. haulimus*, 226; *A. hortensis*, 226; *A. patula*, 19, 226; *A. pat* var. *erecta*, 226; *A. pat.* var. *hastata*, 226; *A. pat.* var. *prostata*, 226; *A. portulacoides*, 226; *A. repens*, 226
Attalea, 164; *A.* (*Bornoa*) *amygdalina*, 164; *A.* (*Orbignya*) *cohune*, 164; *A. compta*, 164; *A. excelsa* (*Scheelea martiana*), 164
Aulospermum: A. longipes, 89; *A. purpureum*, 89
Avena, 291–5; *A. brevis*, 291; *A. byzantina*, 291; *A. fatua*, 291; *A. nuda*, 291; *A. orientalis*, 291; *A. sativa*, 291; *A. sterilis*, 291; *A. strigosa*, 291
Averrhoa bilimbi, 108–9
Aydendron firmulum, 323

Babiana plicata, 143
Bactris utilis (*Guilielma gasipaes*), 303
Basella: B. alba, 229; *B. condifolia*, 229; *B. japonica*, 229; *B. lucida*, 229; *B. rubra*, 229
Batatas edulis (*Ipomea batatas*), 230
Batis maritima, 203
Bauhinia, 58; *B. esculenta*, 58; *B. malabarica*, 59; *B. reticulata*, 58; *B. variegata*, 59
Bambusa, 39–41; *B. arundinacea*, 41; *B. cornuta*, 41; *B. multiplex*, 41; *B. oldhami*, 41; *B. spinosa*, 41; *B. tulda*, 41; *B. vulgaris*, 41
Barbarea: B. praecox, 102; *B. vulgaris*, 102
Benincasa cerifera, 121–2
Bertholettia excelsa, 273
Beta, 60–1; *B. maritima*, 61; *B. vulgaris*,

Beta—cont.
60; *B. vulg. cicla*, 60; *B. vulg. macrorrhiza*, 60; *B. vulg. rapacea*, 60; *B. vulg. saccharifera*, 60
Betonica (Stachys) officinalis, 32
Bidens pilosa, 229
Blighia sapida, 25
Blitum (Chenopodium) capitatum, 224
Bomarea: B. acutifolia, 141; *B. edulis*, 141; *B. glaucescens*, 141; *B. ovata*, 141; *B. salsilla*, 141
Borago officinalis, 61
Borassus flabellifer, 164
Borrichia frutescens, 203
Boscia, 319; *B. angustifolia*, 319; *B. caffra*, 319; *B. senegalensis*, 319
Boussingaltia baselloides, 230
Bowenia spectabilis, 303
Brachychiton populneum, 190
Brachystelma: B. bingere, 148; *B. lineara*, 148
Brachytrichia quoyi, 213
Brasenia peltata, 245
Brassica, 66–72; *B. adpressa (Sinapis incana)*, 150; *B. (Sinapis) alba*, 149; *B. arvensis*, 150; *B. campestris (asperifolia)*, 149–50, 240; *B. camp.* var. *sarsun*, 149; *B. camp.* var. *toria*, 149–50; *B. chinensis*, 72; *B. juncea*, 150; *B. napobrassica (rutabaga)*, 242–3; *B. nigra*, 149; *B. oleracea*, 66; *B. ol. acephala*, 70; *B. ol. botrytis*, 72; *B. ol. botr. cauliflora*, 71; *B. ol. bullata gemmifera*, 71; *B. ol. bull. major*, 70; *B. ol. capitata*, 69; *B. ol. cap. rubra*, 69; *B. ol. costata*, 72; *B. ol. caulo rapa (gonglyodes)*, 243; *B. ol. gemmifera*, 72; *B. ol. italica*, 72; *B. pekinensis*, 72; *B. rapa*, 241; *B. sinapistrum (Sinapis arvensis)*, 66
Brodiaea capitata, 126
Bromelia: B. karatas, 305; *B. laciniosa*, 305
Brosimum: B. alicastrum, 65; *B. costaricanum*, 65; *B. utile*, 65–6
Bryonia dioica (alba), 20, 36–7
Bultmannia, 17
Bunias: B. cakile (Cakile maritima), 75; *B. erucago*, 75; *B. orientalis*, 75
Bunium: B. (Conopodium) bulbocastanum, 178; *B. (Carum) flexuosum (Conopodium majus)*, 178; *B. (Conopodium) incrassatum*, 178
Buphthalum oleraceum, 92
Butia yatay, 167
Butomus umbellatus, 255

Cajanus indicus, 175–6
Cakile: C. edulenta, 75; *C. maritima (Bunias cakile)*, 75
Caladium: C. sequinum, 268; *C. striatipes*, 97
Calamus, 164–7; *C. scipionum*, 167; *C. zeylanicum*, 164
Calandrina: C. balonensis, 193, *C. caulescens* var. *menziesi*, 193; *C. polyandra*, 193
Calathea: C. allonia, 265; *C. macrosepala*, 265; *C. violacea*, 265
Calendula officinalis, 333–4
Calla palustris, 270
Callirhoe: C. digitata, 146; *C. pedata*, 146
Calochortus, 128–9; *C. elegans*, 129; *C. gunnisonii*, 129; *C. nuttallii*, 129; *C. nutt.* var. *aureum*, 129; *C. pulchellus*, 129
Caloncytion aculeatum (Ipomea bonanox), 233
Caltha: C. leptosepala, 83; *C. palustris*, 20, 83
Calypso bulbosa, 268
Camassia: C. esculenta (scilloides), 129; *C. quamata*, 129
Campanula: C. latifolia, 195; *C. rapunculus*, 195; *C. versicolor*, 195
Canarium: C. commune, 264; *C. ovatum*, 264
Canavallia, 53–4; *C. ensiformis*, 53; *C. ensiformis nana*, 54; *C. gladiate*, 53; *C. obtusifolia*, 54; *C. plagiosperma*, 54
Canella alba, 322
Canna: C. bidentata, 265; *C. coccinea*, 265; *C. discolor*, 265; *C. edulis*, 265; *C. glauca*, 265; *C. iridifolia*, 265; *C. languinosa*, 265; *C. latifolia*, 265; *C. paniculata*, 265
Capparis, 319–20; *C. corymbifera*, 319; *C. sodala*, 319; *C. spinosa*, 319
Capsella bursa-pastoris, 102
Capsicum: C. frutescens, 320; *C. frut.* var. *abbreviatum*, 320; *C. frut.* var. *acuminatum*, 320; *C. frut.* var. *cerasiforme*, 320; *C. frut.* var. *connoides*, 320; *C. frut.* var. *fasciculatum*, 320; *C. frut.* var. *grossum*, 321; *C. frut.* var. *longum*, 321
Caragana: C. arborescens, 177; *C. chamlagu*, 177
Cardamine, 103; *C. hirsuta*, 103; *C. nasturtioides*, 103; *C. pratensis*, 103; *C. yesoensis*, 103
Carduus, 234; *C. (Cirsium) edule*, 234; *C. (Cirsium, Cnicus) undulatum*, 234

362 / Index of Latin Names

Carica papaya, 109
Carlina, 234–5; C. acaulis, 234
Carnegiea gigantea, 79
Carthamus tinctorius, 142
Carum (Bunium) flexuosum (Conopodium majus), 178
Caryoca: C. nuciferum, 274; C. tormentosum, 274
Caryota: C. mitus, 302; C. urens, 302
Cassia tora, 177
Castanea, 279; C. crenata, 270; C. dentata, 279; C. mollissima, 279; C. pumila, 279–80; C. sativa, 270
Castanopsis: C. boisii, 280; C. chrysophylla, 280; C. cuspidata, 280; C. philippensis, 280; C. solerophylla, 280; C. sumatrana, 280; C. tibetiana, 280
Castanospermum australe, 281
Catanella impudiea, 213
Cayratia clematides, 233
Ceiba pentandra, 146–7
Celosia argentea, 28
Centaurea, 235; C. calcitrapa, 235; C. eryngoides, 235
Centranthus macrosiphon, 99
Ceramium, 214
Cerasus laurocerasus, 317
Ceratonia siliqua, 51–2
Ceratopteris thalictroides, 115
Cercidium, 177; C. parvifolium, 177; C. torreyanum, 177
Cercis: C. canadensis, 177; C. siliquastrum, 177
Ceretotheca sesamoides, 352
Cetraria islandica, 287
Chaerophyllum: C. bulbosum, 170; C. prescottii, 170
Chaetomorpha, 213
Chamaedorea, 167
Chamaerops humilis, 167
Chelidonium, 82
Chenopodium, 223–5; C. album, 19, 224; C. amaranticolor, 224; C. ambrosoides, 225; C. bonus henricus, 223–4; C. (Blitum) capitatum, 224; C. erosum, 224–5; C. fremontii, 225; C. leptophyllum, 225; C. murale, 224; C. pallidicaule, 225; C. quinoa, 225; C. rhadinostachyum, 225; C. rubrum, 224
Chlorella, 215
Chondrus crispus, 207–8
Chrysanthemum, 91–2; C. coroniarum, 92; C. indicum, 92; C. leucanthemum, 91; C. pyrethrum (Tanacetum balsamita), 92; C. segetum, 91; C. sinense, 92
Cicer arietinum, 174–5

Cichorium: C. endiva, 90; C. end. var. crispa, 90; C. end. var. latifolia, 90; C. intybus, 90
Cinnamomum, 322–3; C. burmannii, 322; C. cassia, 322; C. loureirii, 322; C. massoia, 322; C. oliveri, 322; C. tamala, 322; C. zeylanicum, 322
Circelium (Atractylis) cancellatum, 235
Cirsium: C. drumondii, 234; C. (Carduus) edule, 234; C. occidentale, 234; C. (Cnicus) oleraceum, 234; C. scopulorum, 234; C. (Cnicus) tuberosum, 234; C. (Carduus, Cnicus) undulatum, 234; C. virginianum, 234
Cissus opaca, 233
Citrullus: C. lanatus, 119; C. lan. var. fistulous, 109
Citrus bergamia, 315
Cladophora, 213
Cladosiphon, 213
Claytonia: C. acutifolia, 192; C. caroliniana, 192; C. cubensis, 192; C. (Montia) perfoliata, 192; C. sibirica, 192; C. virginica, 192
Clematis vitalba, 84
Cleome: C. integrifolia, 319–20; C. vicosa, 320
Clethra, 343; C. acuminata, 343; C. alnifolia, 343; C. barbinervis, 343
Clintonia: C. borealis, 129; C. umbellata, 129
Cnicus: C. (Cirsium, Carduus) edulis, 234; C. japonicus, 234; C. (Cirsium) oleraceus, 234; C. (Cirsium) tuberosus, 234; C. (Cirsium, Carduus) undulatus, 234
Coccinea indica, 122
Cochlearia: C. armoracia (Armoracia lapathifolia), 20, 330, 333; C. officinalis, 333
Cocos nucifera, 167
Codium: C. lindenbergii, 208, C. mucronatum, 208; C. muelleri, 208; C. tenus, 208; C. tormentosum, 208
Codonopsis ussuriensis, 195
Coelopleurum (Archangelica) gmelini, 28
Cogeswellia, 17, 170
Coix lachryma jobi, 195–6
Coleus: C. aromaticus, 153; C. rotundifolius, 153
Colocasia, 94–7, 226, 256, 268; C. antiquorum, 97; C. esculenta, 97; C. macrorrhiza, 94
Comarium (Potentilla) palustris, 216
Conopodium: C. (Bunium) bulbocastanum, 178; C. carvi, 178–9; C. denudatum, 178; C. (Bunium) incrassatum, 178;

Conodiumpo—cont.
  C. kelogii, 178; *C. majus* (*Bunium Carum*) *flexuosum*, 178
Corchorus olitorius, 147
Cordeauxia edulis, 58
Cordia dichotoma, 264
Cordyline (*Dracaena*), 19, 34; *C. australis*, 129; *C. terminalis*, 129
Coriandrum sativum, 324
Coronopus, 101
Corylus, 282–5; *C. americana*, 285; *C. amer.* var. *avellana*, 285; *C. avellana* var. *grandis*, 285; *C. californica*, 285; *C. chinensis*, 285; *C. colurna*, 285; *C. cornuta*, 285; *C. ferox*, 285; *C. heterophylla*, 285; *C. maxima*, 285; *C. sieboldiana*, 285
Corynocarpus laevigata, 264
Corypha utan, 302
Costus speciosus, 92 n
Couroupita (*Lecythis*) *guianensis*, 274
Crambe: C.maritima, 72–5; *C. tartarica*, 75
Crataeve macrocarpa, 319
Crinum flaccidum, 141
Crithmum maritimum, 202
Crocosmia aurea, 142
Crocus, 141–2; *C. cancellatus*, 141; *C. damascenus* (*edulis*), 141; *C. sativus*, 141–2; *C. sieberi*, 141
Crotalaria, 177; *C. glauca*, 177; *C. guatemalensis*, 177
Cryptocarya moschata, 340
Cryptonemia decumbens, 213
Cryptotaemia canadensis, 89
Cucumis: *C. anguria*, 108; *C. melo*, 109; *C. metuliferus*, 108; *C. sacleuxii*, 108; *C. sativus*, 107–8, *C. trigonis*, 108
Cuminum cyminum, 327
Curcuma, 265, 324, 327; *C. angustifolia*, 265; *C. leucorrhiza*, 265; *C. pierreans*, 265; *C. xanthorrhiza*, 265; *C. zeodaria*, 327; *C. zerumbet*, 327
Cucurbita, 117–18; *C. ficifolia* (*melanosperma*), 118; *C. foetidissima*, 118; *C. maxima*, 118; *C. max.* var. *maxima*, 118; *C. max.* var. *turbaniformis*, 118; *C. mixta*, 118; *C. moschata*, 118; *C. pepo*, 118; *C. pepo* var. *medullosa*, 118; *C. pepo* var. *melopepo*, 118; *C. pepo* var. *pepo*, 118
Cyamopsis psorialiodes, 58
Cycas, 303; *C. circinalis*, 303; *C. revoluta*, 303; *C. media*, 303
Cyathea: *C. dealbata*, 305; *C. medullaris*, 304; *C. viellardii*, 305
Cyclanthera: *C. edulis*, 108; *C. explodens*, 108; *C. pedata*, 108

Cymbidium: *C. caniculatum*, 268; *C. virescens*, 268
Cymbopogon citratus, 327
Cymopterus fendleri, 170
Cynara: *C. cardunculus*, 29; *C. scolymus*, 30
Cyperus, 196; *C. aristatus*, 196; *C. esculentus*, 196, *C. longus*, 196; *C. papyrus*, 196
Cyphomandra, 240; *C. betacea*, 240; *C. hartwegi*, 240
Cyrtosperma merkusii, 98

Daucus: *D. carota*, 79; *D. carota sativa*, 79–80; *D. pussilus*, 80
Dasyliron wheeleri, 140
Deeringia amaranthoides, 28
Dendrocalamus, 39–41; *D. hamiltonii* var. *edulis*, 41; *D. latifolius*, 41
Dendrolobium salaccense, 268
Dentaria, 333; *D. diphylla*, 333; *D. laciniata*, 333
Descurainia, 150
Dicksonia antarctica, 305
Dictyopteris, 213
Dictyota, 213
Dicypellium caryophyllatum, 322, 324
Digenea simplex, 213
Digitaria: *D. exilis*, 290; *D. iburna*, 290; *D.* (*Syntherisma*) *sanguinalis*, 290
Dillenia, 17; *D. indica*, 327
Dioon edule, 304
Dioscorea, 255–8; *D. aculeata*, 256; *D. alata*, 246; *D. altissima*, 256; *D. atropurpurea*, 256; *D. batatas*, 255, 256; *D. bemandry*, 256; *D. bulbifera*, 256; *D. cayenensis*, 256–7; *D. daemona*, 256; *D. dedecaneura*, 257; *D. dumetorum*, 256; *D. eburnea*, 257; *D. esculenta*, 257; *D. fargesii*, 257, *D. fasciculata*, 257; *D. glandulosa*, 257; *D. globosa*, 257; *D. hastata*, 257, *D. hastifolia*, 257; *D. japonica*, 257; *D. latifolia*, 257; *D. lutea*, 257; *D. luzonensis*, 257; *D. maciba*, 257; *D. oppositifolia*, 257; *D. papuana*, 257; *D. pentaphylla*, 257; *D. praehensis*, 257–8; *D. purpurea*, 258; *D. pyrennaica*, 258; *D. pyrifolia*, 258; *D. rotundata*, 257; *D. rotundifolia*, 258; *D. rubella*, 258; *D. sativa*, 258; *D. spinosa*, 258; *D. toxicaria*, 256; *D. transversa*, 258; *D. trifida*, 258; *D. trifoliata*, 258; *D. tuberosa*, 258
Diplazium: *D. esculentum*, 115; *D. asperum*, 115
Dodecatheon hendersonii, 149

364 / Index of Latin Names

Dolichos, 53; *D. biflorus*, 53; *D. bracteatus*, 53; *D. bulbosus*, 53; *D. hirsutus* (*japonicus*) (*Pueraria thunbergiana*), 53; *D. lablab* (*Lablab niger*), 53; *D. lablab* var. *lablab* (*typicus*) 53; *D. lablab* var. *lignosus*, 53; *D. lubia*, 53; *D. sinensis*, 53
Dondia suffrutescens, 226
Dracaena, 34–5; *D.* (*Cordyline*) *australis*, 34; *D. draco*, 35; *D. mannii*, 34
Drynaria (*Polypodium*): *D. quercifolia*, 115; *D. rigidula*, 115
Durvillea antarctica, 210

Echinocereus pectinatus, 79
Echinochloa: *E. colonum*, 289; *E. crusgalli* var. *frumentaceum*, 289; *E. decompositum*, 289; *E. maximum*, 289; *E. stagnina*, 289
Ecklonia: *E. bicyclis*, 207; *E. kurome*, 207; *E. latifolia*, 207
Edanthe tepejilote, 167
Eichornia (*Piaropus*), 252
Elaeagnus angustifolia, 155
Elaterium ciliatum, 122
Eleocharis, 196–9; *E. esculenta*, 199; *E. platiginea* (*sphacelata*), 199; *E. tuberosa*, 196
Eletaria: *E. cardamomum*, 321; *E. speciosa*, 321
Eleusine: *E. aegyptica*, 290; *E. corocana*, 289–90; *E. indica*, 290; *E. tocussa*, 290
Eleutheropetalum sartoni, 167
Elymus condensatus, 301
Emex australis, 193
Emilia sonchifolia, 237
Encephalartos, 303–4; *E. altensteinii*, 303; *E. caffer*, 303
Enhalus koenigii, 209
Ensete (*Musa ensete*), 46–7; *E. ventricosum*, 47
Entada phaseoloides, 281
Enteromorpha prolifera, 212
Ephedra: *E. distachya*, 117; *E. nevadensis*, 117
Epidendron cochleatum, 268
Epilobium rosmarinifolium, 114
Equisetum, 20, 116–17; *E. arvense*, 117, *E. laevigatum*, 116–17; *E. pratense*, 116
Eragrostis abyssinica, 290
Erechthites hieracifolia, 112
Eremurus, 130; *E. aucherianus*, 130; *E. spectabilis*, 130
Erigenia bulbosa, 89
Eriogonum: *E. corymbosum*, 220; *E. inflatum*, 220
Eruca sativa, 75

Erucaria aleppica, 75
Ervum lens (*Lens culinaris esculenta*), 123–4
Eryngium: *E. campestre*, 235; *E. foetidum*, 235
Erythea edulis, 167
Erythrina: *E. edulis*, 60; *E. rubrinervia*, 60
Erythronium: *E. dens canis*, 130; *E. grandiflorum*, 130
Escobedia, 142
Eucheuma speciosa, 214
Eugenia: *E. aromatica*, 323–4; *E. pimenta*, 343
Eulophus, 170
Euryale ferox, 245
Euterpe edulis, 167
Eutrema wasabi, 333
Evernia: *E. furfuracea*, 286; *E. prunastri*, 286
Exidia auricula-judae, 38
Exogonium (*Ipomea*) *bractatum*, 233

Faba vulgaris (*Vicia faba*), 47–51
Fagopyrum, 274–5; *F. emarginatum*, 274; *F. esculentum*, 274; *F. tartaricum*, 275
Fagus: *F. grandiflora*, 262; *F. sylvatica*, 21, 262
Fedia cornucopiae, 99
Foeniculum: *F. dulce*, 329; *F. vulgare*, 328–9
Freycinettia: *F. banksii*, 66; *F. funicularis*, 66
Fritillaria: *F. camschatensis*, 131; *F. imperialis*, 131; *F. meleagris*, 130; *F. pudica*, 130; *F. roylei*, 130
Fucus: *F. esculentus*, 209; *F. fuscatus*, 209; *F. nodosus*, 209; *F. versiculosus*, 209
Funkia ovata, 131

Gastrodia: *G. cunninghamii*, 268; *G. sesamoides*, 268
Geitonoplesium cymosum, 35
Gelidium divaricatum, 213–14
Gelidopsis, 214
Geonoma binervia, 167
Geum, 38–9; *G. rivale*, 38; *G. urbanum*, 38
Giganthochloa, 39–41; *G. ater*, 41; *G. verticillata*, 41
Gigartina, 214
Ginkgo biloba, 303, 304
Gladiolus: *G. edulis*, 143; *G. quartianus*, 143; *G. spicatus*, 143; *G. zambesiacus*, 143
Glaux maritima, 149

Gleichenia dichotoma, 115
Glyceria fluitans, 199
Glycine (Soja) max, 52–3
Gracilana: G. compressa, 214; *G. corronopifolia*, 214; *G. lichenoides*, 214; *G. verrucosa*, 214
Grateloupia, 213
Guilielma gasipaes (Bactris utilis), 303
Guizotia abyssinica, 352
Gundelia tournefortii, 235
Gymenema: G. lactiferum, 148; *G. sylvestre*, 148
Gymnogongus, 214
Gynandropsis: G. pentaphylla, 103, 319; *G. speciosa*, 103, 319
Gyrophora esculenta, 287

Habenaria sparciflora, 268
Haloxylon salicornicum, 226
Halymenia formosa, 213
Hedysarum, 93
Helianthus: H. annuus, 31; *H. decapetalus*, 31; *H. dronicoides*, 31; *H. giganteus*, 31; *H. maximiliani*, 31; *H. tuberosus*, 20, 30–2
Helicornia bihai, 46
Helminthia echoides, 62
Hemerocallis, 131–2; *H. aurantiaca*, 131; *H. flava*, 131; *H. minor*, 131
Heracleum: H. lanatum, 169; *H. persicum*, 169; *H. sphondylium*, 169
Heterochordaria, 213
Heterospathe elata, 167
Hibiscus: H. cannabinus, 147; *H. (Abelomoschus) esculentus*, 147, 216; *H. ficulneus*, 147; *H. furcatus*, 147; *H. heterophyllus*, 148; *H. sabdariffa*, 147; *H. tiliaceus*, 148
Hijikia fusiforme, 213
Hoffmanseggia densiflora, 190
Hordeum, 270–3; *H. distichon*, 273; *H. hexastichon*, 273; *H. trifurcatum*, 273; *H. vulgare*, 273; *H. zeocriton*, 273
Hovea longipes, 177
Humulus lupulus, 37
Hydrosme (Amorphophallus) rivieri, 269
Hydrodictyon reticulatum, 213
Hydrolea zeylanica, 251
Hydrophyllum: H. appendiculatum, 251; *H. virginicum*, 251
Hypnea: H. cenomyce, 214; *H. cervicornis*, 214; *H. musiformis*, 214; *H. nidifica*, 214
Hyphochoeris maculata, 112

Impatiens: I. flaccida, 327; *I. noli-metangere*, 327

Inga, 59
Inula crithmoides, 203
Ipomea, 230–3; *I. aquatica (reptans)*, 233; *I. batatas (Batatas edulis)*, 230; *I. bonanox (Caloncytion aculeatum)*, 233; *I. (Exogonium) bracteata*, 233; *I. mammosa*, 233; *I. tiliacea (fastigata)*, 233
Iriartrea (Socratea) durissima, 168
Iridaea edulis, 209–10
Iris, 143–5; *I. edulis*, 143; *I. florentina*, 143; *I. foetidissima*, 143; *I. juncea*, 143
Isomeris arborea, 319

Jubaea chilensis, 167
Juglans, 305–6; *J. australis*, 306; *J. boliviana*, 306; *J. californica*, 306; *J. cathayensis*, 306; *J. cinerea*, 306; *J. major*, 306; *J. mandshurica*, 306; *J. regia*, 306; *J. rupestris*, 306; *J. sieboldiana*, 306

Kaempferia galanga, 196
Kerstingiella geocarpe, 282

Lablab niger (Dolichos lablab), 53
Lactuca: L. canadensis, 124; *L. indica*, 124; *L. perennis*, 124; *L. quercina*, 124; *L. saligna*, 124; *L. sativa*, 124–5; *L. sat. var. angustana*, 125; *L. scariola*, 124–5; *L. tsitsa*, 124; *L. virosa*, 124
Lagenaria vulgaris, 121
Lagoecia cumminoides, 327
Laminaria: L. digitata, 210; *L. japonica*, 210; *L. potatorum*, 210; *L. saccharina*, 207, 210
Lamium, 153–4; *L. album*, 154; *L. galeobdelon*, 154; *L. purpureum*, 154
Laportea bulbifera, 153
Lappa major var. edulis, 204
Lapsana, 101
Lasia aculeata, 98
Lathyrus, 21, 173; *L. aphaca*, 174; *L. macrorrhizus*, 189; *L. ochroleucus*, 174; *L. ornatus*, 174; *L. polymorphus*, 174; *L. sativus*, 173–4
Laurelia sempervirens, 340
Laurencia pinnatifolia, 209–10
Laurus nobilis, 316–17
Lavatera plebia, 190
Lecanora esculenta, 286
Lecythis, 273–4; *L. (Couroupita) guianensis*, 274; *L. laevifolia*, 274; *L. lanceolata*, 274; *L. ollaria*, 274; *L. usitata (zabucajo)*, 274
Lens culinaris esculenta (Ervum lens), 123–4

366 / Index of Latin Names

Lepidium, 103–4, 149; *L. campestre*, 104; *L. draba*, 104; *L. fremontii*, 104; *L. mayenii*, 103; *L. oleraceum*, 104; *L. rotundum*, 104; *L. sativum*, 103; *L. virginicum*, 104
Levisticum officinale, 87
Lewisia, 17; *L. rediviva*, 192
Ligularia (Senecio) kaemferi, 112
Ligusticum: *L. canadense*, 87; *L. scoticum*, 87
Lilium, 132–5; *L. auratum*, 133; *L. bulbiferum*, 133; *L. callosum*, 133; *L. chalcedonium*, 133; *L. cordifolium*, 133; *L. cord.* var. *glehni*, 133; *L. dauricum*, 134; *L. elegans*, 134; *L. japonicum*, 134; *L. lancifolium*, 134; *L. longifolium*, 134; *L. martagon*, 134; *L. medeoloides*, 134; *L. parviflorum*, 134; *L. philadelphicum*, 134; *L. pomponium*, 134; *L. sargentiae*, 134; *L. speciosum*, 134; *L. spectabile*, 134; *L. superbum*, 134; *L. tenuifolium*, 134; *L. tigridum*, 134–5
Limnocharis emarginata, 255
Lindera benzoin, 343
Litsea cubeba (*citrata*), 323
Livistonia australis, 167
Lobelia succulenta, 195
Lomatium: *L. ambiguum*, 170; *L. circumdatum*, 170; *L. foeniculaceum*, 170
Lotus: *L. arabicus*, 174; *L. edulis*, 174; *L. purpureus*, 174; *L. tetragonolobus*, 174
Luffa: *L. acutangula*, 122; *L. aegyptiaca*, 122
Lunaria annua, 243–4
Lupinus, 21, 176; *L. albus*, 176; *L. cruickshankii*, 176; *L. littoralis*, 176; *L. luteus*, 176; *L. perennis*, 176; *L. termis*, 176
Lycium: *L. chinense*, 240; *L. europaeum*, 240; *L. humile*, 240
Lycopersicum esculentum (*Solanum lycopersicum*), 237
Lysimachia: *L. clethroides*, 104; *L. fortuneii*, 104; *L. obovata*, 104

Macadamia, 297–8; *M. integrifolia*, 298; *M. ternifolia*, 298; *M. tetraphylla*, 298; *M. whelani*, 298
Macrozamia: *M. miquelii*, 303; *M. reidlei*, 303; *M. spiralis*, 303
Malva, 145–6, 216; *M. rotundifolia*, 145; *M. sylvestris*, 145
Manihot: *M. palmata aipi*, 276–9; *M. utilissima* (*esculenta*), 276–9
Maranta arundinacea, 264–5
Marattia fraxinea, 305

Marjorana hortensis, 335
Marsdenia: *M. australis*, 190; *M. viridifolia*, 190
Marsilea: *M. drummondii*, 115; *M. hirsuta*, 115; *M. quadrifolia*, 115; *M. salvatrix* (*macrocarpa*), 115
Mastocarpus klenzinaus, 214
Mauritia flexuosa, 303
Medeola virginica, 108
Melilotus, 93; *M. altissima*, 93; *M. elegans*, 93; *M. officinalis*, 93; *M. ruthenica*, 93
Melissa officinalis, 315
Melocalamus compactifolius, 41
Mentha, 335–9; *M. aquatica*, 336; *M. arvensis*, 336; *M. australis*, 336; *M. citrata*, 336; *M. gentilis*, 336; *M. gracilis*, 336; *M. hirsuta*, 336; *M. javanica*, 336; *M. laxifolia*, 336; *M. piperita*, 336; *M. pubescens*, 336; *M. pulegium*, 336; *M. rotundifolia*, 336; *M. rubra*, 336; *M. serpyllifolia*, 336; *M. spicata*, 336; *M. subglabrata*, 336; *M. sylvestris*, 336, 339; *M. verticillata* (*sativa*), 339; *M. viridis*, 339
Mertensia maritima, 62
Mesembryanthemum, 229–30; *M. aequilaterale*, 230; *M. cordifolium*, 230; *M. crystallinum*, 230
Mesogloia, 213
Metaplexis stauntoni, 148–9
Metroxylon: *M. rumphii*, 301–2; *M. sagu* (*laeve*), 301–2
Microseris scapigera, 190
Microtis porrifolia, 268
Momordica: *M. balsamina*, 110; *M. charantia*, 109; *M. char.* var. *abreviata*, 109–10; *M. dioica*, 110
Monarda, 315; *M. didyma*, 315; *M. pectinata*, 315
Monolepsis nuttalaina, 225
Monostroma, 212
Monotropa: *M. hypopystis*, 36; *M. uniflora*, 36
Montia: *M. fontana*, 192; *M. (Claytonia) perfoliata*, 192
Moringa pterygosperma, 333
Mucuna gigantea, 58
Murraya koenigii, 327
Musa, 41–7; *M. acuminata*, 42; *M. balbisiana*, 42; *corniculata*, 44; *M. ensete* (*Ensete*), 46; *M. nana* (*cavendishii*), 43; *M. paradisiaca*, 44; *M. sapientum*, 43; *M. sap.* var. *oleracea*, 46; *M. textilis*, 46
Myristica, 339–40; *M. argentea*, 339; *M. fragrans*, 339; *M. insipida*, 339; *M.*

Myristica—cont.
  laurifolia, 339; *M. malabarica*, 339; *M. muelleri*, 339; *M. otoba*, 339
Myrrhis odorata, 88

Najas major, 105
Nasturtium (*Radicula, Roripa*), 104–5; *N. aquaticum* (*officinale*), 105; *N. palustre*, 105
Neanthe elegans, 167
Nectandra cinnamonoides, 322
Nelumbo, 245–9; *N. lutea*, 249; *N. nucifera*, 249; *N. speciosa*, 249
Nereocystis leutkeana, 210
Nerium oleander, 155
Nigella: *N. damascena*, 344; *N. sativa*, 344
Nopalea dejecta, 79
Nostoc commune, 211
Nuphar, 249–50; *N. advena*, 249–50; *N. luteum*, 250; *N. multisepalum*, 250
Nymphaea, 250–1; *N. alba*, 250; *N. ampla*, 250; *N. caerulea*, 250; *N. calliantha*, 250; *N. gigantea*, 250–1; *N. lotus*, 250; *N. magnifica* (*rubra, edulis*), 250; *N. micrantha*, 250; *N. rudgeana*, 250; *N. stellata*, 250
Nymphoides, 251–2; *N. crenatum*, 251; *N. cristatum*, 251; *N. peltatum*, 251

Ocimum, 315–16; *O. basilicum*, 315; *O. canum*, 316; *O. minimum*, 315; *O. sanctum*, 316; *O. suave*, 315
Oenanthe: *O. crocata*, 252; *O. pimpinelloides*, 252; *O. sarmentosa*, 252; *O. stolonifera*, 252
Oenothera: *O. acaulis*, 114; *O. biennis*, 113–14; *O. mollissima*, 114
Olea europaea, 154–5
Omolanthus pedicellatus, 238
Oncosperma filamentosa, 167
Onopordon: *O. acanthium*, 142, 235–6; *O. illyricum*, 236
Ophioglossum reticulatum, 115
Ophiopogon japonicus, 135
Ophrys: *O. apifera*, 266; *O. arachnites*, 266
Opuntia, 76; *O. basilaris*, 79; *O. clavata*, 76; *O. engelmanii*, 76; *O. megacantha*, 76
Orchis: *O. fusca*, 266; *O. latifolia*, 266; *O. maculata*, 267; *O. mascula*, 267; *O. militaris*, 266; *O. morio*, 267; *O. purpurea*, 267; *O. simia*, 266
Oreodoxa oleracea, 302–3
Origanum, 334–5; *O. heracloeticum*, 335; *O. marjorana*, 335; *O. onites*, 335; *O. vulgare*, 334
Ornithogalum, 35; *O. narbonensis*, 35; *O. pyrenaicum*, 35; *O. umbellatum*, 35
Orobanche, 35–6; *O. californica*, 36; *O. cruenta*, 35; *O. fasciculata*, 36; *O. ludoviciana*, 36; *O. major*, 36; *O. tuberosa*, 36
Orontium aquaticum, 270
Oryza: *O. glaberrhima*, 299; *O. sativa*, 298–9
Oryzopsis hymenoides, 299
Osmorrhiza: *O. claytoni*, 89; *O. occidentalis*, 89
Oxalis, 21, 216, 220–2; *O. acetosella*, 220–1; *O. cernua*, 222; *O. compressa*, 220; *O. corniculata*, 220; *O. deppei*, 222; *O. enneaphylla*, 220; *O. frutescens*, 220; *O. stricta*, 220; *O. tetraphylla*, 222; *O. tuberosa*, 222; *O. zonata*, 220
Oxyria: *O. digyna*, 219; *O. reniformis*, 219
Oxystelma esculentum, 148

Pachucereus pringlei, 79
Pachyrrhizus: *P. angulatus*, 190; *P. erosus*, 190; *P. palmatilobus*, 190; *P. tuberosus*, 190
Paeonia, 84–5; *P. alibflora*, 85; *P. officinalis*, 85
Pandanus, 66; *P. ceramicus*, 66; *P. connoides*, 66; *P. edulis*, 66; *P. odorus*, 66; *P. polycephalus*, 66; *P. spiralis*, 66; *P. tectorius* (*odoratissimus*), 66; *O. utilis*, 66
Panicum: *P. fasciculatum*, 289; *P. miliare*, 289; *P. obtusum*, 289; *P. spicatum* (*Pennisetum typhoideum*), 289; *P. turgidum*, 289
Papaver, 21, 179–80; *P. rhoeas*, 179; *P. somniferum*, 179
Parkia, 52; *P. africana*, 52; *P. biglobosa*, 52; *P. filicoides*, 52; *P. speciosa*, 52
Parmelia physodes, 286
Paspalum scrobiculatum, 289
Pastinaca sativa, 168–9
Pedicularis lanata, 32
Peltandra virginica, 270
Pennisetum typhoideum (*Panicum spicatum*), 289
Peperomia, 343
Pereskia: *P. aculeata*, 79; *P. bleo*, 79
Persea, 39; *P. americana*, 39; *P. drymifolia*, 39; *P. schiediana*, 39
Petalostemon oligophyllum, 93
Petasites, 222; *P. frigidus*, 222; *P. japonicus*, 222; *P. vulgaris*, 222

Index of Latin Names

Petroselinum, 340–1; *P. crispum*, 340; *P. crispum filicinum*, 340; *P. crispum neapolitanum*, 340; *P. sativum*, 86; *P. sat.* var. *tuberosum*, 86; *P. segetum*, 89 n
Peucedanum, 169–70; *P. canbyi*, 170; *P. cous*, 170; *P. farinosum*, 170; *P. graveolens*, 329; *P. ostruthium*, 329; *P. palustre*, 329
Phalaris canariensis, 291
Phaseolus, 54–8; *P. aconitifolius*, 56; *P. acutifolius*, 56; *P. adenanthus*, 56; *P. angularis*, 56; *P. aureus*, 56; *P. calcaratus*, 56; *P. coccineus* (*multiflorus*), 55; *P. diversifolius*, 56; *P. lunatus*, 55; *P. lun.* var. *limensis*, 55; *P. mungo*, 56; *P. mun.* var. *radiatus*, 56; *P. polystachys*, 56; *P. tunkiniensis*, 56; *P. vulgaris*, 54
Phellopterus: P. littoralis, 170; *P. montanus*, 170
Phelypaea, 36
Phoenix, 164; *P. farinifera*, 302
Phragmites communis, 199
Phyllitis, 213
Phyllophera rubens, 214
Phyllostachys, 39–41; *P. bambusoides*, 41; *P. edulis*, 41; *P. riga*, 41
Physalis, 240
Phytolacca, 228; *P. abyssinica*, 228; *P. acinosa*, 228; *P. decandra*, 228; *P. esculenta*, 228; *P. octandra*, 228; *P. rivinoides*, 228
Piaropus (*Eichornia*), 252
Pimpinella saxifraga, 319
Pinus, 296–7; *P. albicaulis*, 297; *P. cembra*, 296–7; *P. cem.* var. *sibirica*, 297; *P. cembroides*, 297; *P. cemb.* var. *edulis*, 297; *P. cemb.* var. *monophylla*, 297; *P. cemb.* var. *parryana*, 297; *P. coulteri*, 297; *P. gerardiana*, 297; *P. koraiensis*, 297; *P. lambertiana*, 297; *P. pinea*, 296; *P. sabiana*, 297; *P. torreyana*, 297
Piperaceae, 320–1
Piper, 341–3; *P. bantamense*, 342; *P. betle*, 342; *P. clusii*, 343; *P. cubeba*, 342; *P. excelsum*, 343; *P. guiense*, 342; *P. longum*, 341, 342; *P. methysticum*, 342; *P. nigrum*, 342; *P. officinale*, 342; *P. saigonense*, 342; *P. sanctum*, 343
Pisum, 171–3; *P. arvense*, 171; *P. elatium*, 171; *P. sativum*, 171–2; *P. sativum macrocarpum*, 172
Pithecolobium: P. dulce, 59; *P. flexicaule*, 59; *P. labatum*, 59

Pithophora, 213
Plantago, 345–8; *P. major*, 345; *P. media*, 345; *P. maxima*, 345; *P. lanceolata*, 345
Platanthera bifolia, 266
Plectranthus: P. esculentus, 191; *P. ternalis*, 191; *P. tuberosus*, 191
Pleopeltis longissima, 115
Polakowskia tacaco, 122
Polyanthes tuberosa, 140
Polygala: P. butyracea, 149; *P. vulgaris*, 149
Polygonatum, 21, 135; *P. biflorum*, 135; *P. giganteum*, 135; *P. japonicum*, 135; *P. officinale*, 135
Polygonum, 219–20; *P. bistorta*, 219; *P. cuspidatum*, 220; *P. douglasii*, 220; *P. maximowiczii*, 220; *P. muelenbergi*, 220; *P. sachalinense*, 220; *P. weyrichii*. 220
Polypodium: P. lineare, 131; *P.* (*Drynaria*) *quercifolium*, 115
Pontederia cordata, 252
Porphyra, 211–12; *P. lacianata*, 211; *P. leucostica*, 212; *P. nerocystis*, 212; *P. perforata*, 212; *P. suborbiculata*, 212; *P. vulgaris*, 211
Portulacca: P. grandiflora, 192; *P. intraterranea*, 192; *P. lutea*, 192; *P. napiformis*, 192; *P. oleracea*, 191–2; *P. quadrifida*, 192; *P. retusa*, 192; *tuberosa*, 192
Potentilla: P. anserina, 215–16; *P.* (*Comarium*) *palustris*, 216
Poterium sanguisorba, 319
Primula, 99–101; *P. elatior*, 100; *P. veris*, 99–100; *P. vulgaris*, 100–1
Pringlea antiscorbutea, 76
Prunus amygdalus (*Amygdalus communis*), 263
Psophacarpus tetragonolobus, 58, 174
Psoralea: P. castorea, 191; *P. esculenta*, 190–1; *P. maphitica*, 191
Pteridium: P. aquilinum, 116; *P. esculentum*, 115
Pterocladia, 214
Ptychosperma elegans, 167
Pueraria thunbergiana (*Dolichos hirsutus* (*japonicus*)), 53
Pyrrhopappus catolianus, 112

Quercus, 21, 261–2; *Q. alba*, 261; *Q. ballota*, 262; *Q. emoryi*, 261; *Q. glabra*, 262; *Q. glauca*, 262; *Q. ilex*, 262; *Q. lobata*, 261; *Q. persica*, 262; *Q. primus*, 261; *Q. robur*, 262; *Q. sessifolia*, 262

Radicula, 104–5
Ranunculus: R. ficaria, 20, 80–3; R. pallasi, 80
Raphanus, 193–4; R. acanthiformis, 194; R. caudatus, 194; R. maritimus, 193; R. raphanistrum, 193; R. sativus, 194
Raphia: R. flabelliformis, 302; R. ruffia, 302; R. taedigera, 302; R. vinifera, 302
Raphionacme brownii, 148
Rhodymenia: R. ciliata, 212; R. palmata, 212
Rhopalostylis sapida, 167–8
Rinorea castanaefolia, 244
Robinia pseudacacia, 52
Roripa, 104–5
Rosmarinus officinalis, 348–9
Roystonea: R. oreodoxa, 168; R. regia, 168
Rumex, 21, 216–19, 221; R. abyssinicus, 218–19; R. acetosella, 217; R. acetosa, 217; R. alpinus, 218; R. arcticus, 219; R. berlanderi, 219; R. brasiliensis, 219; R. crispus, 218; R. hymenocephalus, 219; R. mexicanus, 219; R. obtusifolius, 218; R. occidentalis, 219; R. patientia, 218; R. paucifolius, 219; R. roseus, 219; R. sanguineius, 218; R. scutatus, 216; R. vesicarius, 219
Ruscus aculeatus, 35
Ruta graveolens, 349–50

Sabal palmetto, 168
Saccharum spontaneum, 41
Sadlera cyathecides, 305
Sagittaria: S. sagittifolia (chinensis), 252; S. variabilis, 252
Salicornia: S. australis, 202; S. fruticosa, 202; S. herbacea, 202
Salsola: S. asparagoides, 203; S. kali, 202–3; S. soda, 203
Salvadora persica, 151
Salvia, 350–1; S. carduacea, 351; S. columbariae, 351; S. horminum, 351; S. officinalis, 351; S. rutilans, 351; S. sclarea, 351; S. splendens, 351; S. verbenaca, 351
Sambucus, 37–8; S. canadensis, 37; S. nigra, 38
Samuela carnerosa, 139–40
Samolus valerandi, 105
Sanguisorba, 318–19; S. officinalis, 319
Santalum: S. acuminatum, 263; S. murrayanum, 263–4
Saranthe marcgravii, 265
Sargassum: S. echinocarpum, 209; S. enerve, 209; S. vulgare, 209

Sarothamnus scoparius, 21, 176–7
Satureia: S. hortensis, 352; S. montana, S. thymbra, 352
Scabiosa succisa, 345
Scandix: S. grandiflora, 89; S. pecten veneris, 89
Scendesmus, 215
Scheelea martiana (Attalea excelsa), 164
Scilla non-scripta (nutans), 159
Schinus: S. latifolius, 343; S. molle, 343
Scirpus, 199–200; S. lacustris, 199–200; S. maritima, 200; S. nevadensis, 200; S. paludosus, 200; S. tuberosus, 200
Scolymus hispanicus, 236
Scorzonera: S. hispanica, 203–4; S. mollis, 204; S. schweinfurthii, 204
Secale: S. cereale, 300–1; S. montanum, 301; S. sylvestre, 301
Sechium edule, 121
Sedum, 105–6; S. acre, 106; S. album, 105; S. reflexum, 105; S. roseum, 106; S. rupestre, 106; S. telephium, 105–6
Senebiera: S. lepidoides, 106; S. pinnatifida, 106
Senecio: S. (Ligularia) kaemferi, 112; S. palmatus, 112
Serenoa semilata, 168
Serpyllum montanum (Thymus serpyllum), 356
Sesamum: S. alatum, 352; S. calycinum, 352; S. indicum, 352–3; S. radiatum, 352
Sesbanea: S. grandiflora, 177; S. sericea, 177; S. tetraptera, 177
Sesuvium portulacastrum, 192
Setaria italica, 290
Sicaria odorifera, 109
Silene vulgaris (inflata), 227–8
Silybum marianum, 236
Sinapsis: S. alba (Brassica alba), 21, 49; S. arvensis (Brassica sinapistrum), 21, 66; S. incana (Brassica adpressa), 150
Sisymbrium (Descurainia), 150; S. alliara, 150–1; S. halictorum, 150; S. incisa, 150; S. officinale, 150; S. parviflora, 150; S. pinnata, 150
Sium: S. helenianum, 215; S. sisarum, 215
Smilax: S. aspera, 135; S. heracea var. nipponicum, 135; S. megacarpa, 135; S. rotundifolia, 135
Smyrnium olustratum, 86–7
Socratea (Iriartea) durissima, 168
Soja (Glycine) max, 52–3
Solanum: S. aethiopicum, 238; S. agrarium, 238; S. ajanhuiri, 16, 182; S. andigenum, 182, 183; S. anomalum, 238; S. anthropophagum, 238; S.

Solanum—cont.
antipovickii, 16, 182; *S. aviculare (vescum)*, 238; *S. boyacense*, 182; *S. chacoense*, 16, 182; *S. commersoni*, 182; *S. curtilobium*, 182; *S. diversifolium*, 238; *S. ellipticum*, 238; *S. esuriale*, 238; *S. fendleri*, 182; *S. ferox*, 239; *S. gilo*, 238; *S. goniocalyx*, 182; *S. hyperchordium*, 239; *S. hysterix*, 239; *S. inaequilaterale*, 240; *S. incanum*, 239; *S. indicum*, 239; *S. jamesii*, 182; *S. juzepczukii*, 16, 182; *S. laciniatum*, 238; *S. lypopersicum (Lycopersicum esculentum)*, 237; *S. kesselbrenneri*, 16, 182; *S. macrocarpum*, 113, 239; *S. maglia*, 182; *S. mamilliferum*, 16, 182; *S. melongena*, 112–13; *S. muricatum*, 239; *S. nigrum*, 239; *S. nodiflorum*, 239; *S. olivare*, 239; *S. phureja*, 16, 182; *S. pierreanum*, 239; *S. pimpinellifolium*, 237; *S. rybiri*, 182; *S. stenotomum*, 186; *S. torvum*, 240; *S. triflorum*, 239–40; *S. tuberosum*, 16, 180–9; *S. xanthocarpum*, 239
Sonchus: *S. megalocarpus*, 236; *S. oleraceus*, 236
Sophia, 150
Sorghum: *S. durra (vulgare)*, 290; *S. saccharatum*, 290–1
Sparganium eurycarpum, 200
Spergula arvensis, 228
Sphaerococcus gelatinosus, 214
Sphenostylus schweinfurthii, 190
Spilanthes: *S. acmella*, 106; *S. oleracea*, 106
Spinachia oleracea, 222–3
Stachys: *S. officinalis (betonica) (Betonica officinalis)*, 32–3; *S. palustris*, 32; *S. sieboldii*, 32
Stellaria media, 227
Stenochlaena palustris, 115
Sterculia monosperma (nobilis), 281–2
Sticta glomulifera, 286
Stigeoclonium, 213
Stizolobium: *S. niveum*, 58; *S. pruritum*, 229
Symphytum officinale, 62
Symplocarpus (Spathyema) foetidas, 97
Syntherisma (Digitaria) sanguinalis, 290

Tacca: *T. hawaiiensis*, 265–6; *T. pinnatifida*, 265
Talinum: *T. aurantiacum*, 192; *T. crassifolium*, 192; *T. patens*, 192; *T. triangulare*, 192
Tamarindus indicus, 59
Tamus communis, 36

Tanacetum: *T. balsamita (Chrysanthemum pyrethrum)*, 92; *T. vulgare*, 92
Taraxacum officinale, 110–12
Telfairia: *T. occidentalis*, 121; *T. pedata*, 121
Telosma: *T. cordata*, 149; *T. procumbens*, 149
Tetragonia: *T. expansa*, 229; *T. implexicoma*, 229
Thalictrum aquilegifolium, 84
Thelygonum cynocrambe, 75
Thlaspi, 102
Thymus, 355–7; *T. azoricas*, 356; *T. citriodorus*, 356; *T. cit. aureus*, 356; *T. cit. variegatus*, 356; *T. fragrantissimus*, 356; *T. herba-barona*, 356; *T. mastachinus*, 356; *T. serpyllum (Serpyllum montanum)*, 356; *T. virginicus*, 356; *T. vulgaris*, 356
Tigrida pavonia, 143
Tordylium apulum, 89
Torreya, 17; *T. californica*, 340
Trapa: *T. bicornis*, 281; *T. bispinosa*, 281; *T. natans*, 281
Tragopogon: *T. pratensis*, 201; *T. porrifolium*, 201–2
Treculia africana, 66
Trichosanthes: *T. anguina*, 122; *T. cucumeroides*, 122; *T. dioica*, 122; *T. origera*, 122; *T. subvelutina*, 122
Trifolium: *T. amabile*, 93; *T. ciliatum*, 93; *T. gracilientum*, 93; *T. microcephalum*, 93; *T. tridentatum*, 93
Triglochin: *T. maritima*, 200; *T. procerum*, 200
Trigonella: *T. coerulia*, 93; *T. foenum-graecum*, 93; *T. suavissima*, 93–4
Trillium: *T. erectum*, 136; *T. grandifolium*, 136; *T. pendulum*, 136; *T. stylosum*, 136; *T. undulatum*, 136
Triticum, 306–12; *T. aestivum*, 308; *T. abyssinicum*, 308; *T. compactum*, 308; *T. dicoccum*, 307; *T. durum*, 307–8; *T. hybernum*, 308; *T. macha*, 308; *T. monococcum*, 307; *T. persicum*, 308; *T. polonicum*, 308; *T. spelta*, 308; *T. sphaerococcum*, 308; *T. turgidum*, 308; *T. vulgare*, 308
Tropaeolum, 104, 106–7; *T. edule*, 107; *T. majus*, 106; *T. minus*, 106; *T. patagonium*, 107; *T. pentaphyllum*, 107; *T. polyphyllum*, 107; *T. sessifolium*, 107; *T. tuberosum*, 107
Trophis anthropophagum, 238
Tulipa, 21, 136–9; *T. edulis*, 139; *T. gesneriana*, 139; *T. suaveolens*, 139; *T. sylvestris*, 139

Turbinaria, 209
Tylostemon manii, 323
Typha, 200–1; *T. angustifolia*, 201; *T. elephantina*, 201; *T. latifolia*, 201; *T. muelleri*, 201
Typhonium, 98
Typhonodorum lindleyana, 97–8

Ullocus tuberosus, 233
Ulva: *U. fasciata*, 212; *U. lactuca*, 212; *U. latissima*, 212
Umbilicaria, 287
Undaria pinnatifida, 207
Unona, 344
Urtica, 151–2; *U. crenulata*, 153; *U. chamaeroides*, 153; *U. dioica*, 151; *U. gigas*, 153; *U. gracilis*, 153; *U. holosericea*, 153; *U. pilulifera*, 151–2; *U. thunbergiana*, 153; *U. tuberosa*, 153; *U. urens*, 151; *U. urentissima*, 153
Uvularia sessifolia, 139

Valeriana edulis, 99
Valerianella: *V. eriocarpa*, 99; *V. olitoria*, 98–9
Veronica: *V. anagallis*, 107; *V. beccabunga*, 107
Vicia, 173–4; *V. ervilia*, 173; *V. faba* (*Faba vulgaris*), 47–51; *V. faba* subsp. *eu-faba* var. *equina*, 48; *V. faba* subsp. *eu-faba* var. *major*, 48; *V. faba* subsp. *paucijuga*, 48; *V. narbonensis*, 47; *V. sativa*, 21, 173
Victoria regia (*amazonica*), 251
Vigna: *V. lanceolata*, 190; *V. sesqui-pedalis*, 175; *V. sinensis*, 175; *V. unguiculata*, 175; *V. vexillata*, 190
Viola, 244–5; *V. palmata*, 244; *V. sylvatica*, 244–5
Vitis, 233
Voandzeia subterranea, 282

Washingtonia filifera, 168

Xanthorrhea, 140
Xanthosoma: *X. atrovirens*, 97; *X. brasiliense*, 97; *X. caracu*, 97; *X. sagittaefolium*, 97; *X. violaceum*, 97
Xylopia: *X. aromaticum*, 343–4; *X. carminative*, 343; *X. frutescens*, 343; *X. grandiflora*, 343; *X. sericea*, 343
Xysmalobium cordata, 148

Yucca, 139–40; *Y. arborescens*, 139; *Y. baccata*, 139; *Y. elephantipes*, 139; *Y. macrocarpa*, 139; *Y. mohavensis*, 139; *Y. whipplei*, 139

Zamia: *Z. floridana*, 304; *Z. furfuracea*, 304; *Z. integrifolia*, 304; *Z. lindenii*, 304; *Z. pumila*, 304
Zantedeschia, 270
Zanthoxylum: *Z. aromaticum*, 344; *Z. burgei*, 344
Zea mays, 287–8
Zephyranthes atamasco, 141
Zingiber, 329–30; *Z. amaricans*, 330; *Z. cassumnar*, 330; *Z. mioga*, 330; *Z. officinale*, 329–30
Zizenia aquatica, 299
Zygophyllum fabago, 319

INDEX OF COMMON NAMES

Numbers in bold type refer to illustrations

acacia, 59
acorn, 18, 21, 261–2, 307
adder's tongue fern, 115
Afghan thistle, 239
African millet (Kaffir corn), 290
African spinach, 27
agar-agar, 65, 213–14
agati, 177
agave aloe, 140
akee, 25, **66**
alder, white, 343
akaka, 158
Alexander parsley, 86–7
alkanet, 61–2
alligator pepper, 321
allspice, **337**, 343, 344
almond, 162, 262; native, 164; Peruvian, 274
alocasia, 98
aloe, 140–1
alloto, 256–7
alpine dock, 218
amaranth, 25–8, 226
amaryllis, 140–1
American beech, 262
American wood-lily, 135–6
anemone, 84; narcissus, 84; rue, 84; wood, 84
angelica, 28–9, 169; wild, 29
Angola mandubi, 282
aonori, 212
ape orchis, 266
aracacha, 170
arame, 207
archangel, 28, 154
arrowgrass, sea, **197**, 200
arrowleaf, 252, **253**, 255
arrowroot, **231**, 264–5, 303, 304; Brazilian, 276; reed, 264
artichoke: Chinese, 32; globe, **3**, 29–30, 31; Jerusalem, 30, 201, **231**; root, 30, 31
arum: bog, 270, giant, 269; Italian, 269; lily, 132, 270; wild (cuckoo pint), 268, 269
ashleaf fern, 305
asparagus, 15, 33–4, 35, 126
asparagus bean, 175
asparagus bush, 34
asparagus samphire, 303
asphodel, 126
assai, 167

atamasco lily, 141
aubergine, 113
Australian bean tree (Moreton Bay chestnut), **277**, 281
Australian grove-fern, 305
Australian jelly plant, 214
Australian spinach, 224, 229
Australimusa, 43
avens, 38–9; water, 38–9; wood, 38
avocado pear, 39

balm, 315; bee, 315
balsam apple, 110
balsam pear, 109, **119**
bamboo, 39–41, 164
banana, 41–7
Bancoul nut (country walnut), 263
baobab, 146
Barbary oak, 262
barilla plant, 202, 203
barley, 16, 270–3, **293**
basella, 229
basil, 315–16; bush, 315–16; common, 315; hoary, 316; purple-stalked, 316; sweet, 315
bay, 316–18, 322
bead lily, 129
bean, 15, 47–60, 171; asparagus, 175; broad, 47–8; butter, **49**, 55; carob, 51–2; cluster, 58; curl-brush, 59; Dolichos, 53; Goa, 58, 174; haricot, 54–5; horse, 48, 53; jack, **49**, 53–4; kidney, 54; lima, 55; locust, 48; Mexican black, **49**, 54; mung, 56; pea, **49**, 55; pigeon, 48; rice, 56; runner, **49**, 55; soya, **49**, 52–3, 306; string, 54; Texas, 56; velvet, 58
bean caper tree, 319
beaver bread, 191
beavertail, 77, 79
bee balm, 315
bee orchis, 266
beech, 21, 262, 274–5; American, 262; European, 262
beet, 60–1; rhubarb, 61; sea, 61; seakale, 61; sugar, 60; white-rooted, 60
beetroot, 60; red, 60; white, 60
belembe, 97
bell-flower, 195
bellwort, 139
beniseed, 352
benzoin, 343

bergamot, 315, 330
bergamot mint, 336
betel nut, 168
betel pepper, 342
betony, 32–3; marsh, 32; sweet, 32; wood, 32
bihai, 46
bistort, 219, 220
bitter almond, 263
bitter cassava, 276
bitter cress, 103
bitter vetch, tuberous-rooted, 189
bitter wood, 343
black-berried nightshade, 239
black cummin, 344
black pepper, 342–3
bladder campion, 227–8
bladder-wrack, **205**, 209
blade kelp, **205**, 210
bleo, 79
blite, 224; strawberry, 224; white, 224
blood-veined dock, 218
blue dicks, 126
bog arum, 270
Bombay mace, 339
Bombay nutmeg, 339
bopple nut, 298
borage, 61–2, **325**
borecole (kale), 69
box, 163 n
bracken, 20, 116
bracken fern, 116
brank buckwheat, 274
Brazil nut, 273–4, **277**
Brazilian arrowroot, 276
Brazilian pepper, 343
breadfruit, 62, 65, **67**
breadnut, 62
bread-root, 170
brimstonewort, 329
bristle oat, 291
broad bean, 47–8, 51
broad-leaved dock, 218
broccoli, 72; Italian, 72
brook weed, 105
brookline, 107
broom, 21, 35, 176–7
broomrape, 35–6, 126
Brussels sprout, 71, 72, **73**
bryony, 20, 36; black, 36
buckeye, 280
buckwheat, 216, 274–5; brank, **274**; notch-seeded, 275; Tartary, 275
bugloss, 61; sea, 62
bugre, 343
bull-nosed pepper (green, red), 321, **337**

bulrush (lake clubrush), **197**, 199; alkali, 200; Nevada, 200; tuberous, 200
Burdekin sorrel-vine, 233
burdock, 204; great, 204
buri palm, 302
burnet, 318–19; great, 319; salad, 319
burnet saxifrage, 319
bur-reed, giant, 200
bush basil, 316
butcher's broom, 33, 35
butter bean, **49**, 155
butternut, 306
butterbur, 222

cabbage, 15, 66, 69, 217; Chinese, 72, **73**; dog's, 75; Kerguelen, 76; meadow, 97; Portuguese, 72; red, 69; Savoy, 70; wild, 66
cabbage palm, 167, 202–3; Australian, 167; Guadeloupe, 167
cabbage palmetto, 168
cabbage tree, 129
cabbage turnip, 243
cactus, 76, 79; giant, 79
calalu, 27; branched, 239; prickly, 27, 239; Spanish, 228, 239
Californian chia, 351
Californian Washington palm, 168
calla: lily, 132, 270; wild, 270
caltrop, 235
calypso orchid, 268
camass root, 129, 286
camomile, 329
campion: bladder, 227
Canada snakewort, 330
canaigre, 219
Canary dragon tree, 35
Canary grass, 291
cancer root, 36
cane, 41
cane palm, 164, 167
cañihua, 225
canna, 265
cannibals' tomato, 238
cannonball tree, 274
Cape spinach, 193
caper, 319–20; prickly, 319; South African, 319; Timbuctoo, 319
capucine (ysaño), 107, **231**
caracu, 97
cardamom, 321, **331**
cardon, 79
cardoon, 29–30
carline thistle, 234–5
carob, 51, 52
carragheen, **205**, 207

Index of Common Names / 375

carraway, 178–9, 329; edible-rooted, 178
carrot, 79–80, 241; Peruvian, 170; spring, 215
cashew, 275–6, **277**
cassava, **231**, 276, 279; bitter, 276; sweet, 276
cassia (Chinese cinnamon), 322–3; clove, 322–3; Indian, 322; Padang, 322
cassumnar ginger, 330
capsicum, 320–1
cat briars (green), 135
cat's ear, 112
catechu, 168
catjung, 175
cauliflower, 71, 243
cayenne, 320, 343
celandine, **63**, 80–5; great, 82; lesser, 20, 80
celeriac, 86, **95**
celery, 85–6, 340
celtuce, 125
centaury, 235
Ceylon moss, 214
chaff-flower, prickly, 27
chard, 60–1
charlock, 21, 66, 149, 150, 193
chayote, **119**, 121
cherry laurel, 317
cherry pepper, 320
chervil, 87–9; great, 88; root, 170; salad, 88; wild, 88
chestnut, 261, 279–82; China, 291–2; dwarf, 279; Fiji, 281; horse, 280; Moreton Bay, **277**, 281; sea, 281; sweet, 279; water, 281; wild, 279
chestnut dion, 304
chick pea, 171, 174–5
chickweed, 19, 227–8; Chinese, 192
chickling vetch, 173–4
chicory, 90, 110, 201
children's tomato, 238
Chile pine (monkey puzzle), 297
Chile sassafras, 340
chilli, 320, 321, **337**
China chestnut, 281–2
China pea, 175
China squash, 118
Chinese artichoke, 32
Chinese cabbage, 72, **73**
Chinese cinnamon (cassia), 322
Chinese mustard (Indian), 72, 150
Chinese pepper, 344
Chinese spinach, 229
Chinese water lily, 249
Chinese yam, 54
Chinese yellow bean, **54**
chinquapin, 261, 280; evergreen, 280; golden, 280; Philippine, 280; Sumatra, 280; Tibetan, 280; water, 249
chive, 159–60, 220; Chinese, 160
cholas, 76
christophine, 121
chrysanthemum, 91–2
chucklusa, 170
ciceley, 88–9
cinnamon, 321, 322–3, **331**; Chinese (cassia), 322–3; Santa-Fe, 322
cinnamon vine, 258
clammy weed, 320
clary, 351
clematis, 84
clove, 323–4, **337**; Brazil, 324; Madagascar (clove-nutmeg), 323
clove bark (Brazil clove), 324
clove cassia, 322, 324
clove-nutmeg, 340
clover, 93–4, 221
club wheat, 308
clubrush: lake (bulrush), **197**, 199; seaside, **197**, 200
cluster bean (guar), 58
cob nut, 285
cock's spur grass, 346
cockspur, 191
cocona, 239
coconut, **165**, 167
cohune, 164
colocasia, 94
columbine meadow rue, 84
comfort root, 304
comfrey, 62, **63**
common basil, 315
common melilot, 93
common orach, 19
common wheat, 308
cone pepper, 320
Congo bean, 282
convolvulus, 230
coral bean, 60
coriander, 324, **325**, 329
cork-wood, 148
corn (wheat), 306–12
corn-cockle, 228
corn marigold, 91
corn salad, 98–9; Italian, 99
corn spurrey, 228
cos lettuce, 125
costmary, 92
cotton (Scotch) thistle, 142, 235–6
country walnut (Bancoul nut), 263
courgette (zuccini), 118
cous, 170
cow lily, 249–50
cow-parsley, 88

cow's parsnip, 169
cow pea, 175, 190
cow-plant, 148
cow tree, 65–6
cowage, New Zealand, 58, 229
cowitch, 58
cowslip, **63**, 83, 99–100; American, 149
crab-apple, 346
crab-grass, 290
cress: bitter, 103; garden, 103, 149; hairy bitter, 103; para, 106; penny, 102; rock, 102; swine's, 106; winter, 102
crinkleroot, 333
crocus, **127**, 141–2; saffron, 141
crowfoot, Arctic, 80
crown imperial, 131
crowscress, 106
cubeb, 342
cuckoo flower, 103
cuckoo pint (wild arum), 19, 132, 268, 269
cucumber, 107–10, 117
cumin, **325**, 327
cummin; black, 344; wild, 327
curcumas, 265–6
curl-brush bean, 59
curra, 305
custard marrow, 118
cycad (fern palm), 303–4
cyclamen, 152
cypress, sweet, 196, 197

daffodil, 126
dandelion, 110–12
dandelion sundrop, 114
Darling lily, 141
dasheen, **95**, 97
date, 164, 302; shrubby, 302
daikon, 194
datil, 139
day lily, 131
dead nettle, 153–4; purple, 154
deer vetch, 174
desert candle, 130
desert thorn, 240
desert trumpet, 220
dill, **325**, 327–8, 329; East Indian, 329
dillenia, 327
ditch reed, 199
dock, 216, 217, 224; alpine, 218; Arctic, 219; bladder, 219; blood-veined, 218; Brasilian, 219; broad-leaved, 218; rhubarb, 218–19; spinach (patience), 218, 219
dog's cabbage, 75
dog's tooth violet, **127**, 130

Dolichos bean, 53, 54
doubah, 190
dourra (maize), 288; (sorghum), 290–1
dragon plant, 34–5, 129
Drummond thistle, 234
dulse, 209–10, 218; craw, 212; pepper, 209
durmast oak, 262

East Indian dill, 329
earth-nut, 178
egg fruit, 113
egg plant, 15, 112–13
einkorn, 307
elder, 37; ground (dog), 29
elephant grass, 201
emmer, 307–8
emory oak, 261
endive, 90; Batavian, 90
English oak, 262
English sorrel, 217
enhalus, 209
ensete, 46
ervillia, 173
Ethiopian nightshade, 238
Eumusa, 43
European beech, 262
European oak moss, 286
European palm, 167
evening primrose, 113–14

fabirama, 153
false orchis, 266
fameflower, 192
fawn lily, 130; dogtooth, 130; lamb's tongue, 130
feather cockscomb, 28
fennel, **325**, 328–9, 346, 355, Italian (Florence), **95**, 329
fennel-flower, 327, 344
fenugreek, 93, **325**
field cabbage, 240
field mustard, 149
field poppy, 179–80
fig marigold, 229
fikongo, 148
filbert, 285
finger millet, 289, 290, 353
finger poppy, 146
fireweed, American, 112
fish-tail palm, 302, 303
flageolet, 54
flame-nettle, 153
flat sedge, 196
fleece-flower, 220
flixweed, 150
floating heart, 251–2

Index of Common Names | 377

Florence fennel (Italian), **95**, 329
florimer, 27
French millet, 289
French sorrel, 216
French vetch, 47, 173
friar's cowl, 269
fringed water lily, **253**
fritillary, 130–1
frost plant, 230

galingale, 196, 320
garden cress, 103, 149
garlic, **95**, 126, 155, 156, 159, 160–3, 193
garlic mustard, 150
garlic weed, 319
gherkin, 108
giant arum, 269
giant knotweed, 220
gilo, 239
ginger, 321, 324, 329–30, **331**, cassumnar, 330; common, 329
ginkgo, 303, 304
gladiolus, 143
glass-wort, 202
globe artichoke, **3**, 29–30, 31, 234
Goa bean, 58, 174
golden marsh fern, 115
golden samphire, 203
good King Henry (mercury goosefoot), 223–4
goosefoot, 16, 222, 223–5
gourd, **119**, 117–23; bottle, 121; Malabar, 118; scarlet-fruited, 121; snake, **119**, 121, 122; wax, 121–2
gram, 175; green, 56; Madras, 53; Turkish, 56
grape hyacinth, 141
grape vine, 233
great bell-flower, 220
great burnet, 319
green pepper (bull-nosed), 321
green winged orchis, 267
gromwell reed, 195–6
ground cherry, 240
ground elder, 29
ground nut, 178–9, 189, 282; Congo, 282
ground nut pea, 189
Guinea pepper, 343

hacub, 235
haricot bean, 54–5
hart-wort, 89
hazel, 163 n, 282–5
hazelwort, 330
hedge mustard, 150
hellebore, 152
hemlock, 88

herb lily, 140–1
herb-trefoil, 93
hibiscus, 147–8
hijiki, 213
hoary basil, 316
hog fennel, 169–70, 329
hogweed, 169
holm oak (ilex), 262
Holy Ghost, 28
honesty, 243
honewort, 89
hop, 37
horn plant, 207
horse bean, 48
horse chestnut, 280, 348
horse mint, 315
horse-tail, 20, 116–17
horsehair lichen, 286
horseradish, 20, 330, 333
horseradish tree, 333

Iceland moss, 287
iceplant, 229–30
ilex (holm oak), 262
Indian copper leaf, 320
Indian lettuce, 192
Indian morel, 239
Indian (Chinese) mustard, 72, 150
Indian pea, 21
Indian pipe, 36
Indian shot, 265
Indian turnip, 269
iris, **127**, 143–5
Irish moss (carragheen), **205**, 207, 213
Italian arum, 269
Italian broccoli, 72
Italian corn salad, 99
Italian fennel (Florence), **95**, 329
Italian millet, 290, 306
ivy: poison, 275; prickly, 135

jack bean, **49**, 53–4
jackfruit, 65
jaggery palm (sugar palm), **165**, 302
Japanese kelp, 210
Japanese millet, 289, **293**
Japanese oak, 262
Japanese pepper, 344
Japanese spinach, 229
jasmin orange-tree, 327
Jerusalem artichoke, 30, 201, **231**
joyweed, 28
Judas tree, 177
juniper, 226, **325**
jute, 147

Kaffir bread, 303–4

378 / Index of Common Names

Kaffir potato, 153, 191
kale, 70; sea, 72, **73**, 75; wild, 150
kangaroo apple, 238
kapok-tree, 146–7
kava pepper, 342
kelp, 210; Japanese, 210
kemando, 65
Kerguelen cabbage, 76
kidney bean, 54
kie-kie, 66
kiery, 27
king plant, 268
knapweed, 235
knob-tang, 209
knotweed, 219–20; giant, 220
kohl-rabi, 72, **95**, 243
kombu (oarweed), 207
krobonko, 121
kudzu, 53
kurome, 207

lablab, 53
lad's love, 353–4
lady orchis, 267
lakoocha, 65
lamb's cress, 346
laurel, 317; bay, 317; cherry, 317; New Zealand, 264; Roman, 317; sweet, 317
laver, 211–12
leek, **95**, 126, 155, 156, 157, 158–9
lemon grass, 327
lemon scented thyme, 356
lentil, 123–4
lettuce, 98–9, 124–5, 217; Batavian, 125; Indian, 124, 192; least, 124; oak-leaved, 124; Spanish, 192; wild, 90, 124; willow, 124
lichen, 285–7; horsehair, 286; manna, 285–6; puffed shield, 286
lily, 15, 126–36, **127**; arum, 132–5; bead, 129; calla, 132–5; day, 131; fawn, 130; herb, 140–1; martagon, **127**, 133, 134; palm, 129; scrambling, 35; sego, 129; snake's head, 135; sword, 143; tiger, 134
Lima bean, 55
limu: akiaki, 214; eleele, 212; huana, 214; kala, 209; kobu, 213; lipahapaha, 212; lipoa, 213; luan, 212; manauea, 214
locust, 51, 52
long pepper, 321
loosestrife, 104
lotus, 245–6, **247**, 249, 250
lovage, 87; Canadian, 87; Scottish, 87
love-in-a-mist, 344

love-lies-bleeding, 26, 137
lubia, 53
lunary (moonwort), 243–4
lupin, 21, 171, 176, 177
Luzon potato, 257

macambira, 305
Macassar nutmeg, 339
mace, 103, **337**, 339–40; Bombay, 339
macha wheat, 308
maciba, 257
Madeira vine, 233
maidenhair fern, 114
maidenhair tree (ginkgo), 304
maize, 182, 195, 237, 287–8; water, 251
Malabar gourd, 118
Malacca cane, 167
malanga, 97
mallow, 145–8, 216, 226; tree, 190
man orchis, 267
mango, 275
manioc, 53, 276
manna lichen, 285–6
manna oak, 262
mangel-wurzle, 60
marigold, 142, 333–4; corn, 91; fig, 229
mariposa lily, 126, 129
marjoram, 334–5
marrow, 117, 118, 203; custard, 118; Siam, 118; vegetable, 117, 118
marsh betony, 32
marsh marigold, 20, 83; white-flowered, 83
marsh orchis, 266
marsh parsley, 86
marsh woundwort (betony), 32
marsh samphire, 202
marshmallow, 145–6
martagon lily, **127**, 133, 134
masterwort, 329
mastich tree, 343
mayweed, 346
meadow distaff, 234
melon, 109; water, **119**, 233
melon shrub, 239
Mexican black bean, **49**, 54
milfoil, 228, 344
military orchis, 266
milk thistle, 236–7
milk-vetch, 174, 189–90
milkweed, 148–9
milkwort, 149
millet, 289–90; African 290; barnyard, 289; finger, 289, 290, 353; French, 289; great, 290; Italian, 290; Japanese, 289, **293**; koda, 289; little, 289; pearl, 289

Index of Common Names / 379

mint, 335–6, 339; apple, 336; bergamot, 336; corn, 336; country, 336; downy, 336; forest, 336; fringed; 336, hairy, 336; hairy water, 336; horse, 336; lamb, 339; mountain, 315; red, 336; slender, 336; small, 336; smooth water, 336; spear, 336; whorled, 339
mioga, 330
miriti fibre palm, 303
mirume, 208
monkey nut (peanut), 282
monkey pot tree, 273, 274
moon creeper, 233
moonwort, 243–4
Moreton Bay chestnut (Australian bean tree), **277**, 281
morning glory, 230
mougri, 194
mountain rose, 233; white, 192
mugwort, 346, 353, 354–5
mung bean, 56
murlins, 207
mushroom, 155
musk thyme, 356
mustard, 15, **73**, 149–51; black, 149, **325**; field, 149; garlic, 150; hedge, 150; Indian (Chinese), 92, 150; tansy, 150; white, 21, **325**
mustard tree, 151
myrtle, 317, 323

nama dumpa, 251
nardoo plant, 115
nasturtium, 104–5, 106–7
native potato, 190
net fern, 115
nettle, 151–4, 217, 220, 226, 346; dead, 153–4; devil, 153; flame, 153; Greek, 152–3; hoary, 153; Hudson's Bay, 153; Japanese, 153; stinging, 152; weak, 153
Nevada joint-fir, 117
New Zealand laurel, 264
New Zealand spinach, 229
nibong, 167
nigella, 228
Niger seed, 352
nightshade, 16, 183; black-berried, 239; cut-leaf, 239–40; deadly, 112; Ethiopian, 238; red Malabar, 229
nikau, 167–8
nopal, 79
Norfolk Island tree-fern, 305
nostoc, 211
nuphar, 249–50
nut, *see under individual names*
nut-grass, yellow, 196

nutmeg, **337**, 339–40; Bombay, 339; Macassar, 339; native, 339; Queensland, 339; Santa-Fe, 339; wild, 339
nymphaea, 250

oak, 261–2; Barbary, 262; durmast, 262; English, 262; emory, 261; holm, 262; Japanese, 262; manna, 262; swamp chestnut, 261; white, 261
oak moss: European, 286; stag-horn, 286
oarweed (kombu), 207
oat, 220, 291–2, 295, **293**; bristle, 291; red, 291; short, 291; Turkish, 291; wild, 291
oca, 221–2
okra, **67**, 147; long, 122; wild, 244
old man, 353
oleander, 155
oleaster, 155
olive, **67**, 154–5
olive tomato, 239
olombeh, 239
omime root, 191
onion, 15, 20, **95**, 155–8, 193; bog, 269; Canada, 157–8; Egyptian tree, 157; Himalayan, 158; Japanese, 158; old man's, 158; potato, 157; ruddy, 158; Siberian, 158; Spring, 157
opium poppy, 179
orach, 224, 225–6
orchid, 268; adder's mouth, 268; calypso, 268; fringe, 268; potato, 268; tree, 268
orchis, **231**, 266–8; ape, 266; bee, 266; false, 266; green winged, 267; lady, 267; man, 267; marsh, 266; military, 266; spider, 266; tawny, 266
oregano, 335
oriental water fern, 115
orpine, 105–6
Otba wax-tree, 339
ox-eye daisy, 91, 92
oxlip, 100

pacaya, 167
palm, 16, 163–8; Buri, 302; cabbage, 167, 302; Californian Washington, 168; cane, 164; European, 167; date, 164, 302; fern (cycad), 303–4; fish-tail, 302; jaggery (sugar), **165**, 302; jupati, 302; miriti fibre, 303; Piassava, 302; peach, 303; raffia, 302; royal, 168; sago, **165**, 303; shadow, 167; solitaire, 167; toddy, 302
palm cabbage, 46, 164, 305
palm lily, 34, 129
palmetto, 168
palmyra, 164, **165**

palo verde, 177
pandan (screwpine), 66, **67**
paprika, 320, 343
papyrus rush, 196
para cress, 106
paradise nut, 274
parakilja, 193
parsley, 340–1; Alexander, 86–7; Dutch, 86; marsh, 86; sea, 87; turnip-rooted, 86
parsnip, 168–9, 241; wild, 170
pasta, 295–6
patchouli plant, 153
pawpaw, 109
pea, 171–3; chick, 171; China, 175; field, 171; Mediterranean, 171; pigeon, 175–6; scurf, 190; sugar, 172; yellow-flowered, 174
pea bean, **49**, 55
pea-tree, 177
peach, 263
peach palm, 303
peanut (monkey nut), 282
penny cress, 102
penny royal, 336
peony, 84
pepper, 320–1, **337**, 341–4; alligator, 321; Ashanti (West African), 342–3; betel, 342; black, 342–3; Brazilian, 343; cayenne, 320, 343; cherry, 320; Chinese, 344; cone, 320; holy, 343; Japanese, 343; bullnosed (green, red), 321, **337**; kava, 342; long, 321, 341, 342; native, 343; white, 75
pepper-elder, 343
pepper tree, 343
peppermint, 336; Australian, 336
pepperwort, 103–4; Drummond's, 115; four-leaved, 115; hairy, 115
piassava, 164, 302
pickerell weed, 252
pigeon bean, 48
pigeon pea, 175–6
pignut, 32, 178–9
pili nut, 164
pimento, 343
pimiento, 320
pimpernel, 318; scarlet, 318
pindaiba, 343
pindova, 164
pine: Australian, 297; Californian, 297; Chile, 297; Colorado, 297; Coulter, 297; Digger, 297; grey-leaf bull, 297; nut, 297; Swiss stone, 297; Torrey, 297; white bark, 297
pine nut, **277**, 296–7
pinole, 79, 351
pinyon (American nut pine), 296–7

pistachio, 275
plantain, 345–8; (banana), 44, 46; water, 252, **253**
plume thistle, 234
pokeweed (poke), 228
polenta, 279
polypody, 115, 305
pond spice, 323
poppy, 21, 179–80; field, 179–80; finger, 146; opium, 179; Shirley, 179
Port Curtis yellow wood, 177
Portugese cabbage, 72
potato, 16, 19, 31, 169, 180–9, 223, 230, 237; Andean, 182; air, 256; Chinese, 256; country, 191; Cree, 190; Fendler's, 182; Hausa (Kaffir), 191; James, 182; Luzon, 257
potato-bean, 189–90
potato orchid, 268
potato yam, 256, 257
prickly ash, 344
prickly box, 240
prickly caper, 319
prickly ivy, 135
prickly pear, 76, **77**, 219
prickly samphire, 202
primrose, 101
princess-feather, 26, 27
puffed shield lichen, 286
pumpkin, 117, 118, 121; field, 118; musk, 118
purple laver, **205**, 210–11
purple-stalked basil, 316
purslane, 191–3, 226; desert-rock, 193; seaside, 192; winter, 192

quandong, 263
queen palm, 167
Queensland, nut, **277**, 297–8
Queensland nutmeg, 339
Queensland sorrel tree, 148
quelite, 27
quinoa, 225

radish, 193–4; Bavarian, 194; cabbage, 194; mountain, 330; sea, 193; Spanish, 194
raffia palm, 302
rampion, 195
randa, 238
rape, 35, 241; Indian, 149
rape-root, 190
rattan, 167
rattlebox plant, 177
rattlesnake weed, 80
Ravensara nut, 323–4, 340
red-bud, 177

red dead nettle, 32
red oat, 291
red pepper (bull-nosed), 321, **337**
reed: ditch, 199; gromwell, 195–6
reed arrow root, 264
reedmace, **197**, 200–1; broad-leaved, 200; small, 201
rhubarb, 216, 222
rhubarb dock, 218–19
ribweed, 89
rice, **293**, 298–300, 306; Indian, 299; mountain, 299
rice bean, 56
rock cress, 102
rock purslane, 193
rock samphire, 203
rock-tripe, 287
rocket, **73**, 75–6; winter, 102; wound, 102
Roman laurel, 317
Roman pimpernell, 89
root artichoke, 30, 31
root chervil, 170
rosemary, **325**, 348–9
royal water lily, 251
rue, 325, 349–50
runner bean, 55
rush pea, 190
rye, 300–1; mountain, 301; wild, 301

saffron, 141, 142
safflower, 141, 142
sage, 350–2; black, 353; garden, 351; pineapple, 351; red-topped, 351; thistle, 351
sage brush, 353; fringed, 353; large, 353; three-tipped, 353; Wright's, 353
sagisi, 167
sago, 164, **165**, 301–5; cycad, 303; Japanese, 303; smooth, 301; spiny, 301
sago fern, 304
sahuaro (giant cactus), **77**, 79
salad burnet, 319
salep, 266; royal, 158
salep plant, 265
salsify, 201–2; black (scorzonera), 203–4; meadow, 201
salt-bush, 226
saltwort, 202, 226; Australian, 202; black, 149; West-Indian, 203
samphire, 192, 202–3, 218; asparagus, 203; golden, 203; Jamaican, 203; marsh, 202; prickly, 202; rock, 202
sandwort, 227
Santa-Fe cinnamon, 322
Santa-Fe nutmeg, 339
Saracen wheat, 274
Sargasso-weed, 209

savory, 352; Canadian, 352; mountain, 352; winter, 352
Savoy cabbage, 70
saw palmetto, 168
scallion (shallot), 157
scolymus, 236
scorzonera, 203–4, 215; gentle, 204
Scottish thistle (cotton), **3**, 142, 235–6
scrambling lily, 35
screwpine (pandan), 66, **67**
scurf pea, 190
scurvy grass, 333
sea arrowgrass, **197**, 200
sea-beet, 61
sea cabbage, 212
sea chestnut, 281
sea chickweed, 227
sea holly, 235
seakale, 72, **73**, 75
seakale beet, 61
sea lentil, 209
sea lettuce, 212
sea-oak, 209
sea purslane, 226
sea stock, 75–6
seaside clubrush, **197**, 200
seaside purslane, 192
sebastan plum, 264
sedge: bearded flat, 196: flat, 196; spike, 196
seepweed, desert, 226
semolina, 295
senna, 177
sesame, 352–3
shadow palm, 167
shallot (scallion), 157
shamrock, 221; Australian, 93
sheep sorrel, 217
shell dillisk, 212
shepherd's needle, great, 89
shepherd's purse, 102
Shirley poppy, 179
short oat, 291
shot wheat, 308
Siam marrow, 118
silky wormwood, 353
silver fir, 163
silverweed, 215
sisal, 140
skirret, 215–16; wild, 215
skunk top, 191
sloke, 211
snake gourd, **119**, 127
snake root, 235
snake's head lily, 135
snakewort, Canada, 330
snap-weed, 327

sneezewort, 344
solitaire palm, 167
Solomon's seal, 20, **127**, 135
sorghum (dourra), 290–1, 306; sugar, 291
sorrel, 21, 216, 221, 226; English, 217; French, 216; mountain, 219; sheep, 217; wood, 21, 216, 220–1
souari (swarri) nut, 274
soulkhir, 225
South African caper, 319
southernwood, 353, 355
sow thistle, 236
soya bean, **49**, 52–3, 306
Spanish calalu, 228, 239
Spanish needle, 229
Spanish pepper, 343
Spanish salsify (scorzonera), 203–4, 236
spear mint, 336
speedwell, 107
spelt, 308
spicy cedar, 323
spider herb, 103, 319
spider orchis, 266
spiderflower, 319–20
spike sedge, 196
spinach, 27, 203, 222–3, 296; Australian, 224; Cape, 193; Chinese, 229; Cuban, 192; East Indian, 229; Japanese, 229; water, 233
spinach dock (patience), 218
spleenwort, edible, 115
spoonflower, 97
spoonlily, 98
spot flower, 106
spring beauty, 192
spring onion, 157
spring wheat, 308
spurge, 276
squash, 117, **119**; bush, 118; China, 118; summer (crookneck), 117–18
stag-horn oak moss, 186
star of Bethlehem, 35
star thistle, 235
starchwort, 268, 269
stinging nettle, 152
stinking thistle, 235
stone pine, 296
stonecrop, 105–6
strainer vine, 122
string bean, 54
sucrier, 43
sugar beet, 60
sugar palm, 302; (jaggery palm), **165**, 302
sugar pea, 172
sugar wrack, 210
summer squash (crookneck), 117–18, **119**
sunflower, **3**, 20, 30, 31, 334

sunplant, 192
susumber, 333
swamp chestnut oak, 261
swede, 242
sweet Allison, 102
sweet basil, 315
sweet betony, 32
sweet-cane, 41
sweet chestnut, 279
sweet Cicely, 88
sweet corn root, 265
sweet cypress, 196, **197**
sweet flag, 144–5, 268
sweet laurel, 317
sweet potato, 58, 190, 230, **231**, 233, 276
swine's cress, 106
sword lily, 143

tabasco, 320
tacaco, 122
talghouda, 178
tamarind, 59, 321
tangle, 210
tansy, 92
tansy mustard, 150
tapioca plant, 276
tara fern, 115
taro, 94, 256, 268
tarragon, 353
Tartar bread, 75
Tasmanian tree-fern, 305
tassel bur, 234
tawny orchis, 266
telfairia nut, 121
tepejilote, 167
Texas bean, 56
thistle, 16, 234–7; Afghan, 239; carline, 234–5; cotton (Scotch), 3, 142, 235–6; Drummond, 234; milk, 236; pine, 235; plume, 234; sow, 236–7; star, 235; torch, 79; Virginian, 234; wavy-leaf, 234; western, 234
thyme, **325**, 328, 346, 355–7; garden, 356; lemon scented, 356; musk, 356; wild, 356
thymeweed, 227
ti-kouka, 129
ti-pore, 129
tiger lily, 134–5
Timbuctoo caper, 319
tobacco, 237
tomato, 15, 183, 184, 237–40, 316; cannibal's, 238; children's 238; La Paz, 240; New Zealand, tree, 240; olive, 239; vegetable, 240
tongusa, 213
toothwort, 333

Index of Common Names | 383

toria, 149
torch-thistle, 79
tree bean, 58–9
tree-fern, 304–5; Norfolk Island, 305; Tasmanian, 305
tree-mallow, 190
tree orchid, 268
trumpet seaweed, 207
tsu choy, 212
tuhuha, 170
tulip, 21, **127**, 136–9; wild, **127**, 136, 137
turkey berry, 240
Turkish oat, 291
turmeric, 265, 324, **331**
turnip, 15, 72, 169, 240–4; Indian, 269; prairie, 190
turnip fern, 305
two-rowed barley, 273, **293**

ulloco, **231**, 233
urucuri, 164

valerian, 98–9
vanilla, **331**
Venus comb, 89
velvet bean, 58
vetch, 21, 47, 171, 173–7
viha, 97
violet, 63, 244–5; dog's tooth, **127**, 130; wood, 244

wakame, 207
wallowa, 170
walnut, 305–6
warrigal cabbage, 229
wasabi, 333
water caltrop, 281
water chestnut, 281
watercress, 16, **73**, 104–5
water dropwort, 252
water glorybind (water spinach), 233
water hawthorn, 251
water hemlock, 252
water hyacinth, 252
waterleaf, 251
water lily, 245–6, 249, 251; blue, 250; bulb-leaf, 250; Chinese, 249; dot, 250; Egyptian, 250; fringed, **253**; Indian, 250; prickly, 245; red, 250; royal, 251; Rudge, 250; South African, 250; white, 250
water manna grass, **197**, 199
water melon, **119**, 233
water naiad, 105
water plantain, 252, **253**
water shield, 245
wattle (acacia), 59
wavy-leaf thistle, 234

wayaka, 190
West Indian bay tree (pimento), 343
wheat, 219, 306–12; common, 308; club, 308; emmer, 307; einkorn, 307; macha, 308; Saracen, 274; shot, 308; spelt, 308; spring, 308; winter, 308
white alder, 343
white blite, 224
white goosefoot, 19, 224
white mustard, 21, **325**
white oak, 261
white pepper, 75
wild arum (cuckoo pint), 268, 269
wild blite, 27
wild kale, 150
wild longwort, 28
wild oat, 291
wild okra, 244
wild potato, 233
wild thyme, 356
wild tulip, **127**, 136, 137
willow, 163 n, 164
willowherb, rosemary-leaved, 114
Windsor bean, 48
winter cress, 102
winter wheat, 308
wistaria, tuberous-rooted, 189
witloof, 89
wood betony, 32
wood-lily, 135–6
wood sorrel, 21, 216, 220–1; buttercup, 222
wood violet, 244
wormwood, 353, 354, 355; silky, 353

yam, 190, **231**, 255–8, 276; akam, 257; Chinese, 255; common, 258; devil, 256; fancy, 257; Farge's, 257; Japanese, 257; long, 258; Malacca, 256; negro, 258; New Guinea, 257; poison, 256; potato, 256; prickly, 258; Pyrennean, 258; ten months, 256; white, 257; yellow, 257
yam bean, 53, 190; West Indies, 190
yam daisy, 190
yam tree, 190
yarrow, 344
yatay, 167
yebb nut, 58
yew, 163 n
ysaño (capucine), 107, **231**
yucca, 139–40
yuyucho, 211

zamia, 304
Zanzibar oilvine (telfairia nut), 121
zeodary, 327
zuccini, 118

SECTION 2

Miscellaneous Nuts, Grains and Pasta

7. FRUIT II

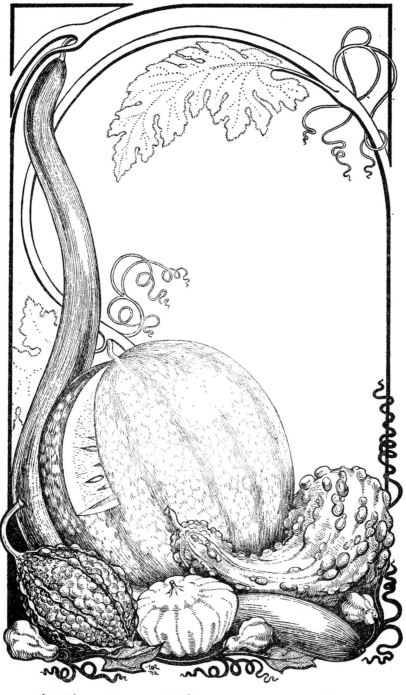

snake gourd
gourd
chayote
water melon squash
summmer crookneck
chayote
balsam pear
chayote

Linda P. Potter 1st Am ed,
B. Farnsworth's ex-lib
5.5.90
$6.50